FOOD AND BEVERAGE CONSUMPTION AND HEALTH

Additional books in this series can be found on Nova's website under the Series tab.

Additional e-books in this series can be found on Nova's website under the e-book tab.

FOOD AND BEVERAGE CONSUMPTION AND HEALTH

WINE

PHENOLIC COMPOSITION, CLASSIFICATION AND HEALTH BENEFITS

FOOD AND BEVERAGE CONSUMPTION AND HEALTH

WINE

PHENOLIC COMPOSITION, CLASSIFICATION AND HEALTH BENEFITS

YOUSSEF EL RAYESS
EDITOR

nova publishers
New York

For permission to use material from this book please contact us:
Telephone 631-231-7269; Fax 631-231-8175
Web Site: http://www.novapublishers.com

Library of Congress Cataloging-in-Publication Data

ISBN: 978-1-63321-048-6

Library of Congress Control Number: 2014939982

Published by Nova Science Publishers, Inc. † New York

CONTENTS

PREFACE

Transformation of grape juice into wine is a natural and spontaneous process known for thousands of years. Wine has importance in many social and religious contexts in our society today. With recent advances in science and technology, wine was shown to be a very complex beverage with countless chemical compounds.

One of the most important groups of compounds in wines is phenolics or polyphenols, which play an important role in the quality of wines. This group contributes to wine organoleptic characteristics such as color, taste, astringency and bitterness. The chemistry of phenolics is complex, but their role in winemaking is well understood, and the application of phenolic extraction in winemaking enables the crafting of many different wine styles. Phenolics are subject to several evolutions during winemaking starting from the grapes till the consumer cup.

Grape and wine phenolic research goes back to the 19th century where Pasteur (1866) recognized the impact of oxygen on the wine color. Since then, a tremendous amount of research on grape and wine phenolics has been published. Nowadays, grape and wine phenolic research can be divided into various categories:

- The chemistry of grape phenolics;
- The development of phenolics in the vineyard and the factors affecting these compounds;
- The chemistry of wine phenolics;
- The development of analytical methods for phenolics identification and quantification;
- The impact of winemaking techniques and processes on wine phenolics;
- The use of phenolics for authenticity assessment of wine based on the geographic origin, variety and/or vintage;
- The role of phenolics in the sensory analysis of wines
- The biological activities and the health benefits of grape and wine polyphenols.

In this book, we attempt to cover all the research categories on wine phenolics. The first two chapters review the chemistry of grape and wine phenolics respectively. The second chapter also aims to present the new devices used in polyphenol analysis by means of sensors and biosensors. The analytical methods used for the assessment of wine polyphenols and their antioxidant activities are presented and discussed in chapter 3. Chapter 4 reviews and discusses the impact of traditional and new processes of winemaking on wine phenolic composition. The health benefits of wine polyphenols are presented in chapter 5. Chapter 6 summarizes facts on possible detrimental effects of *Botrytis* laccase on wine polyphenols but also rules out the enzyme´s positive potential for wine-related applications. In chapter 7, the biosynthesis of resveratrol in grapes, the factors affecting resveratrol content in wines and the current analytical methods for resveratrol determination in wines are revisited. Chapter 8 is an original article aiming to establish relationships between in-mouth sensory attributes perceived in wine and non-volatile composition especially polyphenols. Chapter 9 and 10 present results and evidences about the use of polyphenols for authenticity assessment and wine classification. The recovery of phenolic compounds and the used processes for polyphenols extraction from by-products of the winemaking chain are reviewed in chapter 11 and 12.

It was with great pleasure that I accepted the opportunity offered by Nova publishers to write, assemble and edit this book. I am greatly indebted to the authors and am also delighted that so many of the world's researchers have been willing, so generously, to share their knowledge and experience with others through their contribution to this book.

Youssef El Rayess
Associated professor
Department Food Industry and Technolgy
INP-ENSAT
2 Allée Anita Conti,
31520 Ramonville St Agne
Tel: (33) 06 14 26 37 98
E-mail: youssef.elrayess@ensiacet.fr

In: Wine
Editor: Youssef El Rayess

ISBN: 978-1-63321-048-6

Chapter 1

Grape Phenolic Composition and Antioxidant Capacity

***Fernanda Cosme**[*1] **and António M. Jordão**[2]*

[1]Institute for Biotechnology and Bioengineering, Centre of Genomics and Biotechnology, University of Trás-os-Montes and Alto Douro, School of Life Science and Environment, Vila Real, Portugal

[2]Polytechnic Institute of Viseu (CI&DETS), Agrarian Higher School, Department of Food Industries, Estrada de Nelas, Viseu, Portugal

Abstract

Grapes are one of the most important fruit commodities as economic plants with good agricultural characteristics and they are a significant source of nutritional antioxidants, such as polyphenols. The major C_6-C_3-C_6 flavonoids in grapes include conjugates of flavonols, quercetin and myricetin; flavan-3-ols (+)-catechin and (-)-epicatechin; and malvidin-3-*O*-glucoside and other anthocyanins. Non-flavonoids include C_6-C_1 hydroxy-benzoic acids, and gallic acid, C_6-C_3 hydroxycinnamates caffeic, caftaric, and *p*-coumaric acids; and C_6-C_3-C_6 stilbenes *trans*-resveratrol, *cis*-resveratrol, and *trans*-resveratrol glucoside. Thus, grapes contain a large amount of different phenolic compounds in skins, pulps, seeds and stems, which are responsible for physicochemical and sensory characteristics of the wines.

The composition and content of grape phenolic compounds and consequently their antioxidative or antiradicalar capacity could be affected by many extrinsic and intrinsic factors, such as grape variety, growing regions, vineyard management practices, climate conditions, topography, soil and ripening stage. Phenolic compounds are considered to be free radical scavengers, and their antioxidant properties depend on their chemical structure. Specifically, these properties depend on their ability to donate hydrogen or electron and their ability to delocalize the unpaired electron within the aromatic structure. Flavan-3-ols, including proanthocyanidins, flavonols, and anthocyanins, are the most important phenolic compounds that contribute to grape antioxidant capacity.

* Corresponding Author address: Institute for Biotechnology and Bioengineering – Centre of Genomics and Biotechnology, University of Trás-os-Montes and Alto Douro, School of Life Science and Environment, 5000-911 Vila Real, Portugal. Email: fcosme@utad.pt.

Thus, considering the importance of grape phenolic composition, the purpose of this chapter is to make a review of the most recently publications about the grape phenolic composition and the antioxidant capacities of the main grape phenolic compounds quantified in grapes.

Keywords: Grape variety, phenolic compounds, flavonoids, antioxidant capacity

INTRODUCTION

An important filed of research today is the control of 'redox' status with the properties of food and food components. Natural antioxidants present in the human diet, increase the resistance towards oxidative damages and they may have a substantial impact on human health. The importance of the term 'antioxidant' is evidenced by a search of the PubMed database using the terms 'antioxidant' and 'food', which yielded over 36000 hits.

Antioxidants have a long history of use in the nutrition/health community and food industry. The traditional understanding has been the antioxidant chemicals promote health by removing reactive species that may otherwise exert harmful metabolic effects. By this view, most free radicals and reactive oxygen species (ROS; chemically reactive molecules that contain oxygen) were considered to be harmful (Harman, 1956), implying that maximizing antioxidant concentrations could minimize the risk for chronic disease. However, according to several authors (Finley et al., 2011) as we uncover the complexity of cellular defences and cell signalling pathways, the role of oxidative stress, and the defense system that eliminates it, it has become more complicated.

Dietary antioxidants include ascorbate, tocopherols, carotenoids and bioactive plant phenols. The health benefits of fruit and vegetables are largely due to the antioxidant vitamins supported by the large number of phytochemical, some with greater antioxidant properties, such as phenolic compounds. Phenolics, as secondary plant metabolites and widely distributed in the plant kingdom and abundant in our diet plant phenols, are today among the most talked about classes of phytochemical. In last decade, a lot of works have been presented by the scientific community, which mainly focused on the levels and chemical structures of antioxidant phenols in different plant foods, aromatic plants and various plant materials. The most important natural antioxidant sources and the classes of phenols are sumarized in table 1.

The preparations of food supplements are another field that especially the food industry has developed in last twenty years. Concerning this last field, the use of 'antioxidant' term has been used in advertisements or on package labelling by the market of whole foods or beverages, additives to foods, or dietary supplements. The advertised benefits of antioxidant include slowing the aging process and decreasing the risk of chronic disease. However, the scientific understanding of the functions and health roles of antioxidants is changing, and nutritional messages used to market antioxidant-containing foods often are not supported by contemporary evidence (Aggett, 2009).

The chemical activities of polyphenols in terms of their reducing properties as hydrogen or electron-donating agents predict their potential for action as free-radical scavengers (antioxidants). The activity of an antioxidant is determined by: (i) its reactivity as a hydrogen or electron-donating agent (which relates to its reduction potential); (ii) the fate of the resulting antioxidant-derived radical, which is governed by its ability to stabilize and delocalize the unpaired electron; (iii) its reactivity with other antioxidants; (iv) the transition metal-chelating potential. According to several authors (Rice-Evans et al., 1995; 1996), polyphenols posses ideal structural chemistry for free radical-scavenging activities, and have been shown to be more effective antioxidants *in vitro* than vitamins E and C on a molar basis.

Table 1. Principal sources of antioxidant plant phenols

Sources	Antioxidant compounds	References
Fruits		
Berries	Flavanols, hydroxycinnamic acids, hydroxybenzoic acids, anthocyanins	Wang and Lin (2000); Manach et al. (2004); Céspedes et al. (2010); Denev et al. (2010)
Cherries	Hydroxycinammic acids, anthocyanins	Manach et al. (2004); Ballistreri et al. (2013)
Red grapes	Anthocyanins, flavonols, catechins, flavonoids, resveratrol	Ghiselli et al. (1998); Landrault et al. (2001); Manach et al. (2004); Kedage et al. (2007); Spranger et al. (2008); Breksa III et al. (2010); Darra et al. (2012)
Citrus fruits	Flavanones, flavonols, phenolic acids	Beecher (2003); Manach et al. (2004)
Plums, prunes, apples, pears and kiwi	Hydroxycinnamic acids, catechins, flavonoids	Kim et al. (2003); Manach et al. (2004)
Vegetables		
Aubergines	Anthocyanins, hydroxycinnamic acids	Manach et al. (2004)
Chicory, artichoke	Hydroxycinnamic acids	Manach et al. (2004)
Onion	Flavonoids	Albishi et al. (2013)
Parsley	Flavones	Beecher (2003); Manach et al. (2004)
Rhubarb	Anthocyanins	Manach et al. (2004)
Parsley	Flavones	Manach et al. (2004)
Beans	Flavanols	Manach et al. (2004)
Spinach	Flavonoids, *p*-coumaric acids	Bergman et al. (2001)

Consumption of adequate amounts of fruits and vegetables is considered essential for a healthy life. According to Willett and Stampfer (2003), in order to achieve an optimum health, it's recommended to serves 2 to 3 fruits daily. In addition, a higher intake of fruits and vegetables has been correlated with a lower incidence of major human illnesses, like cardiovascular diseases and cancer (Surh, 2003; Brandt et al., 2004).

During the last two decades, several epidemiological studies have shown that moderate consumption of red wine may contribute to a reduction in the mortality rate due to cardiovascular diseases. This has been associated with designated 'French paradox' theory (Renaud and Lorgeril, 1992; Hertog et al., 1997; Cordova et al., 2005; Lang and Melzer, 2009). This theory was formulated by French epidemiologists in the 1980s (Richard et al., 1981) and is based on the observation that low coronary heart disease death rates despite high intake of dietary cholesterol and saturated fat (Renaud and Lorgeril, 1992; Artaud-Wild et al., 1993). This protective effect seems to be due to their dietary habits that include a high consumption of wine, grapes and grape juice. In this respect, grapes (*Vitis vinifera* L.) and their derived dietary constituents like grape juices and wines have attracted a great deal of attention in recent years this being noticed by countless studies published in last twenty years. Grapes contain large amounts of phenolic compounds (Ricardo-da-Silva et al., 1990; Jordão et al., 1998a; 1998b; 2001a; 2001b; Peña-Neira et al., 2004) which play an important role in human health, such as an important role in the oxidation inhibition of low-density lipoproteins (LDLs) (Ghiselli et al., 1998; Serafini et al., 2000), a decrease of inflammatory and carcinogenic processes (Tapiero et al., 2002) and a inhibition of platelet aggregation (Gryglewski et al., 1987; Escarpa and Gonzalez, 2001). In addition, phenolic compounds are considered to be the most important components of red wine, due to their direct relationship with color, astringency, bitterness, and susceptibility to oxidation reactions. A large number of polyphenols and flavonoids such as *p*-coumaric acid, cinnamic acid, caffeic acid, ferulic acid, vannilic acid, (+)-catechin, (-)-epicatechin, quercetin, proanthocyanidins, besides trihydroxy stilbenes such as resveratrol have been reported from grapes.

In recent years, all these phenolic compounds have been described as playing an important role in the antioxidant capacity of grapes in an isolated form or in combination. These aspects, as well as the different types and distribution of these compounds in grape bunch (including in the different grape varieties) will be covered in this chapter.

Types and Distribution of Polyphenols in Grapes: General Approach

Phenolic compounds, also called polyphenols, constitute a diverse group of secondary metabolites which exist in grapes, mainly in the grape berries skins and seeds (Cheynier and Rigaud, 1986; Souquet et al., 1996; Rodriguez et al., 2006; Lorrain et al., 2011).

The phenolic compounds in *Vitis Vinifera* grapes include two classes of phenolic compounds: non-flavonoids and flavonoids. The non-flavonoid compounds include phenolic acids divided into hydroxybenzoic acids and hydroxycinnamic acids, but also other phenol derivatives such as stilbenes. Flavonoids covered a large number of subclasses, such as flavonols, flavanols, and anthocyanins (Downey et al., 2006; Dopico-Garcia et al., 2008; Novaka et al., 2008; Hernandez-Jimenez et al., 2009). Flavonols are the most abundant phenolic compounds in grape skins (Cheynier and Rigaud, 1986), while grape seeds are rich in flavan-3-ols (Ricardo-da-Silva et al., 1991; Prieur et al., 1994; Jordão et al., 2001a).

In general, the average concentration of total phenolic compounds is around 2178.8 mg/g gallic acid equivalent in seeds, 374.6 mg/g gallic acid equivalent in skins, and 23.8 mg/g gallic acid equivalent in pulps (Pastrana-Bonilla et al., 2003). In a quantitative point of view, total phenolic compounds from grapes are influenced by different factors such as, grape variety (Sun et al., 2001; Monagas et al., 2003; Obreque-Slier et al., 2010; 2012), sunlight exposition (Bergqvist et al., 2001; Spayd et al., 2002), solar radiation (Madeira, 2011), altitude (Mateus et al., 2001; Malinovski, 2009), soil composition (Ubalde et al., 2010), climate (Jackson and Lombard, 1993, Bergqvist et al., 2001; Downey et al., 2006), cultivation practices (Jackson and Lombard, 1993; Downey et al., 2006), exposure to diseases (Bruno et al., 2007), and the degree of grape ripeness (Jordão et al., 2001a; Ó-Marques et al., 2005; Obreque-Slier et al., 2010; 2012). Total phenolic compounds quantified in different grape varieties cultivated in several countries are summarized in table 2.

Table 2. Total phenolic compounds quantified in different grape varieties

Grape variety/Grape fraction	Total phenolic compounds	References
Red grapes		
Cabernet Sauvignon	1111.30 - 1431.67[a]	Madeira (2011)
	1003[a]	Falcão (2007)
	1207.13 - 945.15[a]	Malinovski (2009)
Saint Laurent	3.76[f]	Balík et al. (2008)
Blauer Portugieser	4.55[f]	
Andre	6.54[f]	
Graciano	16.5-29.9[b]	Nuñez et al. (2004)
Tempranillo	18.2-29.1[b]	
Cabernet Sauvignon	11.4-21.2[b]	
Cabernet Franc	424.6[e]	Yang et al. (2009)
Pinot Noir	396.8[e]	
Touriga Nacional	45.4[g]	Jordão and Correia (2012)
Tinta Roriz	60.7[g]	
Touriga Francesa	45.4[g]	Jordão et al. (2001a)
Castelão Francês	39.3[g]	
Touriga Francesa	46.1[f]	Jordão et al. (1998b)
Castelão Francês	19.0[f]	
Cabernet Sauvignon	0.8[c]	Obreque-Slier et al. (2010)
Carménère	1.1[c]	
Seeds		
Cabernet Sauvignon	17.5[d]	
Carménère	16.6[d]	
Stems		
Saint Laurent	24.0[f]	Balík et al. (2008)
Blauer Portugieser	21.8[f]	
Andre	27.2[f]	
White grapes		
Chardonnay	201.1[e]	Yang et al. (2009)
Riesling	255.8[e]	
Chardonnay	3.59[f]	Balík et al. (2008)
Welschriesling	4.02[f]	
Pinot Gris	3.93[f]	
Stems		
Chardonnay	14.5[f]	Balík et al. (2008)
Welschriesling	15.2[f]	
Pinot Gris	13.6[f]	

[a] mg gallic acid equivalents/100g of skin; [b] g gallic acid equivalents /kg on a dried skin basis; [c] mg gallic acid equivalents/g of skin; [d] mg gallic acid equivalents/g of seed; [e] mg of gallic acid equivalents/100 g of grape; [f] mg/g of grape; [g] total polyphenol index.

Non-flavonoid Compounds

Non-flavonoids cover C_6-C_3 hydroxycinnamates acids, C_6-C_1 hydroxybenzoic acids and C_6-C_3-C_6 stilbenes *trans*-resveratrol, *cis*-resveratrol, and *trans*-resveratrol glucoside (Figure 1).

Phenolic Acids

In grapes, phenolic acids are frequently divided in two main groups: hydroxycinnamic and hydroxybenzoic acids. Hydroxycinnamic acids are aromatic compounds characterized by a C_6-C_3 structure as caffeic, ferulic, *p*-coumaric, sinapic, caftaric, coutaric and fertaric acid (Singleton et al., 1978; Lu and Foo, 1999). They are mainly in the *trans* form but also exist in the *cis* form. They are found mainly in the grape skin and pulp as tartaric acid esters (Baderschneider and Winterhalter, 2001). Comparative analysis of hydroxycinnamic esters content showed higher concentrations in the skin fresh material than in the pulp (Romeyer et al., 1985).

Hydroxybenzoic acids are characterized by a C_6-C_1 structure, a benzene ring with one carbone aliphatic chain substituent. The substitution of the benzene ring differentiates the different acids. Grapes mainly contain gallic acid in the pulp (Lu and Foo, 1999), but also *p*-hydroxybenzoic, protocatechuic, vanillic and syringic acids could be found (Bravo, 1998). Table 3 illustrates different phenolic acids values quantified in several white and red grape varieties.

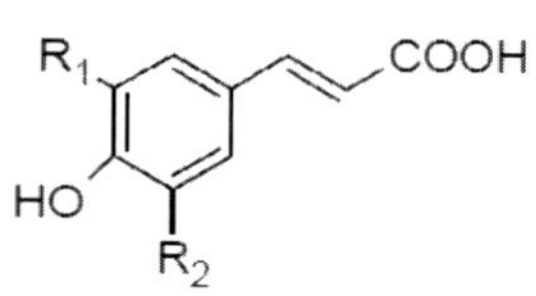

Hydroxycinnamic acids	R_1	R_2
Caffeic acid	OH	H
Ferulic acid	OCH_3	H
p-Coumaric acid	H	H
Sinapic acid	OCH_3	OCH_3

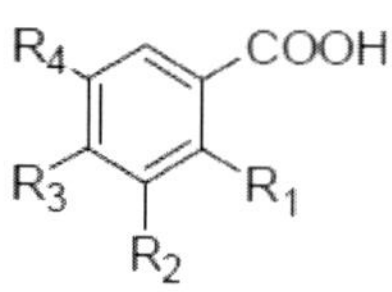

Hydroxybenzoic acids	R_1	R_2	R_3	R_4
Gallic acid	H	OH	OH	OH
p- Hydroxybenzoic acid	H	H	OH	H
Vanillic acid	H	OCH_3	OH	H
Syringic acid	H	OCH_3	OH	OCH_3

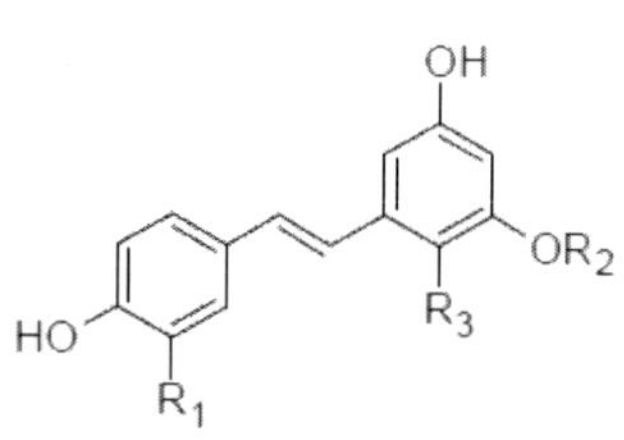

Trans-resveratrol

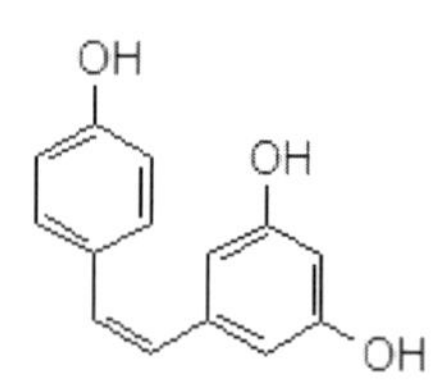

Cis-resveratrol.

Stilbenes	R_1	R_2	R_3
Trans-resveratrol	H	H	H
Piceid	H	OCH_3	OH

Figure 1. Main non-flavonoid compounds found in *Vitis vinifera* grape varieties.

Table 3. Quantification of individual phenolic acids in different white and red grape varieties

Grape variety/Grape fraction	Gallic acid	Protocatechuic acid	Syringic acid	Caftaric acid	*cis*-Caftaric	*trans*-Caftaric	*cis*-Coutaric	*trans*-Coutaric	*trans*-Fertaric	References
Red grapes										
Primitivo		13.4[a]		1.89[a]						Nicoletti et al. (2008)
Negroamaro	7.3[a]	42.0[a]		8.5[a]						
Susumaniello	45.0[a]	8.5[a]		171.7[a]						
Malvasia Nera	77.3[a]	46.0[a]		171.9[a]						
Aglianico	151.9[a]	37.4[a]		320.4[a]						
Cesanese		31.1[a]		28.8[a]						
Alphonse Lavallée				645.0[a]						
Merlot	66.6[a]	328.7[a]		746.3[a]						
Blauer Burgunder	5.2[e]									Mikeš et al. (2008)
Lemberger	6.7[e]									
St. Laurent	13.3[e]									
Stems										
Merlot				40.0[b]						Souquet et al. (2000)
Skins										
Cencibel		1.5[c]			0.25[c]	5.7[c]	2.7[d]	10.0[c]		Montealegre et al. (2006)
Cabernet Sauvignon		2.4[c]			0.44[c]	9.5[d]	2.0[c]	5.8[c]		
Merlot		1.7[c]			0.46[c]	7.0[d]	0.94[c]	3.2[c]		
Shiraz		2.4[c]			0.17[c]	4.9[d]	1.5[c]	6.1[c]		
Carménère	2.8[c]		1.6[c]	0.6[c]						Obreque-Slier et al. (2010)
Cabernet Sauvignon	3.5[c]		1.8[c]	0.7[c]						
Seeds										
Merlot	9.8[d]	8.7[d]								Montealegre et al. (2006)
Cencibel	7.3[d]	3.3[d]								
Cabernet Sauvignon	9.0[d]	7.1[d]								
Shiraz	6.8[d]	6.2[d]								

Table 3. (Continued)

Grape variety/Grape fraction	Gallic acid	Protocatechuic acid	Syringic acid	Caftaric acid	*cis*-Caftaric	*trans*-Caftaric	*cis*-Coutaric	*trans*-Coutaric	*trans*-Fertaric	References
White grapes										
Moscato				48.4[a]						Nicoletti et al. (2008)
Chardonnay					0.85[e]	20.0[e]	2.9[e]	7.9[e]	1.1[e]	Montealegre et al. (2006)
Sauvignon Blanc					0.39[e]	5.2[e]	2.5[e]	4.9[e]	0.30[e]	
Moscatel					0.62[e]	14.0[e]	5.9[e]	18.0[e]	0.61[e]	
Gewürztraminer					0.70[e]	13.0[e]	2.1[e]	6.1[e]	0.87[e]	
Riesling					1.30[e]	31.0[e]	2.5[e]	7.0[e]	2.6[e]	
Viogner						2.4[e]	1.6[e]	1.9[e]		
Grüner Veltliner	3.9[e]									Mikeš et al. (2008)
Hibernal	4.0[e]									
Malverina	3.5[e]									
Müller Thurgau	2.6[e]									
Rheinriesling	2.1[e]									
Welschriesling	1.8[e]									
Neuburger	3.9[e]									
Seeds										
Chardonnay		4.8[c]								Montealegre et al. (2006)
Sauvignon Blanc		4.4[c]								
Moscatel		3.6[c]								
Gewürztraminer		6.0[c]								

[a] mg/kg of berry dry weight; [b] mg/kg of stem; [c] mg/kg of fresh grape skin; [d] mg/kg of fresh grape seed; [e] mg/kg of fresh grape weight.

Stilbenes

Stilbenes are phenolic compounds with high reactivity, including two benzene rings bonded by an ethane bridge. In grapes, a mixture of *trans*- and *cis*-resveratrol forms are present. However the most abundant in grapes are *trans*-resveratrol and glucosyl derivatives: piceid and ε-viniferin (dimeric resveratrol) (Jeandet et al., 1991; Waterhouse and Lamuela-Raventos, 1994; Romero-Pérez et al., 1999). The amounts of stilbenes quantified in different white and red grape varieties are summarized in table 4.

Table 4. Quantification of different stilbenes compounds in several white and red grape varieties

Grape variety/Grape fraction	*trans*-Piceid	*cis*-Piceid	*trans*-Resveratrol	References
Red grapes				
Primitivo	30.7[a]		13.9[a]	Nicoletti et al. (2008)
Negroamaro	4.14[a]		3.6[a]	
Susumaniello	150.3[a]		63.0[a]	
Uva diTroia	15.3[a]		4.6[a]	
Malvasia Nera	98.0[a]		48.5[a]	
Aglianico	75.7[a]		61.1[a]	
Cesanese	12.0[a]		8.1[a]	
Merlot	26.3[a]		9.2[a]	
Alphonse Lavallée	24.1[a]		40.0[a]	
Blauer Burgunder			0.5[d]	Mikeš et al. (2008)
Lemberger			0.3[d]	
Saint Laurent			1.0[d]	
	2.9[d]		2.3[d]	Balík et al. (2008)
Blauer Portugieser	2.5[d]		0.4[d]	
Andre	0.6[d]		0.4[d]	
Skins				
Castelão	67.24[b]	58.9[b]	6.8[b]	Sun et al. (2006)
Syrah	10.43[b]			
Tinta Roriz	11.57[b]			
Stems				
Castelão	61.43[e]	143.85[e]	145.52[e]	Sun et al. (2006)
White grapes				
Moscato			3.89[a]	Nicoletti et al. (2008)
Grüner Veltliner			0.1[d]	Mikeš et al. (2008)
Hibernal			0.3[d]	
Malverina			0.3[d]	
Müller Thurgau			0.3[d]	
Rheinriesling			0.2[d]	
Welschriesling			0.5[d]	
Neuburger			1.5[d]	
Stems				
Chardonnay	1.1[d]		0.3[d]	Balík et al. (2008)
Welschriesling	0.4[d]		1.6[d]	
Pinot Gris	0.6[d]		1.1[d]	

[a] mg/kg of berry dry weight; [b] mg/kg of dry skin; [c] μg/g of dry weight skin; [d] mg/kg of fresh weight; [e] mg/kg of dry stem.

Table 5. Monomeric anthocyanins quantified in different grape varieties

Grape variety	Delphinidin 3-*O*-glucoside	Cyanidin 3-*O*-glucoside	Petunidin-3-*O*-glucoside	Peonidin 3-*O*-glucoside	Malvidin 3-*O*-glucoside	References
Primitivo	44.7[a]	14.7[a]	168.4[a]	207.7[a]	1883.0[a]	Nicoletti et al. (2008)
Negroamaro	112.8[a]	50.6[a]	203.6[a]	110.2[a]	662.6[a]	
Susumaniello	764.0[a]	123.3[a]	881.5[a]	873.0[a]	4131.0[a]	
Uva di Troia	309.2[a]	443.2[a]	354.9[a]	557.6[a]	602.29[a]	
Malvasia Nera	586.6[a]	67.3[a]	625.3[a]	827.5[a]	3396.0[a]	
Aglianico	594.7[a]	52.1[a]	596.5[a]	553.6	3348.0[a]	
Cesanese	1270.4[a]	365.0[a]	1329.2[a]	1333.5[a]	2771.2[a]	
Alphonse Lavallée	219.8[a]	46.4[a]	241.1[a]	52.5[a]	1851.1[a]	
Merlot	974.5[a]	25.93[a]	534.4[a]	580.2[a]	1218.0[a]	
	19.9[b]	9.3[b]	29.0[b]	55.2[b]	357.1[b]	Kallithraka et al. (2005)
Xinomavro	9.2[b]	--	14.6[b]	30.7[b]	399.3[b]	
Mandilaria	29.9[b]	6.9[b]	43.6[b]	137.5[b]	943.5[b]	
Thrapsa	15.7[b]	4.2[b]	22.0[b]	67.2[b]	470.5[b]	
Aidani Mavro	33.2[b]	--	30.0[b]	32.2[b]	368.7[b]	
Sangiovese	73.2[b]	66.3[b]	72.6[b]	67.1[b]	341.1[b]	
Papadiko	34.3[b]	20.9[b]	35.0[b]	68.0[b]	248.0[b]	
Mavrodafni	109.3[b]	11.0[b]	95.0[b]	35.5[b]	694.6[b]	
Grenache Rouge	45.8[b]	4.1[b]	43.4[b]	27.5[b]	513.7[b]	
Negoska	14.9[b]	--	22.5[b]	73.8[b]	651.5[b]	
Vapsa	50.3[b]	9.1[b]	69.4[b]	192.7[b]	1202.6[b]	
Cabernet Sauvignon	43.9[b]	7.3[b]	36.2[b]	58.3[b]	552.5[b]	
Castelão Francês	6.2[c]	2.6[c]	8.5[c]	11.7[c]	59.2[c]	Jordão et al. (1998a)
Touriga Francesa	0.9[c]	0.1[c]	2.5[c]	3.6[c]	46.3[c]	
Tannat	11.35-13.41[d]	3.17-4.56[d]	15.86-17.43[d]	5.87-7.67[d]	56.93-64.13[d]	González-Neves et al. (2007)
Merlot	8.05-8.46[d]	3.75-3.95[d]	10.06-10.49[d]	11.83-11.46[d]	65.84-66.11[d]	
Cabernet Sauvignon	6.54-8.92[d]	1.90-1.97[d]	7.19-8.67[d]	8.79-9.67[d]	71.65-74.73[d]	

Grape variety	Delphinidin 3-*O*-glucoside	Cyanidin 3-*O*-glucoside	Petunidin-3-*O*-glucoside	Peonidin 3-*O*-glucoside	Malvidin 3-*O*-glucoside	References
Cabernet Sauvignon	4.67[d]	0.90[d]	4.21[d]	4.87[d]	41.45[d]	Revilla et al. (2001)
Garnacha	2.26[d]	1.02[d]	3.73[d]	12.69[d]	64.69[d]	
Graciano	6.81[d]	1.28[d]	7.21[d]	12.79[d]	53.69	
Mencia	5.13[d]	2.15[d]	6.68[d]	14.85[d]	47.40[d]	
Merlot	7.53[d]	5.52[d]	7.0[d]	14.27[d]	35.54[d]	
Tempranillo	10.9[d]	3.26[d]	11.11[d]	7.81[d]	46.35[d]	
Cabernet Sauvignon	431.6[e]	53.1[e]	337.4[e]	259.5[e]	2506.3[e]	Ortega-Regules et al. (2006)
Merlot	231.7[e]	48.2[e]	270.9[e]	381.9[e]	1834.7[e]	
Syrah	258.0[e]	27.9[e]	385.2[e]	299.2[e]	2889.7[e]	

[a] mg/kg of berry dry weight; [b] mg/kg of berry fresh weight; [c] % weight of anthocyanins/weight grape; [d] relative amount of anthocyanins (%); [e] μg/g of grape skin.

Flavonoid Compounds

Flavonoids are characterized by a basic structure of 15 carbon atoms comprising two aromatic rings bound through a 3 carbon chain (C_6-C_3-C_6). In grapes, flavonoid compounds are mainly represented by anthocyanins and flavan-3-ols, and by the families of flavonols and flavanols (Bourzeix et al., 1986). The major C_6-C_3-C_6 flavonoids in grapes include conjugates of flavonols (quercetin and myricetin), flavan-3-ols ((+)-catechin and (-)-epicatechin), malvidin-3-*O*-glucoside and other anthocyanins (Figure 2).

Anthocyanins

Anthocyanins are the main compounds responsible for the color of red grapes. They are localized in the skins (Jordão et al., 1998a) and sometimes in the pulp of "teinturier" varieties that have colored flesh (Castillo-Muñoz et al., 2009). Their structure is characterized by a flavylium cation with two benzene rings, linked by an unsaturated cationic oxygenated heterocycle, which is derived from the 2-phenyl-benzopyrylium nucleus and glycosylated at position C_3 (Ribéreau-Gayon et al., 2006). The variation of the degree of hydroxylation, methylation and/or glucosilation leads to five aglycones or anthocyanidins found in *Vitis vinifera*: delphinidin, cyanidin, petunidin, peonidin and malvidin (Figure 2). Anthocyanins are divided in different subclasses, depending on the nature of the substituents on glucose. Glucose can be acylated in position 6 with acetic, *p*-coumaric or caffeic acids. Grapes are characterized by the predominant presence of 3-monoglucosides anthocyanins, especially malvidin 3-*O*-glucoside and its acyl derivatives (Baldi et al., 1995). For the majority of grape varieties, the most abundant individual anthocyanins are malvidin-3-*O*-glucoside while cyanidin-3-*O*-glucoside is the lowest abundant form (Table 5). Nevertheless, it is known that the anthocyanin pattern of each grape variety is under strict genetic control and that their distribution varies considerably among grape varieties (Revilla et al., 2001).

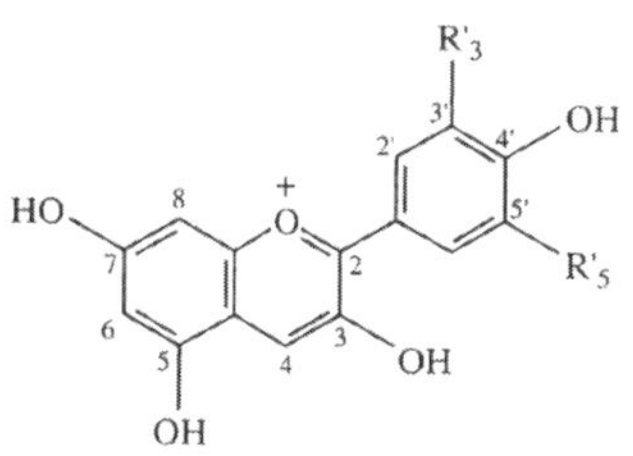

Anthocyanidins	R'$_3$	R'$_5$
Delphinidin	OH	OH
Cyanidin	OH	H
Petunidin	OH	OCH_3
Peonidin	OH	OCH_3
Malvidin	OCH_3	OCH_3

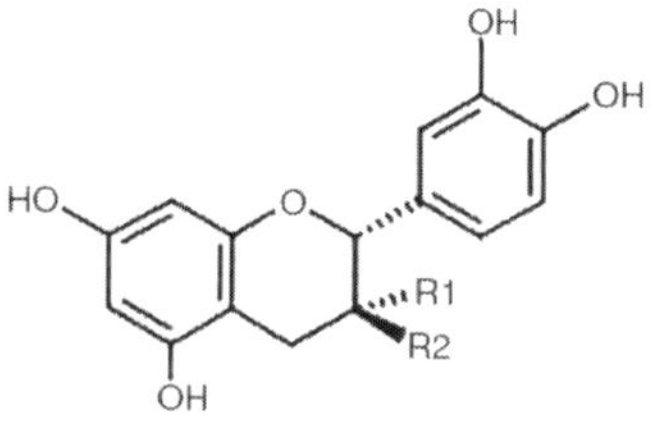

Flavan-3-ols	R_1	R_2
(+) - catechin	H	OH
(+) - epicatechin	OH	H

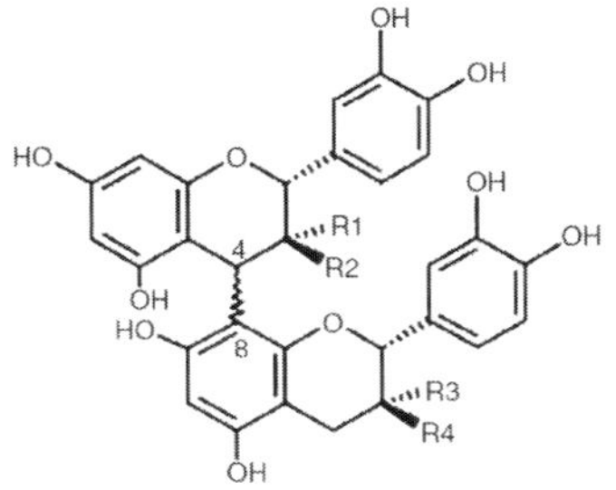

Dimers (C_6-C_8)	R_1	R_2	R_3	R_4
Procyanidin B1	OH	H	H	OH
Procyanidin B2	OH	H	OH	H
Procyanidin B3	H	OH	H	OH
Procyanidin B4	H	OH	OH	H

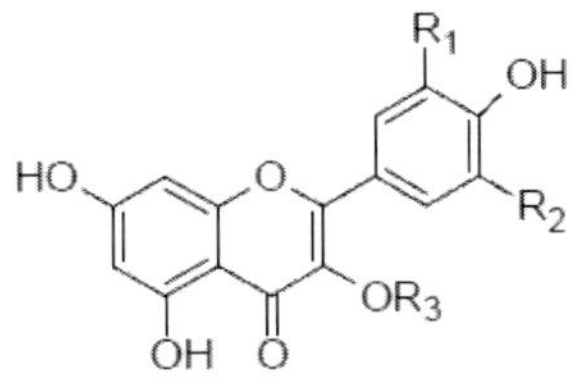

Flavonols	R_1	R_2	R_3
Myricetin	OH	OH	H
Myricetin-3-*O*-glucoside	OH	OH	gluc
Quercetin	OH	H	H
Quercetin-3-*O*-glucoside	OH	H	gluc
Kaempferol	H	H	H
Kaempferol-3-*O*-glucoside	H	H	gluc

Figure 2. Main flavonoid compounds found in *Vitis vinifera* grape varieties.

Flavanols

Flavanols include monomers and condensed tannins (proanthocyanidins), common name for oligomers and polymers of flavan-3-ols (Figure 2). The monomers are (+)-catechin, (-)-epicatechin, (+)-gallocatechin, (-)-epigallocatechin and epicatechin-3-*O*-gallate (Escribano-Bailón et al., 1995; Souquet et al., 1996). There are also dimeric, trimeric, oligomeric, and condensed procyanidins. Dimeric procyanidins are dimers resulting from the condensation of two units of flavan-3-ols linked by a C_4-C_8 (B_1 to B_4) or C_4-C_6 (B_5 to B_6) bond. Trimeric procyanidins are trimers with two interflavan bonds while oligomeric procyanidins are polymers from three to ten flavanol units linked by C_4-C_8 or C_4-C_6 bonds. Condensed procyanidins have more then ten flavan units (Ribéreau-Gayon et al., 2006). Solid parts of grape bunch are the most important source of grape polyphenols, especially proanthocyanidins (Spranger et al., 1998; Sun et al., 1999; Jordão et al., 2001a; 2001b), which are oligomers and polymers of polyhydroxyflavan-3-ol monomer units linked most commonly by acid-labile 4 $\rightarrow$ 8 and in some cases by 4 $\rightarrow$ 6 bonds. Proanthocyanidins that are hydrolyzed to cyanidins are known as procyanidins [polymers of (+)-catechin and (-)-epicatechin]. However, if proanthocyanidin is hydrolyzed to delphinidin it is known as prodelphinidins [polymers of (+)-gallocatechin and (-)-epigallocatechin].

Seed proanthocyanidins are only procyanidins (Prieur et al., 1994; Labarbe et al., 1999), while in skins and stems, proanthocyanidins are procyanidins and prodelphinidins (Souquet et al., 1996; 2000; Cheynier et al., 1997; 2006). The most common class of such compounds are procyanidins. Grape seeds are much richer in these phenolic compounds than grape skins and grape stems (Ricardo-da-Silva et al., 1991; 1992; Escribano-Bailón et al., 1992; 1995; Souquet et al., 1996; 2000; Fuleki and Ricardo-da-Silva, 1997; Kennedy et al., 2000; Sun et al., 2001; Jordão et al., 2001a; 2001b; Negro et al., 2003). Procyanidin B1 has been reported to be the main oligomer in skins (Escribano-Bailón et al., 1995; Jordão et al., 2001a; González-Manzano et al., 2004), while procyanidin B2 is the most abundant in seeds (Bourzeix et al., 1986; Ricardo-da-Silva et al., 1991; Jordão et al., 2001a; González-Manzano et al., 2004).

In seeds and skins the polymeric tannin fractions are present in a greater proportion than the monomeric or dimeric tannins (Cheynier et al., 1997), depending on the grape variety. In addition, polymeric tannins represent 77-85% of total flavanols in seeds and 91-99% of total flavanols in skins (Cosme et al., 2009). The mean degree of polymerization is around 10 in seeds and around 30 in skins (Prieur et al., 1994; Souquet et al., 1996), while the level of galloyled flavan-3-ols is more important in seeds (30%) than in skins (5%) (Moutounet et al., 1996; De Freitas and Glories, 1999; De Freitas et al., 2000; Kennedy et al., 2001; Mateus et al., 2001; Jordão et al., 2001a; Cheynier et al., 2006).

Proanthocyanidins composition and their structural characteristics depend on grape localization (seeds or skins), as well as on the grape varieties as it is shown in some publications (Prieur et al., 1994; Souquet et al., 1996; Sun et al., 1998; Labarbe et al., 1999; Vidal et al., 2003; Perret et al., 2003; Monagas et al., 2003; Cosme et al., 2009). The composition and strututal characteristics of these phenolic compounds quantified in different grape varieties could be observed in Table 6 and 7.

Flavonols

In *Vitis Vinifera* L. grape berry, flavonols are light yellow pigments found in the skins of both red and white grapes (Price et al., 1995), and have also been detected in the pulp of Merlot grape variety (Pereira et al., 2006). They are characterized by the existence of a double bond between C_2 and C_3, and a hydroxyl group in C_3. The main flavonols present in the grape are quercetin, myricetin, kaempferol and isorhamnetin (Mattivi et al., 2006; Cheynier and Rigaud, 1986). Flavonols in grapes are presented only as 3-glycosides (Castillo-Muñoz et al., 2007) and the main were quercetin-3-glucoside, quercetin-3-glucuronide, and myricetin-3-glucoside, while the minor included kaempferol-3-glucoside, kaempferol-3-galactoside, and isorhamnetin-3-glucoside (Cheynier and Rigaud, 1986). The sugar attached to the C-3 position of kaempferol, quercetin, myricetin, and isorhamnetin is commonly glucose. However, glucuronic acid has also been found as the glycosylation sugar of kaempferol, quercetin, and myricetin (Cheynier and Rigaud, 1986). In white and red grapes, kaempferol and quercetin flavonols are presents. Though, myricetin and isorhamnetin flavonols occur merely in red grapes (Mattivi et al., 2006; Castillo-Muñoz et al., 2007). Several examples of individual flavonols quantified in different grapes varieties are presented in Table 8.

Role of Phenolic Compounds in Grape Antioxidant Capacity

Grape contains a large number of different phenolic compounds in the different grape bunch fractions (stem, skin, pulp and seeds). Among the various phenolic antioxidants present especially in red grapes and wines, (+)-catechin, (-)-epicatechin, proanthocyanidins, resveratrol, anthocyanins, quercetin and its glycoside (rutinoside) rutin are the most effective compounds (Sakkiadi et al., 2001). Table 9, summarized the most important class of phenolic compounds with antioxidant capacity quantified in bunch grapes. According to a great number of studies published in last years, these phenolics contribute to antioxidant capacity in grapes in greater deal. Catechins and several dimers and trimer procyanidins from grape seeds showed potent scavenger capacity for superoxide radical and hydroxyl radical (Ricardo-da-Silva et al., 1991).

Table 6. Assessment of different flavan-3-ols and proanthocyanidins in several grape varieties

Grape variety/Grape fraction	(+)-Catechin	(-)-Epicatechin	Procyanidin B3	Procyanidin B1	Procyanidin B4	Procyanidin B2	References
Red Grapes							
Primitivo	307.1[a]	49.8[a]					Nicoletti et al. (2008)
Negroamaro	118.2[a]	27.8[a]					
Susumaniello	147.1[a]	73.5[a]					
Uva diTroia	127.6[a]	62.5[a]					
Malvasia Nera	966.0[a]	734.7[a]					
Aglianico	3214.9[a]	1890.1[a]					
Cesanese	178.8[a]	8.9[a]					
Merlot	601.0[a]	980.7[a]					
Alphonse Lavallée	331.20[a]	32.4[a]					
Blauer Burgunder	515.6[g]	237.2[g]					Mikeš et al. (2008)
Lemberger	197.5[g]	204.8[g]					
Saint Laurent	39.2[g]	467.3[g]					
	118[g]	63.5[g]					Balík et al. (2008)
Blauer Portugieser	467[g]	123[g]					
Andre	533[g]	235[g]					
Baboso Negro	51.61[j]	16.50[j]		15.95[j]		10.39[j]	Pérez-Trujillo et al. (2011)
Listán Negro	54.25[j]	13.77[j]		15.00[j]		5.74[j]	
Negramoll	51.31[j]	15.07[j]		14.69[j]		7.55[j]	
Tintilla	50.10[j]	20.55[j]		13.64[j]		9.92[j]	
Vijariego Negro	49.09[j]	16.13[j]		13.39[j]		7.44[j]	
Stems							
Merlot	60[b]						Souquet et al. (2000)
Saint Laurent	867[g]	81.4[g]					Balík et al. (2008)
Blauer Portugieser	1369[g]	156[g]					
Andre	1582[g]	186[g]					
Touriga Francesa	0.11[i]	1.33[i]	0.89[i]	6.59[i]	0.13[i]	0.27[i]	Jordão et al. (2001b)
Castelão Francês	0.07[i]	1.13[i]	0.04[i]	1.93[i]	0.14[i]	0.10[i]	
Skins							
Merlot	0.57[h]	0.16[h]	0.02[h]	0.08[h]		0.92[h]	De Freitas et al. (2000)
Cabernet Sauvignon	0.31[h]	0.10[h]	0.05[h]	0.11[h]		0.84[h]	

Table 6. (Continued)

Grape variety/Grape fraction	(+)-Catechin	(-)-Epicatechin	Procyanidin B3	Procyanidin B1	Procyanidin B4	Procyanidin B2	References
Merlot	25.0[c]	13.0[c]	35.0[c]	21.0[c]		2.2[c]	Montealegre et al. (2006)
Cencibel	22.0[c]	8.4[c]	39.0[c]	22.0[c]		1.5[c]	
Carménère	1.3[c]		0.60[c]				Obreque-Slier et al. (2010)
Cabernet Sauvignon	0.5[c]		0.70[c]				
	17.0[c]	6.2[c]	27.0[c]	12.0[c]		0.99[c]	Montealegre et al. (2006)
Shiraz	8.5[c]	6.9[c]	16.0[c]	8.4[c]		0.75[c]	
Graciano	4.3[e]						Nuñez et al. (2004)
Tempranillo	8.2-12.3[e]						
Cabernet Sauvignon	1.8-6.2[e]						
Touriga Nacional	0.012-0.021[h]		0.013[h]	0.184-0.260[h]		0.020[h]	Mateus et al. (2001)
Touriga Francesa	0.012[h]	0.010[h]		0.090-0.138[h]		0.011-0.015[h]	
Seeds							
Merlot	240.0[d]	210.0[d]	64.0[d]	170.0[d]	80.0[d]	37[d]	
Cencibel	82.0[d]	60.0[d]	43.0[d]	74.0[d]	39.0[d]	21.0[d]	Montealegre et al. (2006)
Cabernet Sauvignon	270.0[d]	130.0[d]	50.0[d]	150.0[d]	57.0[d]	41.0[d]	
Shiraz	120.0[d]	130.0[d]	55.0[d]	100.0[d]	33.0[d]	23.0[d]	
Merlot	18.5[h]	10.77[h]	7.30[h]	5.92[h]	12.24 [h]	24.0[h]	De Freitas et al. (2000)
Cabernet Sauvignon	27.8[h]	14.6[h]	5.0[h]	5.3[h]	8.0 [h]	30.0[h]	
Touriga Nacional	0.71-0.80[h]	1.39-1.92[h]	0.56-0.96[h]	2.36-3.58[h]	1.54- 2.57[h]	4.86-5.23[h]	Mateus et al. (2001)
Touriga Francesa	0.82-1.89[h]	1.04-2.41[h]	0.88-0.45[h]	1.38-2.29[h]	0.65-1.37[h]	1.75-2.83[h]	
	3.30[i]	2.30[i]	0.33[i]	0.54[i]	0.59[i]	1.35[i]	Jordão et al. (2001a)
Castelão Francês	2.11[i]	4.55[i]	0.52[i]	0.08[i]	2.64[i]	2.02[i]	
White grapes							
Moscato	21.5[a]	4.1[a]					Nicoletti et al. (2008)
Grüner Veltliner	659.1[g]	122.9[g]					Mikeš et al. (2008)
Hibernal	532.9[g]	172.7[g]					
Malverina	242.0[g]	100.0[g]					
Müller Thurgau	70.3[g]	67.1[g]					
Chardonnay	123[g]	144[g]					
Welschriesling	61.0[g]	84.3[g]					Balík et al. (2008)
Pinot Gris	481[g]	251[g]					
Stems							
Chardonnay	899[g]	84.4[g]					Balík et al. (2008)

Grape variety/Grape fraction	(+)-Catechin	(-)-Epicatechin	Procyanidin B3	Procyanidin B1	Procyanidin B4	Procyanidin B2	References
Welschriesling	926^{g}	158^{g}					
Pinot Gris	973^{g}	199^{g}					
Viosinho	1.5^{i}	0.6^{i}	0.1^{i}	1.2^{i}	0.1^{i}	0.4^{i}	Jordão et al. (2001b)
Skins							
Chardonnay	23.0^{c}	5.8^{c}	37.0^{c}	23.0^{c}			Montealegre et al. (2006)
Sauvignon Blanc	9.5^{c}	3.4^{c}	25.0^{c}	16.0^{c}			
Moscatel	16.0^{c}	2.6^{c}	23.0^{c}	21.0^{c}			
Gewürztraminer	19.0^{c}	8.3^{c}	21.0^{c}	48.0^{c}			
Riesling	14.0^{c}		29.0^{c}	12.0^{c}			
Ugni Blanc	2.6-222.0^{h}	0.04-3.0^{h}	0.2-0.3^{h}	1.1-1.9^{h}	0.04^{h}	0.06-0.2^{h}	De Freitas and Glories (1999)
Sémillon	1.2-35.2^{h}	0.03-1.6^{h}	0.01-0.2^{h}	0.02-0.4^{h}			
Seeds							
Chardonnay	390.0^{d}	310.0^{d}	52.0^{d}	380.0^{d}	71.5^{d}	33.0^{d}	Montealegre et al. (2006)
Sauvignon Blanc	200.1^{d}	130.0^{d}	52.0^{d}	250.0^{d}	54.0^{d}	19.0^{d}	
Moscatel	350.0^{d}	120.0^{d}	39.0^{d}	330.1^{d}	40.0^{d}	15.0^{d}	
Gewürztraminer	500.0^{d}	150.0^{d}	56.0^{d}	460.0^{d}	70.0^{d}	22.0^{d}	
Riesling	400.0^{d}	160.0^{d}	43.0^{d}	620.0^{d}	95.0^{d}	33.0^{d}	
Viogner	120.0^{d}	110.0^{d}	51.0^{d}	200.0^{d}	53.0^{d}	19.0^{d}	
Ugni Blanc	10.5-55.8^{h}	2.7-6.4^{h}	0.9-1.8^{h}	0.2-2.0^{h}	2.5-2.9^{h}	3.8-5.6^{h}	De Freitas and Glories (1999)
Sémillon	15.9-18.4^{h}	3.1-6.2^{h}	1.8-2.4^{h}	1.1-1.5^{h}	1.5-3.2^{h}	2.3-4.2^{h}	

[a] mg/kg of berry dry weight; [b] mg/kg of stem; [c] mg/kg of fresh grape skin; [d] mg/kg of fresh grape seed; [e] g (+)-catechin/kg of dried skin; [f] g cyanidin chloride/kg of dried skin; [g] mg/kg of fresh weight; [h] mg/g of dry weight; [i] mg/g; [j] molar percentages.

Table 7. Composition and structural characteristics of proanthocyanidins in several grape varieties

Grape variety/Grape fraction	% Gal	mDP	% Prodelph	% Cat	% Ec	% EcG	% EGC	References
Red Grapes								
Merlot		9.2		14.4	67.7	15.5	2.9	Souquet et al. (2000)
Negrette		10.2		11.7	61.7	21.1	5.4	
Pinot Noir		8.2		15.3	65.1	18.1	1.5	
Tannat		8.7		13.0	65.5	19.8	1.7	
Skins								
Merlot	5.16	28.18		4.96	67.20		22.69	
Carménère	12.5	3.8						Obreque-Slier et al. (2010)
Cabernet Sauvignon	19.0	7.1						
Syrah	4	3.0-19.8						Vidal et al. (2003)
Tempranillo	2.9	72.3	13.3					Monagas et al. (2003)
Graciano	6.5	33.8	10.7					
Cabernet Sauvignon	3.8	85.7	31.2					
Touriga Nacional	3.6	28.4	28.5					Cosme et al. (2009)
Cabernet Sauvignon	4.3	41.7	41.8					
Castelão	4.3	19.8	30.5					
Seeds								
Merlot	32.45	10.35		13.19	54.36			Souquet et al. (2000)
Carménère	27.5	3.2						Obreque-Slier et al. (2010)
Cabernet Sauvignon	16.3	2.7						
Alicante Bouchet	13.2-30.2	2.3-15.1						Prieur et al. (1994)
Tinta Miúda	23.0-26.2	9.8-31.5						Sun et al. (1998)
Cabernet Franc	20.0	4.7-15.7						Labarbe et al. (1999)
Syrah	16.2-22.5	2.8-8.9						Vidal et al. (2003)
Gamay		1.8-19.3						Perret et al. (2003)
Tempranillo		14.3						Monagas et al. (2003)
Graciano		10.9						
Cabernet Sauvignon		12.9						
Touriga Nacional	16.4	6.2						Cosme et al. (2009)

Grape variety/Grape fraction	% Gal	mDP	% Prodelph	% Cat	% Ec	% EcG	% EGC	References
Trincadeira	18.5	6.1						
Cabernet Sauvignon	19.7	5.1						
Castelão	18.6	7.6						
Syrah	18.7	6.5						
Stems								
Merlot	15.55	9.17		14.35	67.66		2.44	Souquet et al. (2000)
White Grapes								
Chardonnay		9.1		14.0	69.4	15.7	0.8	
Clairette		7.7		17.3	68.4	13.4	0.9	

% Gal – percentage of galloylation; mDP – mean degree of polymerization; % Prodeph – percentage of prodelphinidin; Cat – (+)-catechin; Ec – (-)-epicatechin; EcG – (-)-epicatechin gallate; EGC – (-)-epigallocatechin.

Table 8. Flavonols quantified in different white and red grape varieties

Grape variety/Grape fraction	Rutin	Myricetin glucuronide	Myricetin glucoside	Quercetin glucuronide	Quercetin glucoside	Kaempferol glucoside	References
Red grapes							
Primitivo	6.2[a]				6.3[a]	33.9[a]	Nicoletti et al. (2008)
Negroamaro	1.39[a]				1.03[a]	32.0[a]	
Susumaniello	10.6[a]				7.6[a]	22.0[a]	
Uva diTroia	41.4[a]				15.0[a]	45.8[a]	
Malvasia Nera					9.0[a]	31.2[a]	
Aglianico	6.1[a]				6.7[a]	20.9[a]	
Cesanese	54.7[a]				2.87[a]	40.7[a]	
Alphonse Lavallé					4.1[a]	4.3[a]	
Merlot	30.2[a]				45.0[a]	97.7[a]	

Table 8. (Continued)

Grape variety/Grape fraction	Rutin	Myricetin glucuronide	Myricetin glucoside	Quercetin glucuronide	Quercetin glucoside	Kaempferol glucoside	References
Tempranillo		10.70[e]	25.92[e]	22.99[e]	22.00[e]	5.03[e]	
Garnacha		2.17[e]	15.98[e]	25.93[e]	31.18[e]	6.34[e]	Castillo-Muñoz et al. (2007)
Garnacha Tintorera		1.43[e]	29.63[e]	12.98[e]	24.58[e]	3.83[e]	
Syrah		3.49[e]	16.20[e]	17.41[e]	30.07[e]	2.71[e]	
Petit Verdot		10.82[e]	20.85[e]	23.69[e]	20.95[e]	1.80[e]	
Cabernet Sauvignon		6.02[e]	20.88[e]	23.34[e]	26.71[e]	3.17	
Merlot		3.12[e]	18.98[e]	29.94[e]	32.93[e]	2.84[e]	
Skins							
Cencibel		10.0[c]	26.0[c]	29.0[c]	32.0[c]	14.0[c]	Montealegre et al. (2006)
Cabernet Sauvignon		10.1[c]	22.0[c]	59.1[c]	48.1[c]	13.0[c]	
Merlot		5.8[c]	13.0[c]	43.0[c]	31.0[c]	8.0[c]	
Shiraz		7.3[c]	21.0[c]	35.0[c]	55.0[c]	13.0[c]	
Cabernet Sauvignon			3.1[d]	2.2[d]	0.9[d]	0.10[d]	Obreque-Slier et al. (2010)
Carménère			2.4[d]	2.4[d]	6.5[d]	0.50[d]	
Stems							
Merlot				200[b]	18[b]		Souquet et al. (2000)
White grapes							
Moscato	25.2[a]				136.5[a]	101.0[a]	Nicoletti et al. (2008)
Skins							
Chardonnay				25.0[c]	17.0[c]	8.4[c]	Montealegre et al. (2006)
Sauvignon Blanc				12.0[c]	8.9[c]	2.0[c]	
Moscatel				54.0[c]	20.0[c]	17.0[c]	
Gewürztraminer				17.0[c]	24.0[c]	6.7[c]	
Riesling				30.0[c]	22.0[c]	2.9[c]	
Viogner				67.0[c]	66.0[c]	26.0[c]	

[a] mg/kg of berry dry weight; [b] mg/kg of stem; [c] mg/kg of fresh grape skins; [d] mg/kg of skin; [e] molar percentage.

Guendez et al., (2005) analyzed the correlation of low molecular weight polyphenolic constituents of grape seed extracts and *in vitro* antiradical activity employing the stable radical DPPH and obtained a significant correlation with total polyphenol content ($r^2 = 0.649$, $p < 0.01$), but the most effective was shown to be procyanidins from grape seed extracts ($r^2 = 0.793$, $p < 0.002$). In addition, these authors reported that procyanidin B1 may be one of the most important radical scavengers in grape seed extracts, despite its low contribution to overall polyphenol content. These results are in same line with the results published two years later by Villaño et al., (2007), where they analyzed the radical scavenging ability of several flavan-3-ol compounds, and concluded that procyanidin B1 and B2 are among the most potent scavenger compounds. Similar tendency was reported previously by Bors and Michel (1999) that considered that proanthocyanidins are superior radical scavenging agents as compared with the monomeric flavanols and flavones. They propose that these substances represent the antioxidative principle in red grapes, wines and in other beverages, such as green tea.

It is important noted that, several studies indicate that dimeric, trimeric, oligomeric, or polymeric procyanidins account for most of the higher antioxidant capacity of grape seeds. At the same time, it is reported by the same studies that antioxidant activities of grape seed procyanidins are positively related to their degree of polymerization (Yilmaz and Toledo, 2004a; Spranger et al., 2008). Jordão and Correia (2012) analyzed the antioxidant capacity evolution during two Portuguese grape varieties (Touriga Nacional and Tinta Roriz) maturation. The results indicated an important role of the different proanthocyanidin fractions of the skins and seeds in the grape antioxidant capacity during the maturation phase. These authors obtained a correlation between the levels of different proanthocyanidins and the antioxidant capacity from seeds and skins during the maturation of Touriga Nacional and Tinta Roriz grape varieties. For skins, the correlation coefficients ranged from 0.72 to 0.82, from 0.72 to 0.86 and from 0.52 to 0.82 for the monomeric, oligomeric and polymeric fraction of proanthocyanidins, respectively. For seeds, the correlation values ranged from 0.57 to 0.79, from 0.72 to 0.86 and from 0.47 to 0.79 for the monomeric, oligomeric and polymeric fraction of proanthocyanidins, respectively.

Table 9. Main class of phenolic compounds with antioxidant capacity quantified in bunch grapes

Class of phenolic antioxidant compounds	Individual compound or total compound groups/fractions	References
Flavanols	(+)-Catechin and (-)-Epicatechin	Ricardo-da-Silva et al. (1991); Rice-Evans et al. (1996); Revilla and Ryan (2000); Chicón et al. (2002); Kammerer et al. (2004); Yilmaz and Toledo (2004a; 2004b); Yildirim et al. (2005); Muselík et al. (2007); Villaño et al. (2007); Balík et al. (2008); Lacopini et al. (2008); Jordão and Correia (2012)
Hydroxybenzoic acids	Gallic acid	Revilla and Ryan (2000); Chicón et al. (2002); Yilmaz and Toledo (2004a); Yildirim et al. (2005)
	Ellagic acid	Yilmaz and Toledo (2004a; 2004b)
Hydroxycinnamic acids	*p*-Coumaric acid	Chicón et al. (2002); Kammerer et al. (2004)
	o-Coumaric acid	Lafka et al. (2007)

Table 9. (Continued)

Class of phenolic antioxidant compounds	Individual compound or total compound groups/fractions	References
	Caffeic acid	Chicón et al. (2002); Kammerer et al. (2004)
	Caftaric acid	Balík et al. (2008)
	Ferulic acid	Chicón et al. (2002); Kammerer et al. (2004)
Proanthocyanidins	Procyanidins B1 and B2	Jordão et al. (1998b); Jordão et al. (2001a; 2001b); Revilla and Ryan (2000); Guendez et al. (2005); Muselík et al. (2007); Villaño et al. (2007)
	Different proanthocyanidins fractions (oligomeric, polymeric)	Ricardo-da-Silva et al. (1991); Bors and Michel (1999); Jordão et al. (2001a; 2001b); Negro et al. (2003); Muselík et al. (2007); Spranger et al. (2008); Fang et al. (2011); Jordão and Correia (2012); Jordão et al. (2012a,b)
Flavonols	Kaempferol	Breksa III et al. (2010)
	Quercetin	Rice-Evans et al. (1996); Villaño et al. (2007); Lacopini et al. (2008); Darra et al. (2012)
	Kaempferol-3-*O* glucoside	Kammerer et al. (2004)
	Isorhamnetin-3-*O*-glucoside	
	Isoquercitrin	
	Rutin	Rice-Evans et al. (1996); Kammerer et al. (2004); Lacopini et al. (2008)
Anthocyanins	Total anthocyanins	Meyer et al. (1997); Orak (2007); Xu et al. (2010)
	Individual anthocyanins (cyanidin-3-*O*-glucoside, delphinidin-3-*O*-glucoside, peonidin-3-*O*-glucoside, petunidin-3-*O*-glucoside, malvidin-3-*O*-glucoside)	Jordão et al. (1998a); Revilla and Ryan (2000); Chicón et al. (2002); Lafka et al. (2007); Muselík et al. (2007); Jacob et al. (2008), Jordão and Correia (2012); Jordão et al. (2012a, b)
Resveratrols	*cis*-Piceid	Waterhouse and Lamuela-Raventos (1994); Romero-Pérez et al. (2001); Vitrac et al. (2005); Dani et al. (2007); Vincenzi et al. (2013)
	trans-Piceatannol	Bavaresco et al. (2002); Vincenzi et al. (2013)
	trans-ε-Viniferin	Romero-Perez et al. (2001); Vitrac et al. (2005); Rayne et al. (2008)
	α-Viniferin	González-Barrio et al. (2006)
	cis-Resveratrol	Soleas et al. (1995); Melzoch et al. (2001); Moreno et al. (2008); Vincenzi et al. (2013)
	trans-Resveratrol	Rice-Evans et al. (1996); Morales et al. (1997); Frémont (2000); Melzoch et al. (2001); Šulc et al. (2005); Gürbüz et al. (2007); Villaño et al. (2007); Rayne et al. (2008); Balík et al. (2008); Moreno et al. (2008); Vincenzi et al. (2013)
	trans-Piceid	Waterhouse and Lamuela-Raventos (1994); Melzoch et al. (2001); Vitrac et al. (2005); Dani et al. (2007); Balík et al. (2008); Vincenzi et al. (2013)

Moreover, other authors reported also that flavonoids such as proanthocyanidins presented higher antioxidant activities than other well-know antioxidants such as vitamin C and vitamin E, suggesting that grape seed procyanidins might be of interest to be used as alternative antioxidants (Bagchi et al., 1997). Recently, Belviranli et al., (2012) investigated the effects of grape seed extract supplementation on exercise performance and oxidative stress in acutely and chronically exercised rats and concluded that grape seed extracts supplementation prevents exercise-induced oxidative stress by preventing lipid peroxidation and increasing antioxidant enzyme activities. However, it is important to note that proanthocyanidins are either very poorly absorbed or not absorbed at all, and their action is thus restricted to the intestine even if they are very abundant in our diet (Manach et al., 2004).

Another important phenolic compound with high antioxidant activity is resveratrol. Resveratrol was first found in grapevines (*Vitis vinifera*) in 1976 (Langcake and Pryce, 1976) and then reported in wine in 1992 (Siemann and Creasy, 1992). The 'French Paradox' has suggested that resveratrol might be the major bioactive component in red wine (Frankel et al., 1993). Resveratrol is a dietary antioxidant polyphenol, found in grapes, red wine, and peanuts, that has been strongly indicated to have protective effects against cardiovascular disease and cancer, including all stages of carcinogenesis (Bhat and Pezzuto, 2002; Pervaiz, 2003). Several researches have also shown that resveratrol is responsible for enhancing the activity of sirtuin enzymes, which may be responsible for the increased longevity of mice and other unrelated organisms (Baur et al., 2006; Kaeberlein and Rabinovithch, 2006). The chemical structure of resveratrol is basely formed by two phenol rings that are linked by a styrene double bond to generate 3, 4',5-trihydroxystilbene. Resveratrol exists as *cis* and *trans* isomeric forms, with *trans* to *cis* isomerization facilitated by UV exposure (Soleas et al., 1995). Resveratrol is mainly contained in the skins of grapes. The potential of grapevines to synthesize resveratrol and viniferins (oxidation product of resveratrol) has been prooved to be related to the resistance against a variety of plant diseases. According to Frémont (2000), contrary to total phenols the highest *trans*-resveratrol content was found in the grape skins and its levels were higher in comparison with the seeds apparently due to relation with the presence of *Botrytis*.

Recently, Darra et al., (2012) reported that in the pool of red grapes from different grape varieties, there are some secondary compounds that are important for their antioxidant activity, such as for example *trans*-resveratrol. In this study, it became clear that the grape extracts with highest polyphenolic content, including *trans*-resveratrol had the greatest scavenging activity. For example, the grape extracts obtained for Cabernet Franc grape variety represents the lowest scavenging capacity, due to the absence of quercetin, (+)-catechin and low levels of *trans*-resveratrol. Similar results were obtained by Prajitna et al., (2007) in wines made from Chambourcin grape variety, where total resveratrol had a positive correlation with total antioxidant capacity. In addition, it's important to consider that resveratrol is only one compound among several other polyphenols in grapes that play an antioxidant role, including (but not limited to) gallic acid, (+)-catechin, myricetin, quercetin, and caffeic acid. According to Frankel et al., (1995), resveratrol may play only a minor role as an antioxidant. For other authors, it is important to consider not only the effect of a given phenolic compound on the antioxidant capacity, but also the existence of an additive and synergistic effects of the phenolic compounds on the antioxidant capacity (Sun and Ho, 2005). Thus, resveratrol content did not correlate with total antioxidant activity of grapes and the explanation may be attributed to the fact that particular compounds may act additively or synergistically with other compounds and the total expressed antioxidant activity may be dependent on the relative proportions of each compound in the system. Lacopini et al., (2008) analyzed the individual phenolic composition [quercetin, (+)-catechin, (-)-epicatechin, rutin and resveratrol] and antioxidant activity of extracts obtained from skin and seeds of 10 native Tuscan and international *Vitis vinifera* varieties. All the five phenols investigated possessed strong antiradical activity. At the same time, there is a potential antiradical interactive effect among the five analyzed compounds and results indicated possible synergy between quercetin, rutin and resveratrol. These results were also confirmed by Yang et al., (2009) who analyzed the phytochemical profiles and antioxidant activities of 14 different grape varieties from Finger Lakes area of New York State (USA) and by Balík et al., (2008) who studied the relations between polyphenols content and antioxidant activity in vine grapes and leaves from several grape varieties originating from Mikulov region in Czech Republic. In other work, Heo et al., (2007) analyzed the antioxidant capacities of individual and combined phenolics in model systems and concluded that individual compounds showed their characteristic antioxidant capacities, while the mixtures, with two or three phenolics combined revealed that the summation of antioxidant capacities of individual phenolics led to total antioxidant capacity. Moreover, these authors concluded that there was no synergistic effect among the phenolics studied, but only an additive effect of antioxidant capacity was observed.

Prajitna et al., (2007) suggested that the total level of phenolic compounds present in grapes and consequently in wines is a better indicator than resveratrol in determining the antioxidant capacity and health benefits of grape and wines. Thus, the physiological effects attributable to the presence of *trans*-resveratrol in wine could be affected by the content of other resveratrol derivatives, like *cis*-resveratrol and resveratrol glucoside isomers, which have some functional properties similar to *trans*-resveratrol (Jayatilake et al., 1993). According to Soleas et al. (1995), *cis*-resveratrol does not occur naturally in grapes and is generally formed by exposure to UV light whereas both *cis*- and *trans*-isomers of 3-β-glucoside of resveratrol (piceid) have been detected in the grape skin by Waterhouse and Lamuela-Raventos (1994). The contribution of the piceids could be important, as the content of resveratrol glycosides in berry skins is two to four times higher than that of the aglycone forms for the *cis*-and *trans*-isomers, respectively (Romero-Pérez et al., 2001). Hydroxylated resveratrol (piceatannol) was first reported in grapes by Bavaresco et al. (2002), and its anticancer properties, through the inhibition of tyrosine kinase and the induction of apoptosis, are well known (Piver et al. 2004). Other stilbenes, such as astringin and viniferin, could also contribute significantly to the total stilbene dietary intake from grapes and wine (Vitrac et al., 2005).

Grapes are one of the major dietary sources of anthocyanins (Rechkemmer and Pool-Zobel, 1996), which are responsible for the coloring of black, red and purple grapes and are mainly localized in the berry skins (Jordão et al., 1998a; Chicón et al., 2002). Anthocyanin pigments are of noticeable importance in grapes and wines since their dual role: first, they constitute an essential part of the sensory attributes, as their concentration, different forms and derivatives affect directly the color of the final wine; second, they are considered to have diverse biological properties and therefore they are regarded as secondary metabolites with potential nutritional value.

The anthocyanins in grape skins are predominately the 3-*O*-glucosides of malvidin, cyanidin, delphinidin, peonidin, and petunidin. Malvidin, the reddest of all anthocyanins, is the major one in dark red *vinifera* grapes, with higher proportions of cyanidin in red grapes

(Jordão et al., 1998a; 1998b; Costa et al., 2014). However for some grape varieties, such as Concord grapes, cyanidin 3-monoglucoside and delphinidin 3-monoglucoside are the major anthocyanins (Singleton, 1982).

Anthocyanins possess antioxidant activity, which is considered to be an important physiological function. The antioxidant capacity of anthocyanins depends on the aglycone moiety and the glycoside form (Wang et al., 1999), so as of the oxidation state of the C ring (Lapidot et al., 1999), the hydroxylation and methylation pattern (Rice-Evans et al., 1996), as well as the acetylation by phenolic acids (Degenhardt et al., 2000). However, anthocyanins have showed other biological activities, independent of their antioxidant capacities, which also produce health benefits, such as the inhibition of cancer cell growth '*in vitro*' (Zhang et al., 2005), the induction of insulin production in isolated pancreatic cells (Jayaprakasam et al., 2005), the reduction of starch digestion through inhibition of α-glucosidase activity (Matsui et al., 2001; Hou, 2003), the suppression of inflammatory responses (Tall et al., 2004), the protection against age-related declines in cognitive behavior and the dysfunction in the central nervous system (Joseph et al., 1999).

According to several studies, flavan-3-ols, flavonols and anthocyanins are the most important compounds that contribute to red wine antioxidant proprieties (Simonetti et al., 1997; Wang et al., 1997; Ghiselli et al., 1998; Beecher 2003). The high antioxidant activity of the anthocyanins is attributed to their peculiar structure that allows the easy donation of H-atom from aromatic hydroxyl groups, that improves the capacity to support the impaired electron via delocalization around the ??-electron system (Van Acker et al., 1996). Thus, anthocyanins have been described as effective scavengers of reactive oxygen radicals (Tsuda et al., 1996), in inhibiting lipoprotein and platelet aggregation (Ghiselli et al., 1998) and in protecting against cardiovascular disease (Falchi et al., 2006). However, it's also important noted that, in last decade, several authors have reported the anthocyanin content, and their relation with antioxidant capacity of several red and white grapes from different varieties (Muñoz-Espada et al., 2004; Kallithraka et al., 2005; Muselík et al., 2007; Romero et al., 2008; Aguirre et al., 2010; Jordão et al., 2012a; 2012b; Jordão and Correia 2012). Muselík et al., (2007) analyzed the antioxidant activity of different individual anthocyanins extracted from grape skins and the activities of anthocyanin monoglycosides tested decreased in the following order: delphinidin-3-monoglucoside > cyanidin-3-rutinoside > malvidin-3-monoglucoside > petunidin-3-monoglucoside. For these authors, the completely conjugated structure of anthocyanins, which allows electron delocalization, results in very stable radical products, which is favorable for antioxidative capacity. In addition, the degree and the position of hydroxylation and methoxylation in the B ring affect stability and reactivity and, thereby, also antioxidant actions.

Total flavanols contributed to hydroxyl free radical scavenging efficacy and to a lesser extent to antiradical and reducing ability, whereas there was a less significant relationship between the antioxidant properties and the total phenolics and only a weak relationship to total anthocyanin content. Kallithraka et al. (2005) also revealed a low and statistically insignificant correlation between antiradical activity determined by DPPH assay and total anthocyanin content of skin extracts from several Greek grape cultivars at harvest. On the same line with these results, previously Meyer et al. (1997) showed that anthocyanins in grape extracts are rather moderately related to the inhibition of low-density lipoprotein oxidation, which has been attributed mainly to total phenol content. More recently, Jordão and Correia (2012) described a negative relationship between individual anthocyanins and antioxidant capacity during grape skin maturation. For these authors, the results were independent of the antioxidant capacity method used and the grape varieties studied, suggesting that anthocyanins were not the most powerful radical scavengers that occur in grape berry skins during the maturation process. Thus, grape skins contain other molecules, like proanthocyanidins, but perhaps also flavonols and phenolic acids that have a more powerful effect on the antioxidant capacity of grape skins during maturation. For example, Jordão et al. (2012b) described in two Portuguese grape varieties a high linear positive correlation between different proanthocyanidin fractions and the antioxidant activity evolution during the grape maturation.

According to several authors (Arnous et al., 2002; Orak, 2007), the antioxidant capacity of red grape cultivars does not always have a relationship with the presence of their anthocyanin content. For example, Orak, (2007) reported a stronger correlation between antioxidant activity and phenol content than antioxidant activity and anthocyanin content in several red grape varieties. Similar results were obtained by Chiou et al. (2013) when they analyzed the anthocyanins content and antioxidant capacity of Corinthian currants (*Vitis vinifera* L., var. Apyrena). Chiou et al. (2013) support the hypothesis that the contribution of this class of flavonoids may be less pronounced to the grape extracts antioxidant capacity; this behavior may originate from the increased content of malvidin derivatives - containing methylated phenolic hydroxyls in the flavonoid B-ring - and/or the presence of other phenolics with high antioxidant efficiency such as monomeric (+)-catechins and proanthocyanidins.

Thus, there has been conflicting evidence between different studies published in the relationship, especially between anthocyanins and antioxidant activity in grape and wines. It can be said that anthocyanin and phenolic compounds, either alone or in combination, are responsible for the antioxidant activity in grapes. It is possible that the antioxidant capacity of grapes is not a property of single phytochemical compounds but is widely distributed among the phenolic constituents. In addition, for example several authors have demonstrated that phenols and biogenic amines are responsible for the antioxidant activity of wines acting synergistically in a mechanism where the easily oxidized phenols are regenerated by less active phenols (Arnous et al., 2001; López-Vélez et al., 2003).

Finally, it is also necessary to consider that there are various methods to evaluate the antioxidant capacity of different foods, such as grape, which involve different mechanisms. Chemical methods are based on the scavenging of reactive nitrogen and oxygen species such as peroxynitrite (Heijnen et al., 2001), hydroxyl radical and superoxide (Yao et al., 2002). Other methods measure the disappearance of free radicals such as ABTS (2,2-azinobis-(3-ethyl-benzothiazoline-6-sulfonate) cation radical) (Re et al., 1999), DPPH (2,2-diphenyl-1-picrylhydrazyl) (Sroka and Cisowski 2003) or ORAC (oxygen radical absorbance capacity) (Fernandez-Pachon et al., 2004), through spectrophotometric measurement. In order to determine the total antioxidant power, other assays use techniques such as the ferric reducing/antioxidant power method (FRAP method) (Benzie and Strain 1996) or *in situ* electrochemically generated bromine (Abdullin et al., 2001). A photochemiluminescence method (PCL) for the quantification of antiradical properties and nonenzymic antioxidants of polycomponent systems was also developed by Popov and Lewin (1996). The antioxidant capacity values changes according to the method used. For example, according to Lachman et al. (2007), the lack of strong correlation between different methods, e.g., DPPH and ABTS, is probably attributable to the fact that every individual phenol compound causes a different response to the specific radical used in the assay.

Thus, it is evident that the expression of the results of antioxidant power of for example (+)-catechins, procyanidins and of course anthocyanins, as well as the effect of diglycosylation of anthocyanins, depend on the assay used and also on the mechanims of these methods used to evaluate the antioxidant activity. For example, according to Muselík et al. (2007) ABTS, FRAP and inhibition tyrosine nitration methods can be recommended to evaluate the antioxidant capacities of wine polyphenols (procyanidins, anthocyanins and pyranoanthocyanins) while lipid phase antioxidant activity (TBARS assay) is not recommended for anthocyanins and their adducts. Thus, it is important noted that a single method cannot provide a comprehensive prediction of antioxidant efficacy of the different compounds, such as phenolic compounds from grapes.

CONCLUSION

Several thousand molecules have been identified in higher plants with a polyphenol structure. These molecules are secondary metabolites of plants and are generally involved in defense against ultraviolet radiation and plante diseases.

In the last decades, grape phenolic composition has been extensively studied being an important topic of research. It is known that bunch grape phenolic compounds constitute a complex and diverse family of secondary metabolites. Thus, large families of structures have been identifed in the different bunch grape fractions (skin, pulp, seed and stem). The two major classes of phenolic compounds are flavonoids (flavonols, anthocyanins and flavan-3-ols or proanthocyanidins) and non-flavonoids (phenolic acids and stilbenes). The results shown in this chapter demonstrated that there is a significant variation in the concentrations of grape phenolic compounds in the different grape bunch fractions and also between the different grape varieties. In addition, it is important to note that the high phenolic composition variability among the different grape varities is predominantly attributed in a great extent to a varietal genotypic difference.

As mentioned above grapes constitute one of the principal sources of phenolic compounds and the grape heath benefits are related to the bioactivity of these compounds. The importance of these compounds is shown in the high number of publications that have already appeared over the past twenty five years on the analysis of grape phenolic compounds and antioxidant capacity. In a high number of studies, the health aspects of the consumption of plant polyphenols have been examined as well as their incidence in grapes. Polyphenols of grapes are part of complex mixtures of compounds that may react with radicals by different mechanisms and often interact synergistically or inhibitorily. New analytical techniques have unraveled some structures derived from polyphenols in grape and determined how they are formed. Thus, the diversity of methods and experimental procedures reflects the complexity of phenolic analytes in grape.

References

Abdullin, I.F., Turova, E.N., Budnikov, G.K., (2001). Coulometric determination of the antioxidant capacity of tea extracts using electrogenerated bromine. *J. Anal. Chem.* 56, 557-559.

Aggett, J., (2009). The process for the assessment of scientific support for claims on food. *Eur. J. Nutr.* 48 (Suppl. 1), S23-S26.

Aguirre, M.J., Chen, Y.Y., Isaacs, M., Matsuhiro, B., Mendoza, L., Torres, S., (2010). Electrochemical behaviour and antioxidant capacity of anthocyanins from Chilean red wine, grape and raspberry. *Food Chem.* 121, 44-48.

Albishi, T., John, J.A., Al-Khalifa, A.S., Shahidi, F., (2013). Antioxidative phenolic constituents of skins of onion varieties and their activities. *J. Funct. Foods* 5, 1191-1203.

Arnous, A., Makris, D.P., Kefalas, P., (2001). Effect of principal polyphenolic components in relation to antioxidant characteristics of aged red wines. *J. Agric. Food Agric.* 49, 5736-5742.

Arnous, A., Makris, D.P., Kefalas, P., (2002). Correlation of pigment and flavanol content with antioxidant properties in selected aged regional wines from Greece. *J. Food Comp. Anal.* 15, 655-665.

Artaud-Wild, S.M., Connor, S.L., Sexton, G., Connor, W.E., (1993). Differences in coronary mortality can be explained by differences in cholesterol and saturated fat intakes in 40 countries but not in France and Finland. *Circulation* 88, 2771-2779.

Baderschneider, B., Winterhalter, P., (2001). Isolation and characterization of novel benzoates, cinnamates, flavonoids, and lignans from Riesling wine and screening for antioxidant activity. *J. Agric. Food Chem.* 49, 2788-2798.

Bagchi, D., Garg, A., Krohn, R.L., Bagchi, M., Tran, M.X., Stohs, S.J., (1997). Oxygen free radical scavenging abilities of vitamins C and E, and a grape seed proanthocyanidin extract in vitro. *Res. Commun. Mol. Pathol. Pharmacol.* 95, 179-189.

Baldi, A., Romani, A., Mulinacci, N., (1995). HPLC/MS application to anthocyanidins of *Vitis vinifera*. *J. Agric. Food Chem.* 43, 2104-2109.

Balík, J., Kyseláková, M., Vrchotová, N., Tříska, J., Kumšta, M., Veverka, J., Híc, P., Totušek, J., Lefnerová, D., (2008). Relations between polyphenols content and antioxidant activity in vine grapes and leaves. *Czech J. Food Sci.* 26, Special Issue, S25-S32.

Ballistreri, G., Continella, A., Gentile, A., Amenta, M., Fabroni, S., Rapisarda, P., (2013). Fruit quality and bioactive compounds relevant to human health of sweet cherry (*Prunus avium* L.) cultivars grown in Italy. *Food Chem.* 140, 630-638.

Baur, J.A., Pearson, K.J., Price, N.L., Jamieson, H.A., Lerin, C., Kalra, A., Prabhu, V.V., Allard, J.S., Lopez-Lluch, G., Lewis, K., Pistell, P.J., Poosala, S., Becker, K.G., Boss, O., Gwinn, D., Wang, M., Ramaswamy, S., Fishbein, K.W., Spencer, R.G., Lakatta, E.G., Le Couteur, D., Shaw, R.J., Navas, P., Puigserver, P., Ingram, D.K., De Cabo, R., Sinclair, D.A. (2006). Resveratrol improves health and survival of mice on a high-calorie diet. *Nature* 444, 337-342.

Bavaresco, L., Fregoni, M., Trevisan, M., Mattivi, F., Vrhovsek, U., Falchetti, R., (2002). The occurrence of the stilbene piceatannol in grapes. *Vitis* 41, 133-136.

Beecher, G.R., (2003). Overview of dietary flavonoids: Nomenclature, occurrence and intake. *J. Nutr.* 133, 3248S-3254S.

Benzie, I.F.F., Strain, J.J., (1996). The ferric reducing ability of plasma (FRAP) as a measure of 'antioxidant power': the FRAP assay. *Anal. Biochem.* 239, 70-76.

Belviranli, M., Gökbel, H., Okudan, N., Başaral, K., (2012). Effects of grape seed extract supplementation on exercise-induced oxidative stress in rats. *Brit. J. Nutr.*108, 249-256.

Bergman, M., Varshavsky, L., Gottlieb, H.E., Grossman, S., (2001). The antioxidant activity of aqueous spinach extract: chemical identification of active fractions. *Phytochemistry* 58, 143-152.

Bergqvist, J., Dokoozlian, N., Ebisuda, N., (2001). Sunlight exposure and temperature effects on berry growth and composition of Cabernet Sauvignon and Grenache in the Central San Joaquin Valley of California. *Am. J. Enol. Vitic.* 52, 1-7.

Bhat, K.P., Pezzuto, J.M., (2002). Cancer preventive activity of resveratrol. *Ann. NY Acad. Sci.* 957, 210-229.

Bors, W., Michel C., (1999). Antioxidant capacity of flavanols and gallate esters: pulse radiolysis studies. *Free Radic. Biol. Med.* 27, 1413-1426.

Bourzeix, M., Weiland, D., Heredia, N.A., (1986). Study of catechins and procyanidins of grape clusters, the wine and other by-product of the wine. *Bulletin de l'O.I.V* 669-670, 1173-254.

Brandt, K., Christensen, L.P., Hansen-Møller, J., Hansen, S.L., Haraldsdottir, J., Jespersen, L., Purup, S., Kharazmi, A., Barkholt, V., Frøkiær, H., Kobæk-Larsen, M., (2004). Health promoting compounds in vegetables and fruits: A systematic approach for identifying plant components with impact on human health. *Trends Food Sci. Technol.* 15, 384-393.

Bravo, L., (1998). Polyphenols: Chemistry, Dietary Sources, Metabolism, and Nutritional Significance. *Nutr. Res.* 56, 317-333.

Breksa III, A.P., Takeoka, G.R., Hidalgo, M.B., Vilches, A., Vasse, J., Ramming, D.W., (2010). Antioxidant activity and phenolic content of 16 raisin grape (*Vitis vinifera* L.) cultivars and selections. *Food Chem.* 121, 740-745.

Bruno, G., Sparapano, L., (2007). Effects of three esca-associated fungi on *Vitis vinifera* L: V. Changes in the chemical and biological profile of xylem sap from diseased cv. Sangiovese vines. *Physiol. Mol. Plant Pathol.* 71, 210-229.

Castillo-Muñoz, N., Fernández-González, F., Gómez-Alonso, S., García-Romero, E., Hermosín-Gutiérrez, I., (2009). Red-Color related phenolic composition of garnacha tintorera (*Vitis vinifera* L.) grapes and red wines. *J. Agric. Food Chem.*, 57, 7883-7891.

Castillo-Muñoz, N., Gómez-Alonso, S., García-Romero, E., Hermosín-Gutiérrez, I., (2007). Flavonol profiles of *Vitis vinifera* red grapes and their single-cultivar wines. *J. Agric. Food Chem.* 55, 992-1002.

Céspedes, C.L., Valdez-Morales, M., Avila, J.G., El-Hafidi, M., Alarcón, J., Paredes-López, O., (2010). Phytochemical profile and the antioxidant activity of Chilean wild black-berry fruits, *Aristotelia chilensis* (Mol) Stuntz (Elaeocarpaceae). *Food Chem.* 119, 886-895.

Cheynier, V., Dueñas-Paton, M., Salas, E., Maury, C., Souquet, J.M., Pascal, S.M., Fulcrand, H., (2006). Structure and properties of wine pigments and tannins. *Am. J. Enol. Vitic.* 57, 298-305.

Cheynier, V., Prieur, C., Guyot, S., Rigaud, J., Moutounet, M. (1997). The structures of tannins in grapes and wines and their interactions with proteins. In T. R. Watkins (Ed.),

Proceedings of ACS symposium series 661, wine: *Nutritional and therapeutic benefits* (pp. 81–93).

Cheynier, V., Rigaud, J., (1986). HPLC separation and characterization of flavonols in the skins of *Vitis Vinifera* var. Cinsault. *Am. J. Enol. Vitic.* 37, 248-252.

Chicón, R.M., Sanchez-Palomo, E., Cabezudo, M.D., (2002). The colour and polyphenol composition of red wine varieties in Castilla - La Mancha (Spain). *Afinidad.* 59, 435-443.

Chiou, A., Panagopoulou, E.A., Gatzali, F., De Marci, S., Karathanos, V.T., (2013). Anthocyanins content and antioxidant capacity of Corinthian currants (*Vitis Vinifera* L., var. *Apyrena*). *Food Chem.*146, 157-165.

Cordova, A.C., Jackson, L.M., Berke-Schlessel, D.W., Sumpio, B.E., (2005). The cardiovascular protective effect of red wine. *J. Am. Coll. Surgeons* 200, 428-439.

Cosme, F., Ricardo-Da-Silva, J.M., Laureano, O., (2009). Tannin profiles of *Vitis vinifera* L. cv. red grapes growing in Lisbon and from their monovarietal wines. *Food Chem.* 112, 192-204.

Costa, E., Cosme, F., Jordão, A.M., Mendes-Faia, A., (2014). Anthocyanin profile and antioxidante activity from 24 grape varieties cultivated in two Portuguese wine regions. *J. Int. Sci. Vigne Vin*, 48, 51-62.

Clifford, M.N., (2000). Anthocyanins-nature, occurrence and dietary burden. *J. Agric. Food Chem.* 80, 1063-1072.

Dani, C., Oliboni, L.S., Vanderlinde, R., Bonatto, D., Salvador, M., Henriques, J.A.P., (2007). Phenolic content and antioxidant activities of white and purple juices manufactured with organically- or conventionally-produced grapes. *Food Chem. Toxicol.* 45, 2574-2580.

Darra, N.E., Tannous, J., Mouncef, P.B., Palge, J., Yaghi, J., Vorobiev, E., Louka, N., Maroun, R.G., (2012). A comparative study on antiradical and antimicrobial properties of red grapes extracts obtained from different *Vitis vinifera* varieties. *Food Nutr. Sci.* 10, 1420-1432.

De Freitas, V.A.P., Glories, Y., (1999). Concentration and compositional changes of procyanidins in grape seeds and skin of white *Vitis vinifera* varieties. *J. Sci. Food Agric.* 79, 1601-1606.

De Freitas, V.A.P., Glories, Y., Monique, A., (2000). Developmental changes of procyanidins in grapes of red *Vitis vinifera* varieties and their composition in respective wines. *Am. J. Enol. Vitic.* 51, 397-403.

Degenhardt, A., Knapp, K., Winterhalter, P., (2000). Separation and purification of anthocyanins by high-speed countercurrent chromatography and screening for antioxidant activity. *J. Agric. Food Chem.* 48, 338-343.

Denev, P., Ciz, M., Ambrozova, G., Lojek, A., Yanakieva, I., Kratchanova, M., (2010). Solid-phase extraction of berries' anthocyanins and evaluation of their antioxidative properties. *Food Chem.* 123, 1055-1061.

Dopico-Garcia, M.S., Fique, A., Guerra, L., Afonso, J.M., Pereira, O., Valentao, P., Andrade, P.B., Seabra, R.M., (2008). Principal components of phenolics to characterize red Vinho Verde grapes: anthocyanins or non-coloured compounds? *Talanta* 75, 1190-1202.

Downey, M.O., Dokoozlian, N.K., Krstic, M.P., (2006). Cultural practice and environmental impacts on the flavonoid composition of grapes and wine: A review of recent research. *Am. J. Enol. Vitic.* 57, 257-268.

Escarpa, A., Gonzalez, M.C., (2001). An overview of analytical chemistry of phenolic compounds in foods. *Crit. Rev. Anal. Chem.* 31, 57-139.

Escribano-Bailón, M.T., Guerra, M.T., Rivas-Gonzalo, J.C., Santos-Buelga, C., (1995). Proanthocyanidins in skins from different grape varieties. *Z. Lebensm. Unters. For.* 200, 221-224.

Escribano-Bailón, M.T., Gutierrez-Fernandez, Y., Rivas-Gonzalo, J.C., Santos-Buelga, C., (1992). Characterization of procyanidins of *Vitis vinifera* variety Tinta del Pais grape seeds. *J. Agric. Food Chem.* 40, 1794-1799.

Falcão, L. D., (2007). Caracterização analítica e sensorial de vinhos Cabernet Sauvignon de diferentes altitudes de Santa Catarina. Tese (Doutorado em Ciências dos Alimentos) – Universidade Federal de Santa Catarina, Florianópolis, Brasil.

Falchi, M., Bertelli, A., Lo Scalzo, R., Morassut, M., Morelli, R., (2006). Comparison of cardioprotective abilities between the flesh and skin of grapes. *J. Agric. Food Chem.* 54, 6613-6622.

Fang, Y., Meng, J., Zhang, A., Liu, J., Xu, T., Yu, W., Chen, S., Li, H., Zhang, Z., Wang, H., (2011). Influence of shriveling on berry composition and antioxidant activity of Cabernet Sauvignon grapes from Shanxi vineyards. *J. Sci. Food Agric.* 91, 749-757.

Fernandez-Pachon, M.S., Villano, D., Garcia-Parrilla, M.C., Troncoso, A.M., (2004). Antioxidant activity of wines and relation with their polyphenolic composition. *Anal. Chim. Acta* 513, 113-118.

Finley, J.W., Kong, A.-N., Hintze, K.J., Jeffery E.H., Ji, L.L., Lei, X.G., (2011). Antioxidants in foods: State of the science important to the food industry. *J. Agric. Food Chem.* 59, 6837-6846.

Frankel, E.N., Kanner, J., German, J.B., Parks, E., Kinsella, J.E., (1993). Inhibition of oxidation of human low-density lipoprotein by phenolic substances in red wine. *Lancet* 341, 454-457.

Frankel, E.N., Waterhouse, A.L., Teissedre, P.L., (1995). Principal phenolic phytochemicals in selected California wines and their antioxidant activity in inhibiting oxidation of human low-density lipoproteins. *J. Agric. Food Chem.* 43, 890-894.

Frémont, L., (2000). Minireview. Biological effects of resveratrol. *Life Sci.*, 66, 663-673.

Fuleki, T., Ricardo-da-Silva, J.M., (1997). Catechin and procyanidin composition of seeds from grape cultivars grown in Ontario. *J. Agric. Food Chem.* 45, 1156-1160.

Ghiselli, A., Nardini, M., Baldi, A., Scaccini C., (1998). Antioxidant activity of different phenolic fractions separed from an Italian red wine. *J. Agric. Food Chem.* 46, 361-367.

González-Barrio, R., Beltrán, D., Cantos, E., Gil, M.I., Espín, J.C., Tomás-Barberán, F.A. (2006). Comparison of ozone and UV-C treatments on the postharvest stilbenoid monomer, dimer, and trimer induction in var. 'Superior' white table grapes. *J. Agric. Food Chem.* 54, 4222-4228.

González-Manzano, S., Rivas-Gonzalo, J.C., Santos-Buelga, C., (2004). Extraction of flavan-3-ols from grape seed and skin into wine using simulated maceration. *Anal. Chim. Acta* 513, 283-289.

González-Neves, G.,·Franco, J., Barreiro, L., Gil, G., Moutounet, M., Carbonneau, A., (2007). Varietal differentiation of Tannat, Cabernet-Sauvignon and Merlot grapes and wines according to their anthocyanic composition. *Eur. Food Res. Technol.* 225:111–117

Gryglewski, R.J., Korbut, R., Robak, J., Swies, J., (1987). On the mechanism of antithrombotic action of flavonoids. *Biochem. Pharmacol.* 36, 317-322.

Guendez, R., Kallithraka, S., Makris, D.P., Kefalas, P., (2005). Determination of low molecular weight polyphenolic constituents in grape (*Vitis vinifera* sp.) seed extracts: Correlation with antiradical activity. *Food Chem.* 89, 1-9.

Gürbüz, O., Göcmen, D., Dağdelen, F., Gursoy, M., Aydin, S., Sahin, I., Buyukuysal, L., Usta, M., (2007). Determination of flavan-3-ols and trans-resveratrol in grapes and wine using HPLC with fluorescence detection. *Food Chem.* 100, 518-525.

Harman, D., (1956). Aging: a theory based on free radical and radiation chemistry. *J. Gerontol.* 11, 298-300.

Heijnen, C.G.M., Haenen, G.R.M.M., Van Acker, F.A.A., Van Der Vijgh, W.J.F., Bast, A., (2001). Flavanoids as peroxynitrite scavengers: the role of the hydroxyl groups. *Toxicol. In Vitro* 15, 3-6.

Heo, H.J., Kim, Y.J., Chung, D., Kim, D.-O., (2007). Antioxidant capacities of individual and combined phenolics in a model system. *Food Chem.* 104, 87-92.

Hernandez-Jimenez, A., Gomez-Plaza, E., Martinez-Cutillas, A., Kennedy, J.A., (2009). Grape skin and seed proanthocyanidins from Monastrell x Syrah grapes. *J. Agric. Food Chem.* 57, 10798-10803.

Hertog, M.G.L., Feskens, E.J.M., Kromhout, D., (1997). Antioxidant flavonols and coronary heart disease risk. *Lancet* 349:699.

Hou, D.X., (2003). Potential mechanisms of cancer chemoprevention by anthocyanins. *Curr. Mol. Med.* 3, 149-159.

Jackson, D.I., Lombard, P.B., (1993). Environmental and Management Practices Affecting Grape Composition and Wine Quality- A Review. *Am. J. Enol. Vitic.* 44, 409-430.

Jacob, J.K., Hakimuddin, F., Paliyath, G., Fisher, H., (2008). Antioxidant and antiproliferative activity of polyphenols in novel high-polyphenol grape lines. *Food Res. Int.* 41, 419-428.

Jayaprakasam, B., Vareed, S.K., Olson, L.K., Nair, M.G., (2005). Insulin secretion by bioactive anthocyanins and anthocyanidins present in fruits. *J. Agric. Food Chem.* 53, 25-31.

Jayatilake, G.S., Jayasuriya, H., Lee, E.S., Koonchanok, N., Geahlen, R.L., Ashendel, C.L., McLaughlin, J.L., Chang, C.J., (1993). Kinase inhibitors from *Polygonus cuspidatum*. *J. Nat. Prod.* 56, 1805-1810.

Jeandet, P., Bessis, R., Gautheron, B., (1991). The production of resveratrol (3,5,4'-trihydroxystilbene) by grape berries in different developmental stages. *Am. J. Enol. Vitic.* 42, 41-46.

Jordão, A.M., Ricardo-da-Silva, J.M., Laureano, O., (1998a). Evolution of anthocyanins during grape maturation of two varieties (*Vitis vinífera* L.), Castelão Francês and Touriga Francesa. *Vitis* 37, 93-94.

Jordão, A.M., Ricardo-da-Silva, J.M., Laureano, O., (1998b). Influência da rega na composição fenólica das uvas tintas da casta Touriga Francesa (*Vitis vinifera* L.). *Cienc. Tecnol. Aliment.* 2, 60-73.

Jordão, A.M., Ricardo-da-Silva, J.M., Laureano, O., (2001a). Evolution of catechins and oligomeric procyanidins during grape maturation of Castelao Francês and Touriga Francesa. *Am. J. Enol. Vitic.* 53, 231-234.

Jordão, A.M., Ricardo-da-Silva, J.M., Laureano, O., (2001b). Evolution of proanthocyanidins in bunch stems during berry development (*Vitis vinifera* L.). *Vitis* 40, 17-22.

Jordão, A.M., Correia, A.C., (2012). Relationship between antioxidant capacity, proanthocyanidin and anthocyanin content during grape maturation of Touriga Nacional and Tinta Roriz grape varieties. *S. Afr. J. Enol. Vitic.* 33, 214-224.

Jordão, A.M., Simões, S., Correia, A.C., Gonçalves, F.J., (2012a). Antioxidant activity evolution during Portuguese red wine vinification and their relation with the proanthocyanidin and anthocyanin composition. *J. Food Process. Preserv.* 36, 298-309.

Jordão, A.M., Correia, A.C., Gonçalves, F.J., (2012b). Evolution of antioxidant capacity in seeds and skins during grape maturation and their association with proanthocyanidin and anthocyanin content. *Vitis* 51, 137-139.

Joseph, J.A., Shukitt-Hale, B., Denisnova, N.A., Bielinski, D., Martin, A.P.C., (1999). Reversals of age-related declines in neuronal signal transduction, cognitive and motor behavioural deficits with blueberry, spinach, or strawberry dietary supplementation. *J. Neuro. Sci.* 19, 8114-8121.

Kaeberlein, M., Rabinovitch, P.S., (2006). Medicine: grapes versus gluttony. *Nature* 444, 280-281.

Kallithraka, S., Mohdaly, A.A., Makris, D.P., Kefalas, P., (2005). Determination of major anthocyanin pigments in Hellenic native grape varieties (*Vitis vinifera* sp.): association with antiradical activity. *J. Food Comp. Anal.* 18, 375-386.

Kammerer, D., Claus, A., Carle, R., Schieber, A., (2004). Polyphenol screening of pomace from red and white grape varieties (*Vitis vinifera* L.) by HPLCDAD- MS/MS. *J. Agric. Food Chem.* 52, 4360-4367.

Kedage, V.V., Tilak, J.C., Dixit, G.B., Devasagayam, T.P.A., Mhatre, M., (2007). A study of antioxidant properties of some varieties of grapes (*Vitis vinifera* L.). *Crit. Rev. Food Sci. Nutr.* 47, 175-85.

Kennedy, J.A., Hayasaka, Y., Vidal, S., Waters, E.J., Jones, G.P., (2001). Composition of grape skin proanthocyanidins at different stages of berry development. *J. Agric. Food Chem.* 49, 5348–5355.

Kennedy, J.A., Matthews, M.A., Waterhouse, A.L., (2000). Changes in grape seed polyphenols during fruit ripening. *Phytochemistry* 55, 77-85.

Kim, D-O., Jeong, S.W., Lee, C.Y., (2003). Antioxidant capacity of phenolic phytochemicals from various cultivars of plums. *Food Chem.* 81, 321-326.

Labarbe, B., Cheynier, V., Brossaud, F., Souquet, J.M., Moutounet, M. (1999). Quantitative fractionation of grape proanthocyanidins according to their degree of polymerization. *J. Agric. Food Chem.* 47, 2719–2723.

Lachman, J., Šulc, M., Schilla, M., (2007). Comparison of the total antioxidant status of Bohemian wines during the wine-making process. *Food Chem.* 103, 802-807.

Lacopini, P., Baldi, M., Storchi, P., Sebastiani, L., (2008). Catechin, epicatechin, quercetin, rutin and resveratrol in red grape: Content, *in vitro* antioxidant activity and interactions. *J. Food Comp. Anal.* 21, 589-598.

Lafka, T.I., Sinanoglou, V., Lazos, E.S., (2007). On the extraction and antioxidant activity of phenolic compounds from winery wastes. *Food Chem.* 104, 1206-1214.

Landrault, N., Poucheret, P., Ravel, P., Gasc, F., Cros, G., Teissedre, P., (2001). Antioxidant capacities and phenolics levels of french wines from different varieties and vintages. *J. Agric. Food Chem.* 49, 3341-3348.

Lang, I.A., Melzer, D., (2009). Moderate alcohol consumption in later life: time for a trial? *J. Am. Geriatr. Soc.*, 57, 1110-1112.

Langcake, P., Pryce, R.J., (1976). The production of resveratrol by *Vitis vinifera* and other members of the Vitaceae as a response to infection or injury. *Physiol. Plant Pathol.* 9, 77-86.

Lapidot, T., Hard, S., Akiri, B., Granit, R., Kanner, J., (1999). pH-Dependent forms of red wine anthocyanins as antioxidants. *J. Agric. Food Chem.* 47, 67-70.

López-Vélez, M., Martinez-Martinez, F., Del Valle-Ribes, C., (2003). The study of phenolic compounds as natural antioxidants in wine. *Crit. Rev. Food Sci. Nutr.* 43, 233-244.

Lorrain, B., Chira, K., Teissedre, P.-L., (2011). Phenolic composition of Merlot and Cabernet Sauvignon grapes from Bordeaux vineyard for the 2009 vintage: Comparison to 2006, 2007 and 2008 vintages. *Food Chem.* 126, 1991-1999.

Lu, Y., Foo, L.Y., (1999). The polyphenol constituents of grape pomace. *Food Chem.* 65, 1-8.

Madeira, F.C., (2011). Efeito da radiação ultravioleta b no comportamento vitícola da variedade (*Vitis Vinífera* L.) Cabernet Sauvignon em São Joaquim, Santa Catarina. Dissertação (Mestrado em Recursos Genéticos Vegetais), Universidade Federal de Santa Catarina, Florianópolis, Brasil.

Malinovski, L. I., (2009). Comportamento vitícola da videira (*Vitis vinifera* L.) variedade Cabernet Sauvignon nos municípios catarinenses de Campo Alegre, Campo Belo do Sul e Bom Retiro. Dissertação (Mestrado em Recursos Genéticos Vegetais), Universidade Federal de Santa Catarina, Florianópolis, Brasil.

Manach, C., Scalbert, A., Morand, C., Jimenez, L., (2004). Polyphenols: Food sources and bioavailability. *Am. J. Clin. Nutr.* 79, 727-747.

Mateus, N., Marques, S., Gonçalves, A.C., Machado, J.M., De Freitas, V., (2001). Proanthocyanidin composition of red *Vitis vinifera* varieties from the Douro Valley during ripening: Influence of cultivation altitude. *Am. J. Enol. Vitic.* 52, 115-121.

Matsui, T., Veda, T., Oki, T., Sugitak, M., Terahara, N., Matsumoto, K., (2001). α-Glucosidase inhibitory action of natural acylated anthocyanins. 1. Survey of natural pigments with potent inhibitory activity. *J. Agric. Food Chem.* 49, 1948-1951.

Mattivi, F., Guzzon, R., Vrhovsek, U., Stefanini, M., Velasco, R., (2006). Metabolite Profiling of Grape: Flavonols and Anthocyanins. *J. Agric. Food Chem.* 54, 7692-7702.

Melzoch, K., Hanzlikova, I., Filip, V., Buckiova, D., Šmidrkal, J., (2001). Resveratrol in parts of vine and wine originating from Bohemian and Moravian vineyard regions. *Agric. Conspec. Sci.* 66, 53-57.

Meyer, A.S., Yi, O.S., Pearson, D.A., Waterhouse, A.L., Frankel, E.N., (1997). Inhibition of human low-density lipoprotein oxidation in relation to composition of phenolic antioxidants in grapes (*Vitis vinifera*). *J. Agric. Food. Chem.* 45, 1638-1643.

Mikeš, O., Vrchotová, N., Tříska, J., Kyseláková, M., Šmidrkal, J., (2008). Distribution of major polyphenolic compounds in vine grapes of different cultivars growing in South Moravian vineyards. *Czech J. Food Sci.* 26, 182-189.

Monagas, M., Gómez-Cordovés, C., Bartolomé, B., Laureano, O., Ricardo-da-Silva, J. M. (2003). Monomeric, oligomeric, and polymeric flavan-3-ol composition of wines and grapes from *Vitis vinifera* L. Cv. Graciano, Tempranillo, and Cabernet Sauvignon. *J. Agric. Food Chem.* 51, 6475-6481.

Montealegre, R.R., Peces, R.R., Vozmediano, J.L. C., Gascueña, J.M., Romero, E.G., (2006). Phenolic compounds in skins and seeds of ten grape *Vitis vinifera* varieties grown in a warm climate. *J. Food Comp. Anal.* 19, 687-693.

Morales, M., Alcantara, J., Barcelo, A.R., (1997). Oxidation of *trans*-resveratrol by a hypodermal peroxidase isoenzyme from Gamay rouge grape (*Vttis vinifera*) berries. *Am. J. Enol. Vitic*. 48, 33-39.

Moreno, A., Astro, M., Falque, E., (2008). Evolution of *trans*- and *cis*-resveratrol content in red grapes (*Vitis vinifera* L. cv Mencia, Albarello and Merenzao) during ripening. *Eur. Food Res. Technol.* 227, 667-674.

Moutounet, M., Rigaud, J., Souquet, J.M., Cheynier, V., (1996). Caractérisation structurale des tanins de la baie de raisin. Quelques exemples de l'incidence du cépage, du terroir et du mode de conduite de la vigne. *Bulletin OIV*, 433–443.

Muñoz-Espada, A.C., Wood, K.V., Bordelon, B., Watkins, B.A., (2004). Anthocyanin quantification and radical scavenging capacity of Concord, Norton, and Marechal Foch grapes and wines. *J. Agric. Food Chem.* 52, 6779-6786.

Muselík, J., Garcia-Alonso, M., Martin-Lopez, M.P., Zĕmlička, M., Rivas-Gonzalo, J.C., (2007). Measurement of antioxidant activity of wine catechins, procyanidins, anthocyanins and pyranoanthocyanins. *Int. J. Mol. Sci.* 8, 797-809.

Negro, C., Tommasi, L., Miceli, A., (2003). Phenolic compounds and antioxidant activity from red grape marc extracts. *Biores. Technol.* 87, 41-44.

Nicoletti, I., Bello, C., De Rossi, A., Corradini, D., (2008). Identification and quantification of phenolic compounds in grapes by HPLC-PDA-ESI-MS on a semimicro separation scale. *J. Agric. Food Chem.* 56, 8801-8808.

Novaka, I., Janeiroa, P., Serugab, M., Oliveira-Brett, A.M., (2008). Ultrasound extracted flavonoids from four varieties of Portuguese red grape skins determined by reverse-phase high-performance liquid chromatography with electrochemical detection. *Anal. Chim. Acta* 630, 107-115.

Nuñez, V., Monagas, M., Gomez-Cordovés, M.C., Bartolomé, B., (2004). *Vitis vinifera* L. cv. Graciano grapes characterized by its anthocyanin profile. *Postharvest Biol. Tec.* 31, 69-79.

Obreque-Slier, E., López-Solís, R., Castro-Ulloa, L., Romero-Díaz, C., Peña-Neira, A., (2012). Phenolic composition and physicochemical parameters of Carménère, Cabernet Sauvignon, Merlot and Cabernet Franc grape seeds (*Vitis vinifera* L.) during ripening. *LWT-Food Sci. Technol.* 48, 134-141.

Obreque-Slier, E., Peña-Neira, A., Lopez-Solís, R., Zamora-Marín, F., Ricardo-da-Silva, J. M., Laureano, O., (2010). Comparative study of the phenolic composition of seeds and skins from Carménère and Cabernet Sauvignon grape varieties (*Vitis vinifera* L.) during ripening. *J. Agric. Food Chem.* 58, 3591-3599.

Ó-Marques, J., Reguinga, R., Laureano, O., Ricardo-da-Silva, J.M., (2005). Changes in grape seed, skin and pulp condensed tannins during berry ripening: Effect of fruit pruning. *Ciência Téc. Vitiv.* 20, 35-42.

Orak, H.H., (2007). Total antioxidant activities, phenolics, anthocyanins, polyphenoloxidase activities of selected red grape cultivars and their correlations. *Sci. Hortic.* 111, 235-241.

Ortega-Regules, A., Romero-Cascales, I., López-Roca, J.M., Ros-García, J.M., Gómez-Plaza, E., (2006). Anthocyanin fingerprint of grapes: environmental and genetic variations. *J. Sci. Food Agric.* 86, 1460–1467.

Pastrana-Bonilla, E., Akoh, C.C., Sellappan, S., Krewer, G., (2003). Phenolic content and antioxidant capacity of muscadine grapes. *J. Agric. Food Chem.* 51, 5497-5503.

Peña-Neira, A., Dueñas, M., Duarte, A., Hernandez, T., Estrella, I., Loyola, E., (2004). Effects of ripening stages and of plant vegetative vigor on the phenolic composition of grapes (*Vitis vinifera* L.) cv. Cabernet Sauvignon in the Maipo Valley (Chile). *Vitis* 43, 51-57.

Pereira, G.E., Gaudillere, J.P., Pieri, P., Hilbert, G., Maucourt, M., Deborde, C., Moing, A., Rolin, D., (2006). Microclimate influence on mineral and metabolic profiles of grape berries. *J. Agric. Food Chem.* 54, 6765-6775.

Pérez-Trujillo, J.P., Hernández, Z., López-Bellido, F.J., Hermosín-Gutiérrez, I., (2011). Characteristic phenolic composition of single-cultivar red wines of the Canary Islands (Spain). *J. Agric. Food Chem.* 59, 6150–6164.

Perret, C., Pezet, R., Tabacchi, R., (2003). Fractionation of grape tannins and analysis by matrix-assisted laser desorption/ionization time-of-flight mass spectrometry. *Phytochem. Anal.* 14, 202–208.

Pervaiz, S., (2003). Resveratrol: from grapevines to mammalian biology. *Faseb J.* 17, 1975-1985.

Piver, B., Fer, M., Vitrac, X., Mérillon, J.M., Dreano, Y., Berthou, F., Lucas, D., (2004). Involvement of cytochrome P450 1A2 in the biotransformation of *trans* resveratrol in human liver microsomes. *Biochem. Pharmacol.* 48, 268-2686.

Popov, I.N., Lewin, G., (1996). Photochemiluminescent detection of antiradical activity; IV: testing of lipidsoluble antioxidants. *J. Biochem. Bioph. Meth.* 31, 1-8.

Prajitna, A., Dami I.E., Steiner T.E., Ferree D.C., Scheerens J.C., Schwartz, S.J., (2007). Influence of cluster thinning on phenolic composition, resveratrol, and antioxidant capacity in Chambourcin wine. *Am. J. Enol. Vitic.* 58, 346-350.

Price, S.F., Breen, P.J., Valladao, M., Watson, B.T., (1995). Cluster sun exposure and quercetin in Pinot Noir grapes and wine. *Am. J. Enol. Vitic.* 46, 187-194.

Prieur, C., Rigaud, J., Cheynier, V., Moutounet, M., (1994). Oligomeric and polymeric procyanidins from grape seeds. *Phytochemistry* 36, 781-784.

Rayne, S., Karacabey, E, Mazza, G., (2008). Grape cane waste as a source of *trans*-resveratrol and *trans*-viniferin: High-value phytochemicals with medicinal and anti-phytopathogenic applications. *Ind. Crops Prod.* 27, 335-340.

Re, R., Pellegrini, N., Proteggente, A., Pannala, A., Yang, M., Rice-Evans, C., (1999). Antioxidant activity applying an improved ABTS radical cation decolorization assay. *Free Radic. Biol. Med.* 26, 1231-1237.

Rechkemmer, G., Pool-Zobel, B., (1996). Estimation of beneficial health effects of anthocyanin/anthocyanidins. *Vortragstagung* 31, 219-232.

Renaud, S., De Lorgeril, M., (1992). Wine, alcohol, platelets, and the French paradox for coronary heart disease. *Lancet* 339, 1523-1526.

Revilla, E., García-Beneytez, E., Cabello, F., Martín-Ortega, G., Ryan, J.M., (2001). Value of high-performance liquid chromatographic analysis of anthocyanins in the differentiation of red grape cultivars and red wines made from them. *J. Chromatogr. A* 915, 53-60.

Revilla, E., Ryan, J.M., (2000). Analysis of several phenolic compounds with potential antioxidant properties in grape extracts and wines by high performance liquid chromatography - photodiode array detection without sample preparation. *J. Chromatog. A* 881, 461-469.

Ribéreau-Gayon, P., Glories, Y., Maujean, A., Dubourdieu, D., (2006). Handbook of Enology: volume 2. The chemistry of wine stabilization and tratments. John Wiley & Sons, LTD, England.

Ricardo-da-Silva, J.M., Belchior, A.P., Spranger, M.I., Bourzeix, M., (1992). Oligomeric procyanidins of three grapevine varieties and wines from Portugal. *Sci. Alim.* 12, 223-37.

Ricardo-da-Silva, J.M., Darmon, N., Fernandez, Y., Mitjavila S., (1991). Oxygen free radical scavenger capacity in aqueous models of different procyanidins from grape seeds. *J. Agric. Food Chem.* 39, 1549-1552.

Ricardo-da-Silva, J.M., Rosec, J.P.H., Bourzeix, M., Heredia, N., (1990). Separation and quantitative determination of grape and wine procyanidins by high performance reversed fase liquid chromatography. *J. Sci. Food Agric.* 53, 85-92.

Rice-Evans, C., Miller, N.J., Bolwell, P.G., Bramley, P.M., Pridham, J.B., (1995). The relative antioxidant activities of plant-derived polyphenolic flavonoids. *Free Radic. Res.* 22, 375-383.

Rice-Evans, C.A., Miller, N.J., Paganga, G., (1996). Structure-antioxidant activity relationships of flavonoides and phenolic-acids. *Free Radic. Biol. Med.* 20, 933-956.

Richard, J.L., Cambien, F., Ducimetière, P., (1981). Epidemiologic characteristics of coronary disease in France. *Nouv. Presse Med.* 10, 1111-1114.

Rodriguez, M.R., Romero Peces, R., Chacon Vozmediano, J.L., Martinez Gascuena, J., Garcia Romero, E., (2006). Phenolic compounds in skins and seeds of ten grape *Vitis vinifera* varieties grown in a warm climate. *J. Food Comp. Anal.* 19, 687-693.

Romero, I., Sanchez-Ballesta, M.T., Escribano, M.I., Merodio, C., (2008). Individual anthocyanins and their contribution to total antioxidant capacity in response to low temperature and high CO_2 in stored Cardinal table grapes. *Post. Biol. Technol.,* 49, 1-9.

Romero-Pérez, A.I., Ibern-Gòmez, M., Lamuela-Raventos, R.M., De la Torre-Boronat, M.C., (1999). Piceid, the major resveratrol derivate in grape juice. *J. Agric. Food Chem.* 47, 1533-1536.

Romero-Pérez, A.I., Lamuela-Raventos, R.M., Andres-Lacueva, C., Torre-Boronat, M.C., (2001). Method for the quantitative extraction of resveratrol and piceid isomers in grape berry skins. Effect of powdery mildew on the stilbene content. *J. Agric. Food Chem.* 49, 210-215.

Romeyer, F.M., Sapis, J.C., Macheix, J.J., (1985). Hydroxycinnamic esters and browning potential in mature berries of some grape varieties. *J. Sci. Food Agric.* 36, 728-732.

Sakkiadi, A.V., Haroutounian, S.A., Stavrakakis, M.N., (2001). Direct HPLC assay of five biologically interesting phenolic antioxidants in varietal Greek red wines. *Lebensm-Wiss Technol.* 34, 410-413.

Serafini, M., Laranjinha, J.A.N., Almeida, L.M., Maiani, G., (2000). Inhibition of human LDL lipid peroxidation by phenol-rich beverages and their impact on plasma total antioxidant in humans. *J. Nutr. Biochem.* 11, 585-590.

Siemann, E.H., Creasy, L.L., (1992). Concentration of phytoalexin resveratrol in wine. *Am. J. Enol. Vitic.* 43, 49-52.

Simonetti, P., Pietta, P., Testolin, G., (1997). Polyphenolic content and total antioxidant potential of selected Italian wines. *J. Agr. Food Chem.* 45, 1152-1155.

Singleton, V.L., (1982). Grape and wine phenolics: background and prospects, in: Singleton, V.L. (ed) Proceedings of *Symposium: Grape and Wine Centennial*, U.C. Davis, Berkeley, CA: University of California Press, 215-227.

Singleton, V.L., Timberlake, C.F., Lea, A.G.H., (1978). The phenolic cinnamates of white grapes and wine. *J. Sci. Food Agric.* 29, 403-10.

Soleas, G.J., Goldberg, D.M., Diamandis, E.P., Karumanchiri, A., Yan, J., Ng, E., (1995). A derivatized gas chromatographic-mass spectrometric method for the analysis of both isomers of resveratrol in juice and wine. *Am. J. Enol. Vitic.* 46, 346-352.

Souquet, J.M., Cheynier, V., Brossaud, F., Moutounet, M., (1996). Polymeric proanthocyanidins from grape skins. *Phytochemistry* 43, 509-512.

Souquet, J.M., Labarbe, B., Le Guerneve, C., Cheynier, V., Moutounet, M., (2000). Phenolic composition of grape stems. *J. Agric. Food Chem.* 48, 1076-1080.

Spayd, S.E., Tarara, J.M., Mee, D.L., Ferguson, J.C., (2002). Separation of sunlight and temperature effects on the composition of *Vitis vinifera* cv. Merlot berries. *Am. J. Enol. Vitic.* 53, 171-182

Spranger, M.I., Sun, B.S., Leandro, M.C., Cavalho, E.C., Bechior, A.P., (1998). Changes in anthocyanins, catechins and proanthocyanidins during fermentation and early post-fermentation of red grapes. In: Proceeding of *XXIII Word Congress on Vine and Wine*, Lisbon, Portugal, Vol. II, 183-189.

Spranger, M.I., Sun, B.S., Mateus, A.M., De Freitas, V., Ricardo-da-Silva, J.M., (2008). Chemical characterization and antioxidant activities of oligomeric and polymeric procyanidin fractions from grape seeds. *Food Chem.* 108, 519-532.

Sroka, Z., Cisowski, W., (2003). Hydrogen peroxide scavenging, antioxidant and anti-radical activity of some phenolic acids. *Food Chem. Toxicol.* 41, 753-758.

Šulc, M., Lachman, J., Hejtmankova, A., Orsak, M., (2005). Relationship between antiradical activity, polyphenolic antioxidants and free *trans*-resveratrol in grapes (*Vitis vinifera* L). *Hortic. Sci. (Prague)* 32, 154-162.

Sun, B., Leandro, C., Ricardo-da-Silva, J. M., Spranger, I., (1998). Separation of grape and wine proanthocyanidins according to their degree of polymerization. *J. Agric. Food Chem.* 46, 1390–1396.

Sun, B., Pinto, T., Leandro, M., Ricardo-da-Silva, J.M., Spranger, M.I., (1999). Transfer of catechins and proanthocyanidins from solid parts of the grape cluster into wine. *Am. J. Enol. Vitic.* 50, 179-184.

Sun, B., Ribes, A.M., Conceição, M.L., Belchior, A.P., Spranger, M.I., (2006). Stilbenes: quantitative extraction from grape skins, contribution of grape solids to wine and variation during wine maturation. *Anal. Chim. Acta* 563, 382-390.

Sun, B.S., Ricardo-da-Silva, J.M., Spranger, I., (2001). Quantification of catechins and proanthocyanidins in several Portuguese grapevine varieties and red wines. *Ciência Téc. Vitiv.* 16, 23-34.

Sun, T., Ho, C.-T., (2005). Antioxidant activities of buckwheat extracts. *Food Chem.* 90, 743-749.

Surh, Y.J., (2003). Cancer chemoprevention with dietary phytochemicals. *Nat. Rev. Cancer.* 3, 768-80.

Tall, J.M., Seeram, N.P., Zhao, C.S., Nair, M.G., Meyer, R.A., (2004). Tart cherry anthocyanins suppress inflammation-induced pain behavior in rat. *Behav. Brain Res.* 153, 181-192.

Tapiero, H., Tew, K.D., Nguyen, B.A.G., Mathe, G., (2002). Polyphenols: do they play a role in prevention of human pathologies? *Biomed. Pharmacother* 56, 200-207.

Tsuda, T., Shiga, K., Ohshima, K., Kawakishi, Osawa, T., (1996). Inhibition of lipid peroxidation and the active oxygen radical scavenging effect of anthocyanin pigments isolated from Phaseolus vulgaris. *Biochem. Pharmacol.* 52, 1033-1039.

Ubalde, J.M., Sort, X., Zayas, A.A., Poch, R.M., (2010). Effects of soil and climatic conditions on grape ripening and wine quality of Cabernet Sauvignon. *J. Wine Res.* 21, 1-17.

Van Acker, S.A.B.E., Van de Berg, D.J., Tromp, M.N.J.L., Griffioen, D.H., Van Bennekom, W.P., (1996). Structural aspects of antioxidant activity of flavonoids. *Free Radic. Biol. Med.* 20, 331-342.

Vidal, S., Francis, L., Guyot, S., Marnet, N., Kwiatkowski, M., Gawel, R., Cheynier, V., Waters, E., (2003). The mouth-feel properties of grape and apple proanthocyanidins in a wine-like medium. *J. Sci. Food Agric.* 83, 564–573.

Villaño, D., Fernandez-Pachon, M.S., Moya, M.L., Troncoso, A.M., Garcia-Parrilla, M.C., (2007). Radical scavenging ability of polyphenolic compounds towards DPPH free radical. *Talanta* 71, 230-235.

Vincenzi, S., Tomasi, D., Gaiotti, F., Lovat, L., Giacosa, S., Torchio, F., Segade, S.R., Rolle, L., (2013). Comparative study of the resveratrol content of twenty-one Italian red grape varieties. *S. Afr. J. Enol. Vitic.* 34, 30-35.

Vitrac, X., Bornet, A., Vanderlinde, R., Valls, J., Richard, T., Delaunay, J.C., Merillon, J.M., Teissedre, P.L., (2005). Determination of stilbenes (δ-viniferin, trans-astringin, trans-piceid, cis and trans-resveratrol, ε-viniferin) in Brazilian wines. *J. Agric. Food Chem.* 53, 5664-5669.

Wang, H., Cao, G., Prior, R., (1997). Oxygen radical absorbing capacity of anthocyanins. *J. Agric. Food Chem.* 45, 304-309.

Wang, H., Nair, M.G., Strasburg, G.M., Chang, Y.C., Booren, A.M., Gray, I.J., DeWitt, D.L., (1999). Antioxidant and anti-inflammatory activities of anthocyanins and their aglycon, cyanidin, from tart cherries. *J. Nat. Prod.* 62, 294-296.

Wang, S.Y., Lin, H.-S., (2000). Antioxidant activity in fruits and leaves of blackberry, raspberry, and strawberry varies with cultivar and development stage. *J. Agric. Food Chem.* 48, 140-146.

Waterhouse, A.L., Lamuela-Raventos, R.M., (1994). The occurrence of piceid, a stilbene glucoside in grape berries. *Phytochemistry* 37, 571-573.

Willett, W.C., Stampfer, M.J., (2003). Rebuilding the Food Pyramid. *Sci. Am.,* 288, 64-69.

Xu, C., Zhang, Y., Cao, L., Lu, J., (2010). Phenolic compounds and antioxidant properties of different grape cultivars grown in China. *Food Chem.* 119, 1557-1565.

Yang, J., Martinson, T.E., Liu, R.H., (2009). Phytochemical profiles and antioxidant activities of wine grapes. *Food Chem.* 116, 332-339.

Yao, D., Vlessidis, A.G., Evmiridis, N.P., Zhoub, Y., Xub, S., Zhoud, H., (2002). Novel chemiluminescence method for detection of superoxide anions and its application to dry-cured meat. *Anal. Chim. Acta* 467, 145-153.

Yildirim, H.K., Akcay, Y.D., Guvenc, U., Altindisli, A., Sozmen, E.Y., (2005). Antioxidant activities of organic grape, pomace, juice, must, wine and their correlation with phenolic content. *Int. J. Food. Sci. Technol.* 40, 133-142.

Yilmaz, Y., Toledo, R.T., (2004a). Major flavonoids in grape seeds and skins: Antioxidant capacity of catechin, epicatechin, and gallic acid. *J. Agric. Food Chem.* 52, 255-260.

Yilmaz, Y., Toledo, R.T., (2004b). Health aspects of functional grape seed constituents. *Trends Food Sci. Technol.* 15, 422-433.

Zhang, Y., Vareed, S.K., Nair, M.G., (2005). Human tumor cell growth inhibition by non-toxic anthocyanidins, the pigments in fruits and vegetables. *Life Sci.* 76, 1465-1472.

In: Wine
Editor: Youssef El Rayess

ISBN: 978-1-63321-048-6

Chapter 2

Characterization of Red Wines Polyphenolics Employing Sensors and Biosensors

***Irina Mirela Apetrei*[1], *Constantin Apetrei*[2,*] *and Youssef El Rayess*[3,4,5]**

[1]Faculty of Medicine and Pharmacy, "Dunarea de Jos" University of Galati
[2]Faculty of Sciences and Environment, "Dunarea de Jos" University of Galati, Galati, Romania
[3]Faculty of Agricultural and Food Sciences, Holy Spirit University of Kaslik, Jounieh, Lebanon
[4]Université de Toulouse, INPT, UPS, Laboratoire de Génie Chimique, Toulouse, France
[5]Centre de Viticulture et d'Œnologie de Midi-Pyrénées, Castanet-Tolosan, France

Abstract

Phenolic compounds are a very complex group of natural compounds that contribute directly to the flavor, namely to the bitterness and astringency perceived in a variety of food and beverages such as unripe fruits, wines, teas and beers. The phenolic compounds are major components in wines, which are generally recognized as an essential component of wine quality. They provide color, astringency and bitterness to wine. The phenolic composition of the wines depends on the grape and winemaking practices. Polyphenols are implied in different reactions starting as soon as the grape is crushed, continuing throughout winemaking and aging process. These reactions influence the aroma release and the perception sensory properties, thus modifying the quality of the wines. The production of the wine is completely dependent on the long-lasting experience of wine makers and quality of input products. For these reasons, it is necessary to monitor the quality of wine. The adequate control involves continuous, simultaneous and selective monitoring of several key compounds that determine the

* E-mail address: apetreic@ugal.ro.

quality and the flavor of wine. At present, the entire analysis of the sample can be carried out either through well-established techniques of analytical chemistry, or using a large number of calibrated sensors or/and biosensors. Individual sensors and biosensors provide basic information about sample usually limited to a single analyte, hence these individual devices cannot be used for comprehensive characterization of the complex sample. The possible way how to monitor the quality of wines represents the multisensor systems.

Keywords: Polyphenol, red wine, sensor, biosensor, e-tongue

INTRODUCTION

The analysis of wine is an important analytical field, especially focused on novel methodologies for its characterization and elaboration control. One of the mainly important wine parameter is its polyphenol content. Polyphenolic compounds may contribute to the color, flavor (bitterness, astringency etc.), odor and antioxidant properties of wine. Additionally, to their health-protecting properties the polyphenolic compounds are of great importance to both consumers and producers. Numerous methods to quantify total phenols and polyphenols have been described in the scientific literature. The Folin-Ciocalteu method is usually employed in the wine industry. As alternative to this method, the use of polyphenol index (I_{280}) is now arising; in this case wine absorbance is measured directly at 280 nm and straightly correlated with phenolic content. However, these methods yield a total phenol content value, and therefore do not allow for the discrimination between individual constituents.

On the other hand, there are some spectrophotometric methods, developed for quantification of phenolics in plants, which bring some specificity; these assays are based on differential reactivity principles and are used to determine different structural groups simultaneous in phenolic compounds. On the other hand, there is also the use of chromatographic techniques such as HPLC or GC, which are able to perform individual determination of phenolic compounds; however these methods need complex and time-consuming sample pre-treatment procedures, and are not suitable for on-site, on-line or real-time analyses.

Sensors and biosensors are considered as alternatives to the traditional techniques given their low cost and their ease of use to carry out on field analyses. Therefore, sensors and biosensors could represent attractive alternatives for the detection of polyphenolic compounds. Furthermore, a promising approach to analyze wines consists in the use of e-tongues. The e-tongue is a recent development from the sensory field, which involves the use of an array of sensors and/or biosensors capable of giving a global and complete response of the analyzed samples, coupled with chemometric processing tools able to interpret the chemical signals and extract meaningful data from the complex data. The objective of this work is to presents the principal approaches carried out in the field of polyphenolic analyses from wine by means of sensory devices.

POLYPHENOLS IN RED WINES

Polyphenolic compounds are a complex group of substances that are synthesized during plant growth and in response to stress. The main groups of polyphenolics are flavonoids and phenolic acids. The sub-classes of flavonoids are flavanols, flavonols, flavones, flavanones and isoflavones. Phenolic acids include benzoic acid and cinnamic acid and their derivatives (Lorrain et al., 2013; Winkel-Shirley, 2002). These organic compounds, to a greater or lesser extent, have a significant antioxidant capacity (Frankel et al., 1993; O'Byrne, 2009). Among the compounds in wine have emphasized the function exerted by grapes flavonoids, because they are essential to the quality of wine that will be for aging in oak barrel (Singleton, 1995; Chasagne et al., 2003).

White wines are essentially composed of phenols found in flesh of the grapes such as gallic acid, hydroxycinnamic acid esters, catechins isomers and procyanidin. The content varies and normally found in few $mg \times L^{-1}$ units. On the other hand, red wines are mainly composed of several closely related chemical groups: flavonol-3 group, flavane-(3,4)-diol group, flavonol-3, anthocyanins and tannins, in addition to those phenolic compounds found in white wines (Peynaud, 1984; Ignat et al., 2011).

Research on this class of compounds received an added impulse with the discovery of the French paradox (low cardiovascular mortality associated to red wine consumption). This is not the only health benefits of polyphenols. Because of their antioxidative properties, polyphenols have a major influence on the vascular system and protect against coronary heart disease. Furthermore, this antioxidant capacity protects body cells and tissues against the damage cause by free radicals and reactive oxygen species. Polyphenols have also anti-inflammatory effects by inhibiting the cyclooxygenase and 5-lipoxygenase pathways reducing the release of arachidonic acid which is a starting point for a general inflammatory response. Also, it has been stated that polyphenols can inhibit carcinogenesis and angiogenesis but these effects still a point of discussion and need more research to be proven.

Structures of phenolic compounds include simple aromatic ring with low molecular weight to complex high molecular weight tannins. Two groups of phenolic compounds are classically distinguished: flavonoids based on common C6-C3-C6 skeleton and non flavonoids (Lorrain et al., 2013; Moreno-Arribas & Polo, 2009).

Flavonoids. Flavonoids can be divided into various classes on the basis of their molecular structure. The 4 main groups of flavonoids are: anthocyanins, flavones, flavonols, and flavan-3-ols.

Anthocyanins. The anthocyanidins (aglycons form, figure 1.I) are the basic structure of the anthocyanins. When the anthocyanidins are found in their glycoside form (bonded to a sugar moiety) they are known as anthocyanins (figure 1.II). They are mainly located in the vacuoles of grape skins and sometimes in the pulp like teinturier varieties. The chemical structures are presented in Figure 1. The anthocyanins identified in grape skins and wines from *Vitis vinifera* are the 3-O-monoglucosides and the 3-O-acylated monoglucosides of five most important anthocyanidins - cyanidin, delphinidin, petunidin, peonidin and malvidin (the differences are the number and position of the hydroxyl and methoxyl groups located in the B-ring of the molecule).

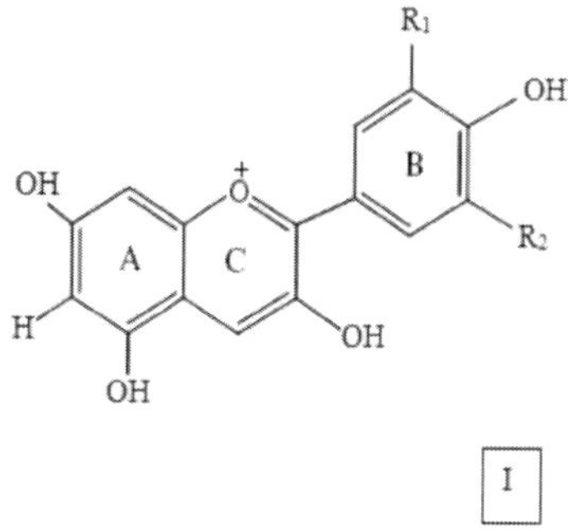

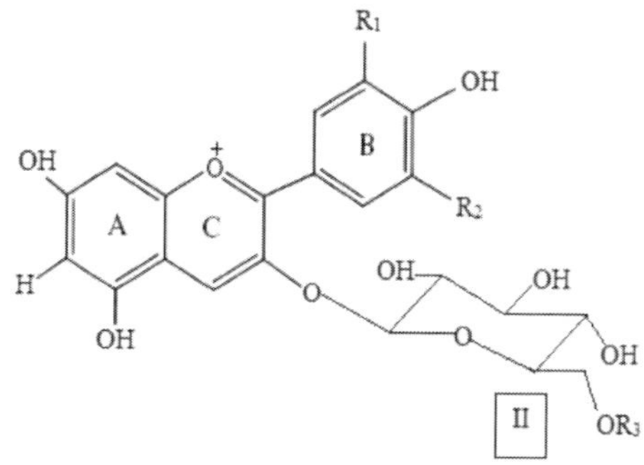

Anthocyanidin	R_1	R_2
Cyanidin	OH	H
Delphinidin	OH	OH
Peonidin	OCH_3	H
Petunidin	OCH_3	OH
Malvidin	OCH_3	OCH_3

Figure 1. I. Anthocyanidin chemical structure. II. Anthocyanin chemical structure.

Acylation (R_3, figure 1.II) occurs at the C-6 position of the glucose molecule by esterification with acetic, lactic, p-coumaric and caffeic acids (Mazza & Miniati, 1993; Kähkönen et al., 2003; Bakowska-Barczak, A., 2005).

The wine anthocyanin composition depends on the original grape profile but also on the extraction and winemaking techniques employed. After reaching a maximum level after a few days of fermentation, the concentration of anthocyanins decreases as a consequence of their adsorption on yeast cell walls, precipitation in the form of colloidal material together with tartaric salts and elimination during filtration and fining (Castillo-Sánchez et al., 2006; Moreno Arribas et al., 2008; Ortega-Regules et al., 2008). Hydrolysis reactions, as well as condensation reactions with other phenols during winemaking also modify the anthocyanin composition of wines. Despite all these changes, the wine anthocyanin profile has also been used as chemotaxonomy criteria to establish differences between grape varieties, vineyard localization and yield, vintages and winemaking techniques (Pomar et al., 2005; Zhao et al., 2010). Anthocyanins are the red pigments of grapes and responsible for the color of red wines whereas flavanols contribute to taste (especially astringency and bitterness) and are also involved in the development of oxidative browning and precipitates (Lesschaeve & Noble, 2005).

In acidic medium as wine, anthocyanins exist in four different structures depending on wine pH (figure 2, figure 3). At pH 1, the red flavylium cation (A^+) is the predominant species and contributes to purple and red colors. At pH range between 2 and 4, the blue quinoidal bases (AO) are the predominant. The colorless hemiketal pseudo-base (AOH) and the colorless chalcone (C) can be observed at pH ranging between 5 and 6.

Anthocyanins are frequently represented as their red flavylium cation, but in aqueous media this form suffers rapid proton transfer reactions, leading to blue quinonoidal bases. By the other hand, the hydration generates colorless hemiketals in equilibrium with chalcone structures (figure 3).

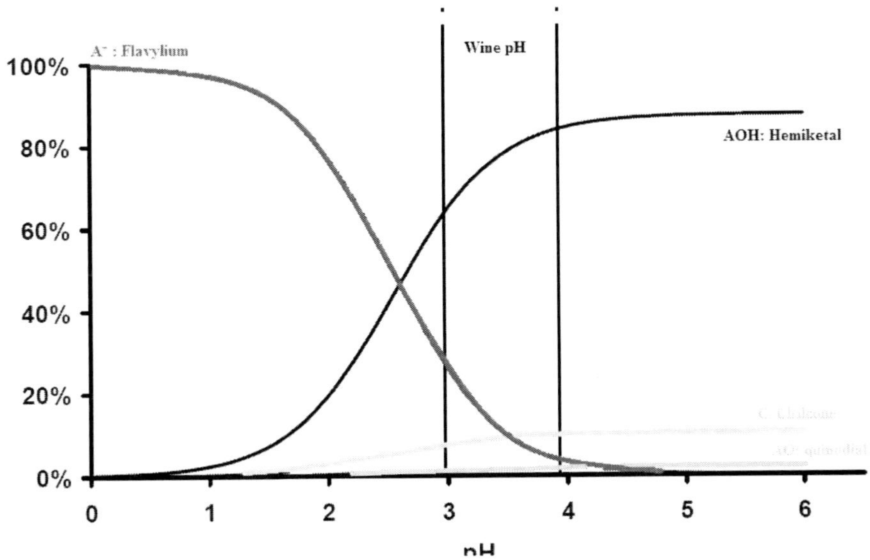

Figure 2. Anthocyanins structures depending on pH (adapted from Brouillard & Dubois, 1977).

The anthocyanins are structurally dependent on the conditions and composition of the media where they are dissolved. The anthocyanins can suffer interactions among them and with other compounds that influence their structural equilibria and modify their color.

The phloroglucinol A-ring of the anthocyanin hemiketal is nucleophilic whereas the C-ring of the flavylium form, having a cationic charge in C2 or C4, reacts as an electrophile. Classical examples of nucleophilic addition reactions onto the flavylium cation are those of water and bisulfite that have long been known to result in anthocyanin bleaching. The contribution of SO_2 during vinification may influence the color of anthocyanins. Indeed, the addition of sodium or potassium metabisulfite to the carbon 4 of the anthocyanin results in discoloration by a 4-sulfonate colorless formation. This reaction is reversible and the red flavylium form may be regenerated by acidification or addition of acetaldehyde which combines bisulfite.

Anthocyanins can undergo condensation reaction with o-quinones (species generated by enzymatic oxidation of caftaric acid and cutaric acid). The nucleophilic hemiketal form of anthocyanin can react with the electrophilic o-quinones species producing a colorless adduct.

Different forms of anthocyanins and their color can be stabilized by physicochemical phenomena such as self-association or intermolecular and intramolecular copigmentation resulting from the hydrophobic interaction of anthocyanins among themselves or with other molecules of the medium. The role of the copigmentation in wine is to bind the free anthocyanins into copigmented forms enabling more pigment to be retained and resulting in higher color intensity (Boutlon, 2001).

Anthocyanins can react with both hydroxycinnamic acids and 4-vinylphenols leading to the formation of pyranoanthocyanins (figure 4). The adduct of malvidin-3-glucoside with pyruvic acid is known as vitisin A. The result from the condensation between anthocyanins and acetaldehyde is known as vitisin B. Portisin is obtained from the reaction between malvidin-3-glucoside–pyruvic acid derivative and (+)-catechin in the presence of acetaldehyde (Mateus et al., 2004).

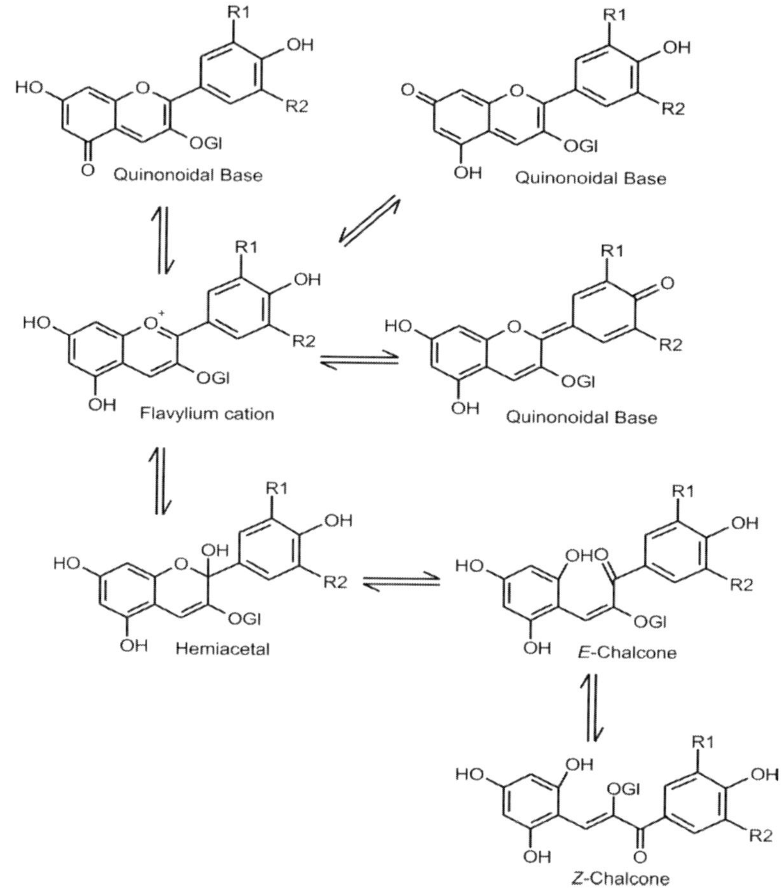

Figure 3. State equilibria of anthocyanins.

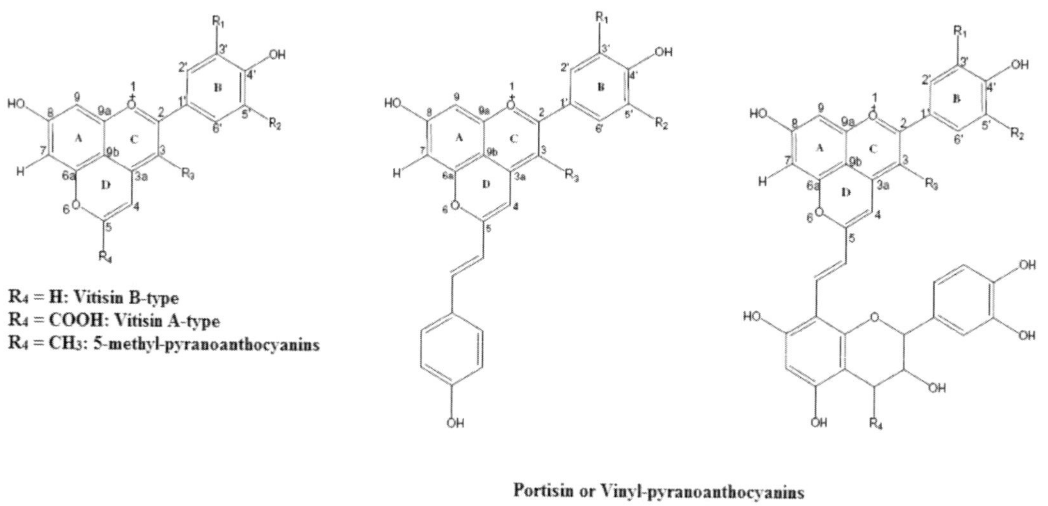

Figure 4. Chemical structures of wine pyranoanthocyanins.

In products that are rich in anthocyanins and possible reaction partners as wine, the formation of pyranoanthocyanins is likely to proceed with increasing storage time, although they may be absent in the initial products.

Pyranoanthocyanins differ from anthocyanins in many analytical aspects, especially the color. The presence of pyranoanthocyanins results in a more orange color. Therefore, they are suspected about their possible contribution to the color of aged red wines. Pyranoanthocyanins are more stable than anthocyanins at wine pH.

Flavonols and Flavones: These compounds are phenolic compounds sharing a common C6-C3-C6 skeleton that consist of two phenolic rings (A and B) linked together by a heterocyclic pyran ring (C), as shown in figure 5.

Flavonols (figure 5) constitute a group of flavonoids that vary in color from white to yellow and are closely related in structure to the flavones. Flavones are represented mainly by kaempferol, quercetin and myricetin. The major flavonols in wines are 3-glycosides and 3-glucuronide of quercetin and myricetin. It is known that the biosynthetic pathways involved in flavonoid production in grapes are greatly influenced by sunlight.

Flavonols, particularly when they occur in their deglycosylated form, are labile molecules and may be degraded upon exposure to heat, enzymes, and oxidative chemical species, such as free radicals (Makris et al., 2006). Therefore, grapes and wine processing and treatments might afford prominent alteration in the flavonol profile.

In wines, common winemaking practices, including maceration, fermentation, stabilization processes and ageing, are responsible for significant changes in flavonols. Factors that may profoundly differentiate flavonol composition are those associated with aging and storage conditions. It was shown that oxygen seems to play a central role, as supplementation with oxygen during storage decreased quercetin levels by more than 50% over a period of 6 months (Castellari et al., 2000).

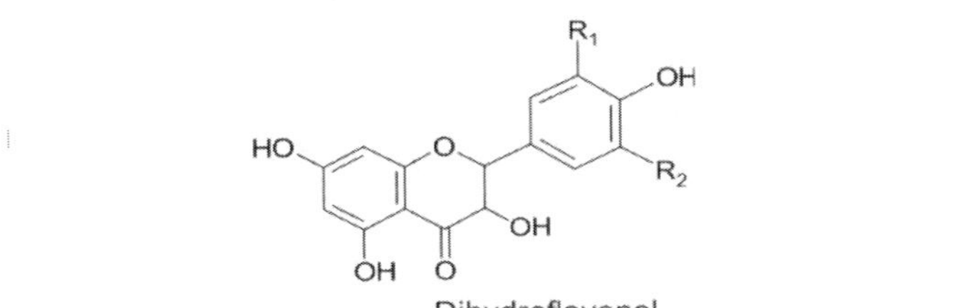

Figure 5. Chemical structures of flavonol and flavone. R_1 and R_2 could be H, OH or OCH_3.

Flavanols (flavan-3-ols) are found as monomers but also as oligomers and polymers. The major flavan-3-ol monomers (figure 6) in grape are (+)-catechin and its isomer, (-)-epicatechin, and, to a lesser amount (-)-epicatechin 3-gallate. Gallocatechin has also been reported in *Vitis vinifera* and catechin 3-gallate and gallocatechin 3-gallate have been detected in some non-*Vinifera* varieties (Piñeiro et al., 2004; Mattivi et al., 2009).

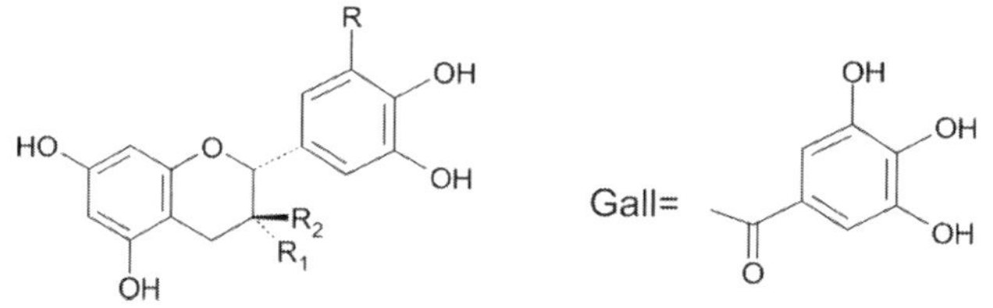

Flavanol	R	R_1	R_2
(+)-Catechin	H	OH	H
(-)-Epicatechin	H	H	OH
(-)-Epicatechin-3-gallate	H	H	OGall
(+)-Gallocatechin	OH	OH	H
(-)-Epigallocatechin	OH	H	OH
(+)-Gallocatechin-3-gallate	OH	H	OGall
(-)-Epigallocatechin-3-gallate	OH	OGall	H

Figure 6. Structures of flavanol monomers.

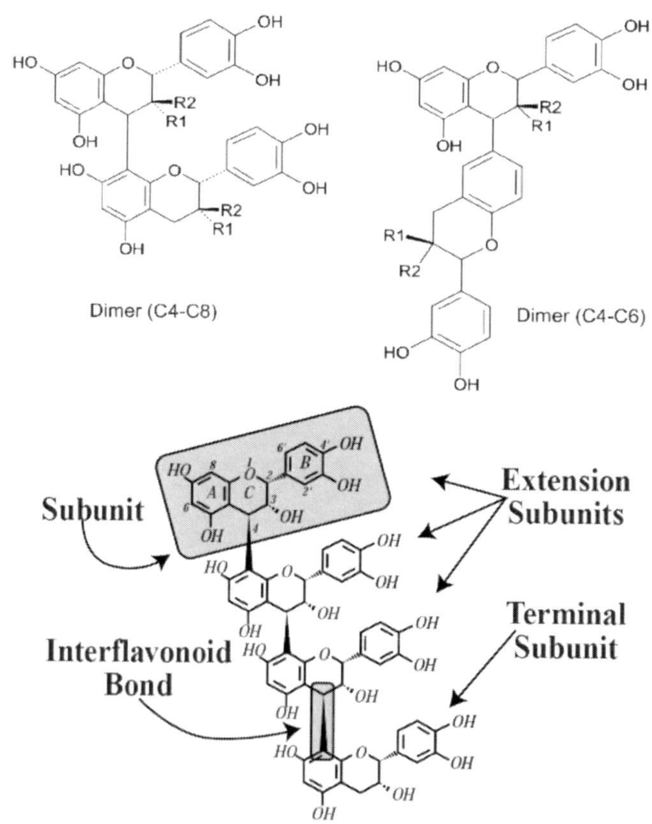

Figure 7. Chemical structures of flavanol dimers and polymers.

The polymers of flavan-3-ols are known as tannins or proanthocyanidins and are formed by flavan-3-ol monomers linked by C4-C6 or C6-C8 bonds as shown in figure 7. The number of flavan-3-ol monomer subunits that make up a proanthocyanidin molecule is referred to as the degree of polymerization. Another class of tannins is known as hydrolysable tannins (gallotannins and ellagitannins). The chemical structures of the hydrolysable tannins are basically composed of a central sugar core, typically a glucose unit, to which gallic acid and gallic acid derived moieties are esterified. These hydrolysable tannins are found in wines after aging in oak barrels.

In wine, proanthocyanidins and monomeric flavan-3-ols undergo condensation reactions to form large polymers, commonly referred to as condensed tannins. These can have high degrees of polymerization and eventually become insoluble in wine.

Anthocyanin units may also be included in the polymerization of tannins, forming polymeric pigments that contribute significantly to red wine color. Three types of condensation reactions between anthocyanin and tannin have been identified leading to the formation of anthocyanin-tannin adduct.

The first reaction is described by a direct condensation between anthocyanin – tannin leading to an A-T adduct. The mechanism proposed for this reaction is that the electrophilic C-4 position of anthocyanin cation can bond with the nucleophilic C-8 or C-6 position of the tannins.

The second reaction is characterized by a direct condensation between tannin – anthocyanin leading to a T-A adduct. After acid cleavage of the interflavanic bond, procyanidins form a carbocation (electrophilic) which can react with the C8 or C6 nucleophilic sites of anthocyanin molecules.

The third reaction is represented as an indirect reaction between tannin and anthocyanin. The mechanism proposed for this reaction starts with the protonation of the ethanol leading to a carbocation that initially undergo nucleophilic addition from the A ring of the flavanols (C6 or C-8 position). After dehydratation, the flavanol-aldehyde adduct attacks the anthocyanin (hemiketal form). Therefore, the new structure is represented by an association of flavanol and anthocyanin through an ethyl bridge. Apparently, the bond between two C8 positions takes place.

The self-association of flavanols and their aggregation have been demonstrated in the literature (Poncet-Legrand et al., 2003, Pianet el al., 2008). It was demonstrated that the hydrophobic interactions are the major driving forces to the flavanols self-association. Flavanols may react also with other wine macromolecules as proteins through both Van der Waals-London interaction and hydrogen bonding. Polysaccharides can also interact with tannins. The reaction of some polysaccharides like mannoproteins and arabinogalacturonan proteins with tannins prevent the aggregation of these latter and limits the precipitation of tannin-proteins complexes.

Non-flavonoids. This group of phenolics is constituted by phenolic acids, volatile phenols and stilbenes.

Phenolic Acids: The name "phenolic acids", in general, describes phenols that possess one carboxylic acid functionality. Phenolic acids are present in their free form or as glycosylated derivatives and esters of tartaric, quinic and shikimic acid in both red and white wines. They are divided into 2 groups: Hydroxybenzoic acids and hydroxycinnamic acids.

Hydroxybenzoic Acids

Hydroxybenzoic acids are derived from benzoic acid being characterized by a C6-C1 skeleton (figure 8).

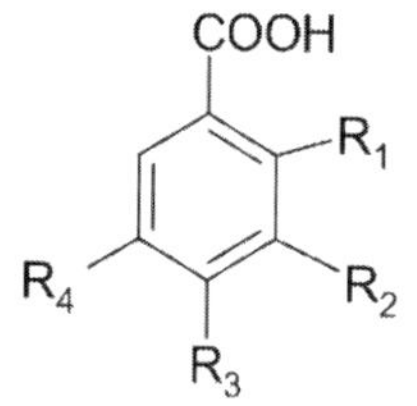

Hydroxybenzoic acids	R_1	R_2	R_3	R_4
Gallic acid	H	OH	OH	OH
Gentisic acid	OH	H	H	OH
p-Hydroxybenzoic acid	H	H	OH	H
Protocatechuic	H	OH	OH	H
Salicylic acid	OH	H	H	H
Syringic acid	H	OCH_3	OH	OCH_3
Vanillic acid	H	OCH_3	OH	H

Figure 8. Chemical structures of hydroxybenzoic acids.

The most common derivatives found in wine are gallic acid, gentisic acid, p-hydroxybenzoic acid, protocatechuic acid, syringic acid, salicylic acid, and vanillic acid. Gallic acid is the most abundant hydroxybenzoic acids in wine. It not only originates from the grape itself but is also formed by hydrolysis of hydrolyzable and condensed tannins (gallic acid esters of flavan-3-ols). The levels of hydroxybenzoic acids in wine show great variability depending on grape variety and growing conditions. The levels of hydroxybenzoic acids and their derivatives are commonly low in wine, compared to the levels of hydroxycinnamic acids (Ribéreau-Gayon et al., 2006; Kelebek et al., 2009).

Hydroxycinnamic Acids

Hydroxycinnamic acids have a C6-C3 skeleton. The different compounds present in wine are mainly derived from the hydroxycinnamic acids such as ferulic acid, p-coumaric acid, caffeic acid, and sinapic acid (figure 9). In wine, hydroxycinnamic acids are present in low amounts in their free form. The most abundant of the hydroxycinnamic acids found in grapes and wine is caftaric acid. Other important substances are the tartaric esters of p-coumaric acid and ferulic acid, and the trans-p-coumaric glucoside (Winter & Herrmann, 1986).

Hydroxycinnamic acids are potent antioxidants but have no sensory impact except when oxidized they may form brown pigments that eventually, precipitate. Hydroxycinnamic acids are more effective antioxidants than hydroxybenzoic acids due to increased possibilities for delocalization of the phenoxyl radical.

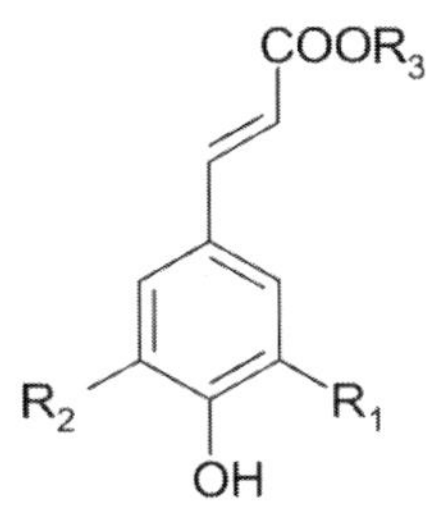

Hydroxycinnamic acids	R_1	R_2	R_3
Caffeic acid	OH	H	H
Caftaric acid	OH	H	Tartaric acid
p-Coumaric acid	H	H	H
p-Coutaric acid	H	H	Tartaric acid
Ferulic acid	OCH_3	H	H
Fertaric acid	OCH_3	H	Tartaric acid
Sinaptic acid	OCH_3	OCH_3	H

Figure 9. Chemical structures of hydroxycinnamic acids.

Volatile Phenols

The volatile phenols have the lowest concentrations among the phenolics in wine. The accumulation of volatile phenols in wine has been a cause of great concern in modern enology being now a key point in the control of wine quality.

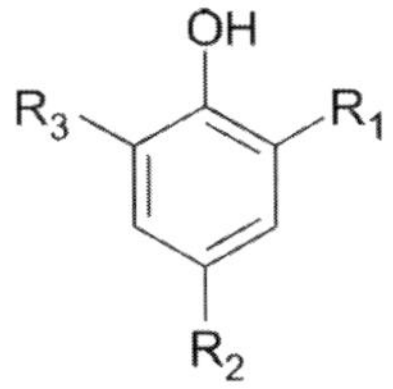

Volatile phenols	R_1	R_2	R_3
Ethyl phenol	H	C_2H_5	H
Vinyl phenol	H	C_2H_3	H
Guaiacol	OCH_3	OH	H
Methyl guaiacol	OCH_3	CH3	H
Ethyl guaiacol	OCH_3	C_2H_5	H
Vinyl guaiacol	OCH_3	C_2H_3	H
Propyl guaiacol	OCH_3	C_3H_7	H
Allyl guaiacol	OCH_3	C_3H_5	H
Syringol	OCH_3	H	OCH_3

Figure 10. Chemical structures of volatile phenols in red wine.

The precise origin(s) of volatile phenols has been under discussion for some time, but most likely they are made from hydroxycinnamic acids (p-coumaric and ferulic acids) or their corresponding tartaric esters under microbial activity. The transformation of p-coumaric and ferulic acids into their volatile phenols is predominantly associated with the activity of the *Brettanomyces* yeast genus. The metabolism of hydroxycinnamic acids by *Brettanomyces/Dekkera* yeasts is due to a sequential action of two enzymes: a cinnamate decarboxylase, which cleaves the phenolic acid directly into the corresponding vinylphenol and a vinylphenol reductase, which transforms the vinyl into the corresponding ethylphenol.

Another source of volatile phenols is the oak wood of barrel. These phenols are characterized by smoked and toasted aromas (Etievant, 1981; Larcher et al., 2007).

Among the volatile phenols, the vinyl and ethyl phenols play the most important role. Once above its limit of detection, 4-ethyl-phenol gives 'animal', 'leather' and 'horse sweat' notes to a wine.

Important volatile phenols in wine are presented in Figure 10.

Stilbenes

Although stilbenes are found in trace quantities in grapes and wine they have drawn attention because of their purported anti-carcinogenic properties. The most famous of the stilbenes is resveratrol (Buiarelli et al., 2007). The grapes and wines are considered one of the most important dietary sources of these compounds.

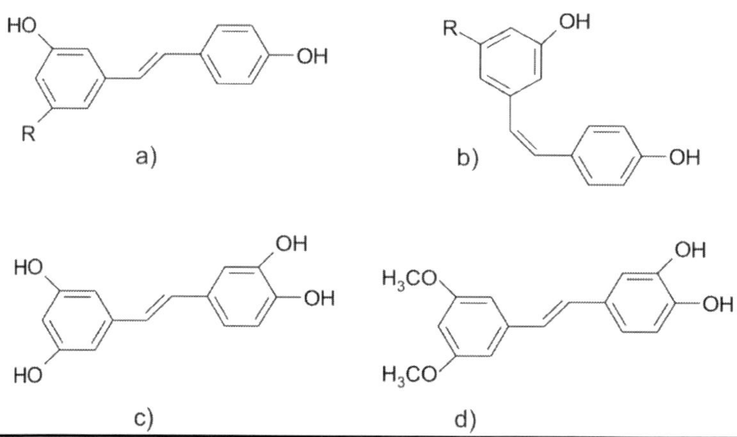

Stilbenes	R
trans-resveratrol (a)	OH
cis-resveratrol (b)	OH
trans-piceid (a)	OGlucose
cis-piceid (a)	OGlucose
trans-piceatannol	-
trans-pterostilbene	-

Figure 11. Chemical structures of stilbenes.

In grapes, resveratrol is synthesized almost entirely in the skin and the synthesis peaks just before the grapes reach maturity. The terminal enzyme in the biosynthesis of resveratrol is stilbene synthase, which is activated in response to exogenous stress factors, such as injury, ultra violet irradiation and chemical signals from pathogen fungi. The levels of resveratrol peak approximately 24 h after stress exposure, and decline after 42–72 h as result of activation of stilbene oxidase (Stervbo et al., 2007). Stilbenes are transferred during the winemaking process into the must and wine. One of the most important and widely studied stilbene is trans-resveratrol (3,5,4'-trihydroxystilbene), a phytoalexin produced by grapevines in response to fungal infection, particularly *Botrytis cinerea* (Vitaglione et al., 2005; Ratola et al., 2004). The chemical structures of resveratrol and further stilbenes are presented in Figure 11. There is a great interest in resveratrol research because there is growing evidence that resveratrol can prevent or delay the onset of cancer, heart disease, ischaemic and chemically induced injuries, diabetes, pathological inflammation and viral infection.

INFLUENCE OF PHENOLICS ON RED WINE ORGANOLEPTIC PROPERTIES

Among polyphenolics, two types of flavonoids, the anthocyanins and flavanols (i.e., catechins, proanthocyanidins, condensed tannins), are important to the quality of red wines, as they are key compounds for color and astringency (Li et al., 2008). Other flavonoids such as flavonols may have some influence on color and bitterness. Phenolic acids and hydrolysable tannins, released from barrel wood, have an influence on wine taste and color.

Wine phenolic composition depends on the original grape and on oenological technologies and storage conditions. Anthocyanins and tannins are located in the solid parts of the cluster from where they are released during winemaking. Additional, they are implied in various enzymatic and chemical reactions as the wine is made and aged. The new compounds formed often exhibit sensory properties, hence modifying the quality of the wines (Cheynier et al., 2006). Among the phenolic compounds, tannins are of particular importance in red wine. Tannins contribute to perceived astringency, an important attribute in wine influencing consumer acceptability. It is often described as the drying, roughing and puckering feeling attributed to the precipitation of a non-covalent complex between salivary-mucoproteins (proline-rich proteins) and tannins in the mouth (McRae & Kennedy, 2011). The sensory properties of tannins dependent upon their structure, molecular size and concentration and they may be bitter and/or astringent. Bitterness is restricted to small molecules with particular structural features enabling them to enter the receptor and activate the signal transduction process. Astringency depends on the number of protein interaction sites in the molecule (Cheynier et al., 2006). The flavanols having low molecular weight (oligomers) are both bitter and astringent. Tannins become gradually less bitter but more astringent as the molecular weight increases. The perception of astringency was also observed to significantly increase as tannin concentration increased, with a tendency to mask perceived bitterness (Cheynier, 2005).

The Color of Red Wines

The color of a red wine is determined by the proportions of the different anthocyanins forms: red flavylium cation, violet quinonoidal bases, yellow chalcones and the colorless hemiketal form. The initial purple-red color of young red wines are due to the anthocyanins extracted from grape skins. But these anthocyanins are unstable and can undergo chemical changes and cleavages. In order to conserve wine color, the anthocyanins pigments must be stabilized by self-association, intramolecular copigmentation and intermolecular reactions. The conversion of anthocyanins to the various pigments and adducts mentioned in the earlier sections increases the range of available colors over time. Pyranoanthocyanins are orange pigments and more stable over time than anthocyanins themselves. Therefore, their contribution to wine color is expected to increase with wine aging

The color of wine is affected by the numerous factors that influence wine phenolic composition such as the own reactive nature of phenolic compounds, the anthocyanin equilibria, and the external and internal conditions of the wines (temperature, oxygen access, pH, acetaldehyde, or SO_2 content) (Liao et al., 1992; Carew et al., 2013).

Astringency

While bitterness is a taste felt by specific receptors in the tongue, astringency is usually defined as the array of tactile sensations felt around the mouth transduced by the trigeminal nerve. Astringent primary descriptors include dryness (lack of lubrication or moisture resulting in friction between oral surfaces), roughness (harsh texture of the oral cavity marked by edges and projections that are felt when oral surfaces contact with each other) and constriction (feeling of puckering and contraction felt in the mouth lips and cheeks) (Ishikawa & Noble, 1995).

In contrast to other foods, astringency in red wine is one of the major attributes that the most exigent consumers of red wines learn to like. The levels of astringency in high quality red wines should be balanced: low levels of astringency could lead to the wine being flat and uninteresting, and an excess could mask other wine characteristics and make it harsh and dry (McRae et al., 2010).

Electroanalytical Techniques Employed in Analysis of Polyphenols in Wine

There are two traditional methods to detect and to quantify the polyphenolic content in foods and therefore more specifically in wines:

a) the determination of a total index by spectrophotometric detection, usually carried out using the Folin-Ciocalteu method and/or the absorbance measurement at 280 nm, which is commonly used in winery practices because of its simplicity (Sánchez-Rangel et al., 2013)
b) a separation of the individual polyphenolic species and their subsequent detection, the most commonly used techniques being high-performance liquid chromatography or capillary electrophoresis with UV, fluorimetric, mass spectrometry or electrochemical detection, with liquid chromatography-based methods extensively used because phenolic acids have high melting points and decompose when heated

above about 200°C (Lorrain et al., 2013; Sun et al., 2006; Aznar et al., 2011; Nicolaou et al., 2010).

Electroanalytical techniques, more specifically voltammetric techniques, are appropriate to the analysis of polyphenols. The antioxidant properties of these compounds are related to their ability to donate electrons. Therefore, the electroactivity and the electrochemical oxidation at low potentials have been exploited for their detection. These characteristics allow selective detection of polyphenols with good sensitivity, even in very complex samples, and responses are independent of the sample turbidity or optical path-length. These advantages have led to consideration of voltammetric techniques as attractive alternatives to the spectrophotometric methods for the polyphenol analysis in such a complex matrix (Sánchez Arribas et al., 2012; Dobes et al., 2013).

Cyclic voltammetry (CV) is a versatile electrochemical technique used for the characterization of redox systems in solution. CV can give information about the stability of reaction products, the number of oxidation states, the electron transfer kinetics, the number of electrons involved, and the reversibility of the reaction. In CV the potential of working electrode is scanned linearly versus time recording the current. The current recorded is proportional to the concentration of redox species present in the solution. Usually, the unmodified electrodes (carbon, inert metals) are suitable for the study of redox compounds such as polyphenols from wine (Nicholson, 1965; Compton & Banks, 2007; Bard & Faulkner, 2001).

Typical electrochemical sensing of polyphenols in wines has been carried out using their direct electrochemical oxidation at carbon-based or metallic electrodes (Kilmartin et al., 2002; Souza et al., 2011; Borato et al., 2006). On the other hand, their participation in enzymatic processes at electrochemical biosensors involving polyphenol oxidases (e.g., polyphenol oxidase, tyrosinase, and laccase) or peroxidases, and monitoring the electrochemical response of the enzymatically-generated products, have been exploited (Bănică, 2012).

Characterization of antioxidant properties and evaluation of total polyphenols in wine samples have been successfully carried out by cyclic voltammetry (CV), square wave voltammetry (SWV) and differential pulse voltammetry (DPV) (Kilmartin et al., 2002; Parra et al., 2006; Souza et al., 2011). These techniques have become alternatives to traditional spectrophotometric tests, and their extensive use is due to not only the relevance of the information provided but also the simplicity of the assay and the ability to process colored and/or turbid samples. These sweep techniques can be considered the electrochemical equivalent of a spectrophotometric scan, as the anodic voltammetric signals obtained in established potential intervals are related to the electrochemical oxidation of different polyphenol classes. In this way, the presence of voltammetric signals at low potentials correlates with the presence of polyphenolics of high antioxidant activity, whereas those compounds with low antioxidant power show electrochemical activity at more positive potentials. Further to the identification of peak potentials or wave potentials, the estimation of antioxidant capacity is performed based on measurements of peak current, or the area under voltammograms, at fixed potential ranges.

Kilmartin et al. (2001) proposed a method to estimate the antioxidant capacity based on CV at glassy carbon electrodes in an acid medium. Under these conditions, complex mixtures were analyzed, and it was found that the most effective antioxidants were those compounds whose electrochemical oxidation occurred at lower potentials. In other researches, total phenolic contents of different wines were calculated from the area under voltammetric recordings up to +0.50 V in model wine solutions, and were expressed as (+)-catechin equivalents (figure 12). These results corresponded to electrochemically easily-oxidizable polyphenols and correlated with the standard Folin-Ciocalteu colorimetric assay for red wines (Kilmartin et al., 2002; Makhotkina & Kilmartin, 2013).

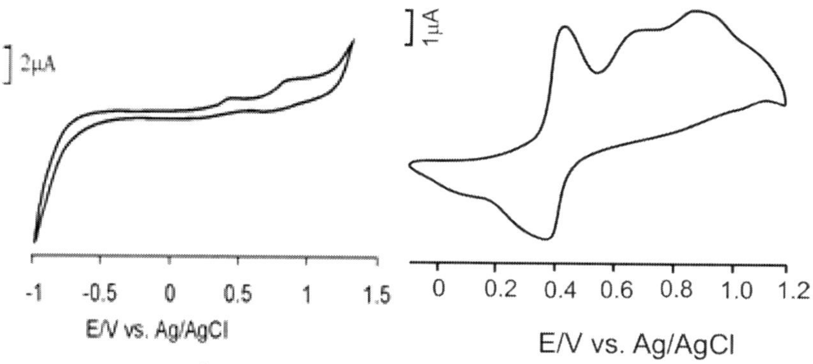

a) b)

Figure 12. a) Cyclic voltammogram of unmodified carbon paste electrode immersed in red wine; b) Cyclic voltammogram taken at a glassy carbon electrode for red wine (Kilmartin et al. (2001).

In general, good correlations of cyclic voltammetry results with the spectrophotometric assay (absorbance at 280 nm) for total polyphenols, and with HPLC data were found. A great number of researches report the "redox spectra of wines" (Ksenzhek et al., 2007) or the "spectral distribution" of the phenolic content (Dhroso et al., 2010). Relationships of these peak-current distributions at such potentials can characterize each wine analyzed. These methods were considered to be a promising tool for the evaluation of the authenticity of wines.

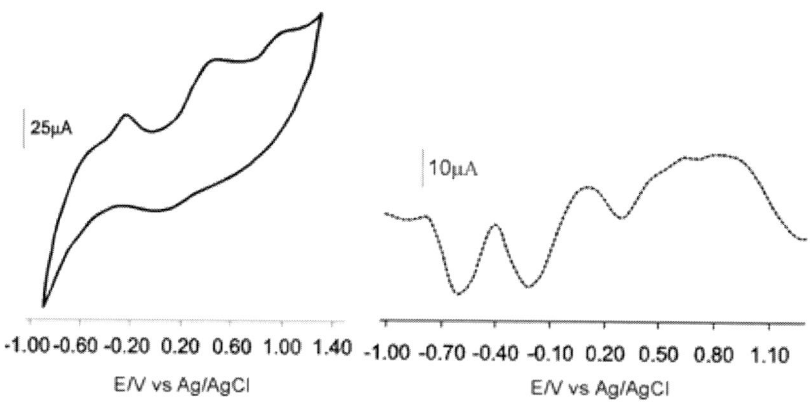

a) b)

Figure 13. a) Cyclic voltammogram of $GdPc_2$ - modified carbon paste electrode immersed in red wine. b) Square wave voltammogram of $GdPc_2$ - modified carbon paste electrode immersed in red wine (Apetrei et al., 2007).

Numerous voltammetric assays for antioxidant capacity and total polyphenolics have been presented where glass carbon electrodes, carbon-polyvinyl chloride composites, modified carbon paste electrodes (figure 13) and screen-printed electrodes were employed in connection with CV, DPV, and SWV (Jakubec et al., 2012; Rebelo et al., 2013; Rodriguez-Mendez et al., 2008).

The results of these studies showed good correlations between total polyphenol contents obtained by Folin-Ciocalteu, HPLC and voltammetric methods (figure 14), and significant variations were observed in the total polyphenol contents and the polyphenol compositions among wines made from grapes grown in different geographical and wine producing regions (Apetrei et al., 2007).

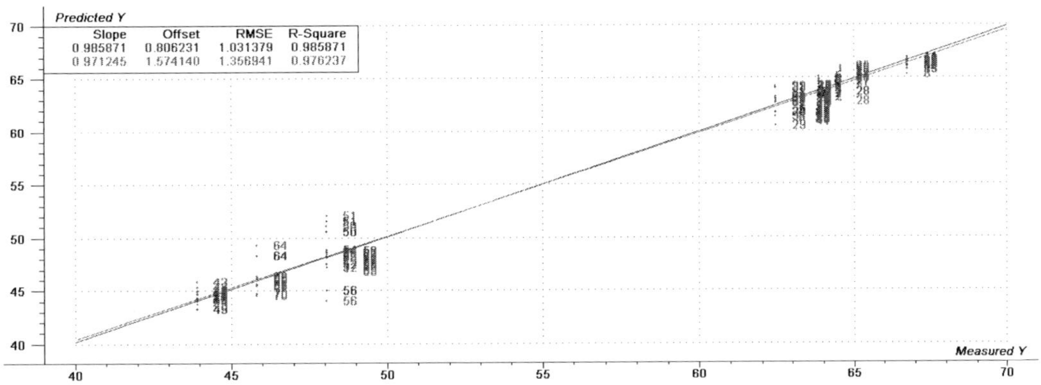

Figure 14. Plots of predicted tannin content obtained with voltammetric method vs. the values of tannin content obtained by HPLC (Apetrei et al., 2012).

Biosensors have been proposed as an efficient analytical tool for the determination of phenolic compounds, exhibiting advantages (e.g., minimal preparation of sample, sensitivity, selectivity, rapid time of response and simple use for continuous on-site analysis) (Apetrei & Ghasemi-Varnamkhasti, 2013).

Electrochemical biosensors that have been employed in the polyphenol analysis of wines were based on the activity of polyphenol oxidases or peroxidases (Carralero Sanz et al., 2005; Gil & Rebelo 2010; Pigani et al., 2011). The former catalyzes the oxidation of selected phenols to quinone derivatives with the concomitant reduction of oxygen to water (figure 15). The latter catalyzes the reduction of hydrogen peroxide to water where phenols act as electron donors. In both cases, quinone derivatives can be electrochemically detected at mild negative potentials. One of the first applications of a polyphenol oxidase-based electrochemical biosensor for wine analysis report a biosensor based on laccase immobilized on derivatized polyethersulfone membranes. The biosensor was applied to red wine samples. The same bioelectrochemical system was used for the determination of the polyphenolic content, as caffeic acid equivalents, in Portuguese wines. Good correlations with total phenolic (Folin-Ciocalteu) and antioxidant capacity (ABTS) assays were shown (Gomes et al., 2004).

A tyrosinase-based biosensor was developed to introduce the bioelectrochemical polyphenol index for wines. The sensing platform included gold nanoparticles electrodeposited onto a GCE, which enhanced both the enzyme stability and the electrochemical response (Carralero Sanz et al., 2005).

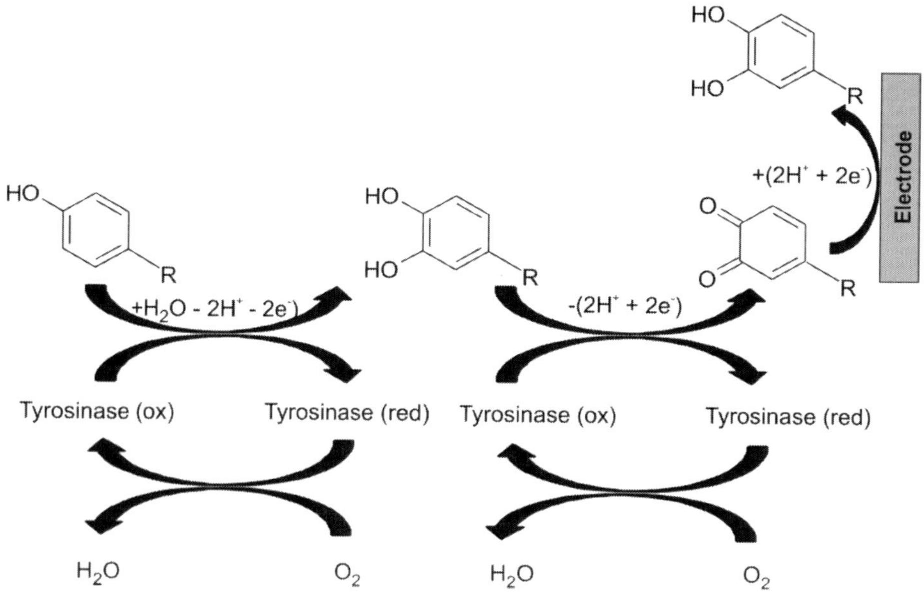

Figure 15. Functioning scheme of tyrosinase based biosensors (Apetrei et al., 2011).

One of the current trends in the development of electrochemical biosensors is the immobilization of enzymes on carbon nanotubes (CNT). This increases the amount of loading enzyme and can biocatalyze the electrochemical processes involved in the detection. Kim et al. have applied a biosensor for sensing phenolics in wine, including one in which tyrosinase is immobilized in multi walled carbon nanotubes (MWCNT) (Kim et al., 2010). Under optimal conditions, the total phenolic contents of commercial red wines were determined.

A more reliable recent example of the integration of enzymes with CNTs in electrochemical devices was provided by Di Fusco et al., who immobilized two laccase oxidases onto screen-printed carbon electrodes modified with CNTs using polyazetidine prepolymer as a binder. Red and white wines were analyzed and the results were compared (recovery study) with those obtained with the Folin-Ciocalteu reference method. Recoveries close to 100% were obtained for all samples when laccase oxidase from *Trametes hirsuta* was employed (Di Fusco et al., 2010).

The application of a peroxidase-based electrochemical biosensor to measure total polyphenols in wine was shown (Imayabashi et al., 2001). They immobilized horseradish peroxidase on a self-assembled monolayer of mercaptopropionic acid deposited on a gold electrode, and used the bio-electrode for polyphenol analysis of red wine. The results obtained correlated well with the Folin-Ciocalteu method.

One recent approach was introduced by Granero et al., who integrated peroxidases from *Brassica napus* hairy roots within a composite matrix made of ferrocene, MWCNTs and mineral oil. Red and white wines were analyzed using this method (*trans*-resveratrol and caffeic acid used as reference compounds) for total polyphenolics and the results correlated well with Folin-Ciocalteu assays. This was despite the reported concentrations being lower for the biosensor when compared with the spectrophotometric method (Granero et al., 2008).

Electronic Tongues Designed for the Characterization of Wines

The perception of wine quality is generally easier to recognize than it is to define. Wine quality is related to the intrinsic aspects of wine, which are consequences of the grape properties and winemaking conditions. The numerous applications of chemical and analytical sensory methods geared towards establishing quality parameters of wine have brought valuable knowledge to understanding perceived quality. In general, the sensory approach includes the evaluation of visual appeal, personality, subtlety and complexity, harmony and balance amongst all the components (aroma, taste, flavor and mouthfeel), all of which are inevitably linked to wine chemical composition (Amerine et al., 1965; Harrington, 2008).

Analysis of odor and taste in food has traditionally been addressed either by a trained sensory panel or using physicochemical methods (i.e., chromatography, spectrometry). In general, these techniques are time-consuming, costly and labor-intensive and the food industry is looking for new methodologies for objective and automated measurements that can characterize odors and tastes.

During the last decades, the concept of an "electronic nose" has been developed (Röck et al., 2008). These instruments provide a signal which is characteristic of the mixture of volatile compounds that form the headspace of a certain sample. For this reason, their main use is not the identification and quantification of the individual components of a sample. Instead, they perform a comparison of volatile patterns as the human nose does. Electronic noses are used in the food industry for odor analysis to evaluate the quality of a variety of products including wine, fish, meat, beer, milk, water, etc. (Wilson & Baietto, 2009; Röck et al., 2008).

More recent, the electronic tongues are defined as an array of non-specific chemical sensors with partial sensitivity (cross-sensitivity) to different components, able to analyze complex liquids. The basic principle behind an electronic tongue is to analyze the signals obtained from an array of non-specific overlapping sensors with pattern recognition routines. Therefore, it can be considered that electronic tongues consist of two different blocks: the array of sensors and the data-processing techniques (Riul Jr. et al., 2010; Ciosek & Wroblewski, 2007; del Valle, 2010).

The most promising approach to find applications for non-specific chemical sensors described in the previous paragraphs consists in the use of arrays of sensors. According to the recent IUPAC definition (Vlasov et al., 2005), an electronic tongue is a multisensor system consisting by a number of low-selective sensors and uses advanced mathematical procedures for signal processing based on pattern recognition and/or multivariate data analysis. In all cases, the sensing elements must have partial specificity so that they respond broadly to a range of liquids, rather than to a specific chemical compound. Therefore, the electronic tongue is an analytical system devised that global information of the sample. Using Principal Component Analysis (PCA) it is possible to discriminate between samples with different characteristics. Classification techniques can also be used, that address the problem of identifying an unknown sample and to assign it to a certain set of previously learned categorized samples. Typical classification models used in electronic tongues are Linear Discriminant Analysis (LDA), Soft Independent Modeling of Class Analogy (SIMCA) or Artificial Neural Networks (ANN). Electrochemical sensors (potentiometric, amperometric, voltammetric or impedimetric) are the most widely used sensing units in electronic tongues.

Potentiometric Electronic Tongues

The researches in this field involved potentiometric sensors based on the measurement of the difference of potential on the interface between the working electrode and the reference electrode. The interface composition determines sensitivity and selectivity of the sensor (Escuder-Gilabert & Peris 2010; Bratov et al., 2010; Ciosek & Wroblewski, 2011).

The interface mostly consists of different kinds of electrode materials, different types of polymeric membranes or crystalline compositions. These membranes can be prepared from different materials, which provide sensors with different selectivity. The development of potentiometric electronic tongue is based on the combination of various types of membranes with different selectivity, sensitivity and cross-reactivity (Gutierrez et al., 2008; Ciosek et al., 2007). The electronic tongue including ion selective sensors (based on solid-state and plasticized organic and chalcogenide glass membranes) served for discrimination of different kinds of wines (Rudnitskaya et al., 2010).

Arrays of potentiometric sensors based on six porphyrin / PVC based electrodes have demonstrated to be efficient ionophores to detect a variety of ions and alcoholic beverages such as wines (Legin et al., 2003; Lvova et al., 2006). The electronic tongue based on platinum coated with cobalt-porphyrin / PVC membranes discriminated Italian wines (Paolesse et al., 2008).

The light-addressable potentiometric sensor are also interesting for the electronic tongue concept. The light-addressable potentiometric sensor methodology was derived from ion-selective field effect transistors. The opportunity of simultaneous detection of several compounds provided light-addressable potentiometric sensor as an effective platform for electronic tongue applications. The ion-selective field effect transistors-based array was suitable to distinguish grape types and vintage of wine samples (Zeravik et al., 2009; Moreno i Codinachs et al., 2008).

Electronic Tongues Based on Voltammetric Sensors

Voltammetry is used in analytical chemistry due to the high sensitivity, versatility, and simplicity. The principle is based on measurement of current flowing between the working and counter electrodes when a pulse of potential is applied between them. Depending on the range of potential and type of the working electrode, electrochemically active compounds become either oxidized or reduced. The disadvantage is low selectivity, because all compounds in sample which are electrochemically active below the applied potential, contribute to the measured signal (Bard & Faulkner, 2001; Compton & Banks, 2007).

The working principle of a voltammetric electronic tongue is illustrated in Figure 16.

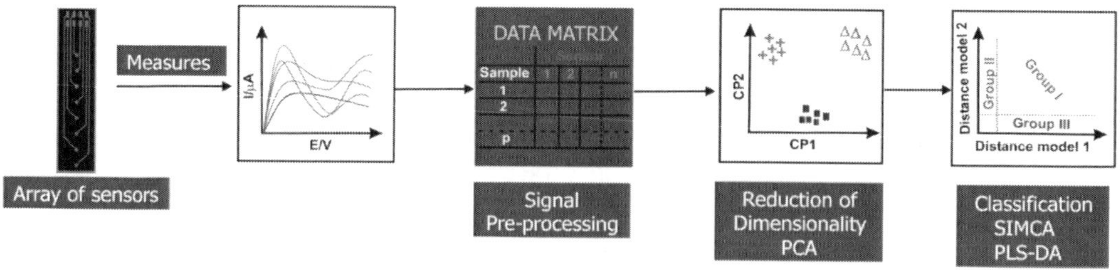

Figure 16. Scheme of principle of an electronic tongue based on voltammetric sensors.

An electronic tongue based on voltammetry consisting of a number of different working electrodes made from different metals in a standard three-electrode configuration was described. Voltammetric measurements are carried out using pulsed voltammetry. The array has been improved by including electrodes fabricated with other materials (stainless steel or glassy carbon) or by combining voltammetric electrodes with potentiometric electrodes (Ivarsson et al., 2001; Winquist et al., 2005). Metallic voltammetric electrodes have been used to discriminate phenolic derivatives and wines (Buratti et al., 2004; Latha & Lakshmi, 2012). Arrays have also been constructed using screen-printed electrodes (Apetrei et al., 2013; Matemadombo et al., 2012).

The chemically modified electrodes, which can be built by either deposition of electroactive compounds over the conductive surface or using chemically modified carbon paste electrodes appeared. Different families of materials have been used as modifiers, including phthalocyanines, conducting polymers deposited electrochemically and doped with a variety of doping agents and perylenes (Apetrei et al., 2004; Arrieta et al., 2004; Parra et al., 2006). The use of different sensitive materials with complementary reactivity greatly improves the cross-selectivity of the array and hence its capability of discrimination. Hybrid arrays formed by phthalocyanines, perylenes and conducting polymers have been successfully used to discriminate between wines with different organoleptic characteristics (Parra et al., 2006).

Voltammetric sensors provide complex voltammograms (variety of peaks at different potentials and with different currents). The intrinsic complexity and cross-selectivity of the signals generated by the array of voltammetric electrodes are advantages because the data set contains large amount of information about the sample. But, the fact that the whole data set contains meaningful information can difficult the data processing. In consequence, a pre-treatment step to reduce of the number of variables (without loss of information) is required.

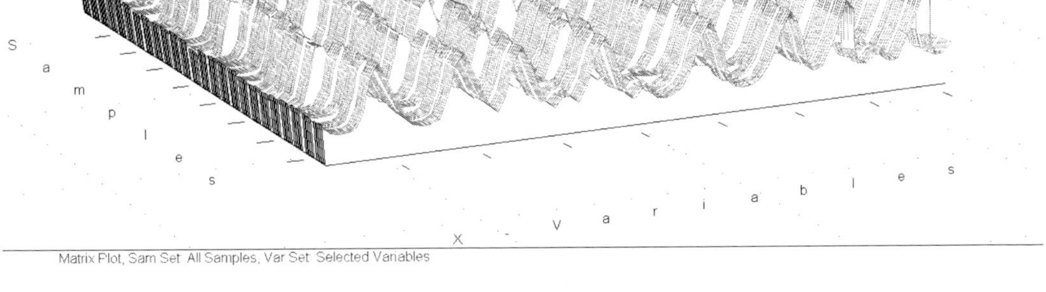

Figure 17. Square wave voltammograms reconstructed from kernel coefficients.

One solution is to simplify the high dimensionality, which in part may be redundant or just be related to noise, employing a feature extraction step such as the wavelet transformation (Moreno-Barón et al., 2006; Holmin et al., 2001). Another procedure to reduce the number of variables consists in pre-processing the voltammograms by the adaptation of a data reduction technique based on predefined response "bell-shaped-windowing" curves called kernels (Parra et al., 2006). Using this method, SWV curve is multiplied by a number of 10 kernel functions, and further integrated with respect to potential. Reconstruction of SWV curves from kernel coefficients in presented in Figure 17. Once the voltammograms have been pre-processed and the number of variables reduced, such variables are used as the input for multivariate data analysis using for instance Principal Component Analysis or Partial Least Squares Discriminant Analysis as discrimination and classification methods (figure 18).

Arrays of voltammetric sensors based on electroactive compounds have been extensively used to analyze foods and beverages. The system has been particularly successful in the field of wines because voltammetric sensors are sensitive to pH, ions (cations and anions) and electroactive compounds (i.e., phenolic compounds) and these compounds play an important role in the organoleptic characteristics of wines.

One of the main interests in the field of e-tongues is to establish correlations between the results obtained by means of these electronic systems and the results obtained by traditional physicochemical analysis or sensorial analysis. This is a highly demanding goal since these electronic systems provide global information of the samples whereas the chemical analysis, provide information about specific compounds.

One of the main successes of voltammetric electronic tongues based on electroactive compounds is the good correlations obtained between the signals provided by the array of sensors and chemical parameters measured using traditional techniques (i.e., chromatography). For instance, using partial least squares regression analysis (PLS2), excellent correlations have been found between the output of the electronic tongue and the polyphenol content or their acidity of red wines (Parra et al., 2006; Prieto et al., 2013). This means that electronic tongues based on electroactive compounds can be used as analytical tools to assess the content of these compounds present in wines with good accuracy.

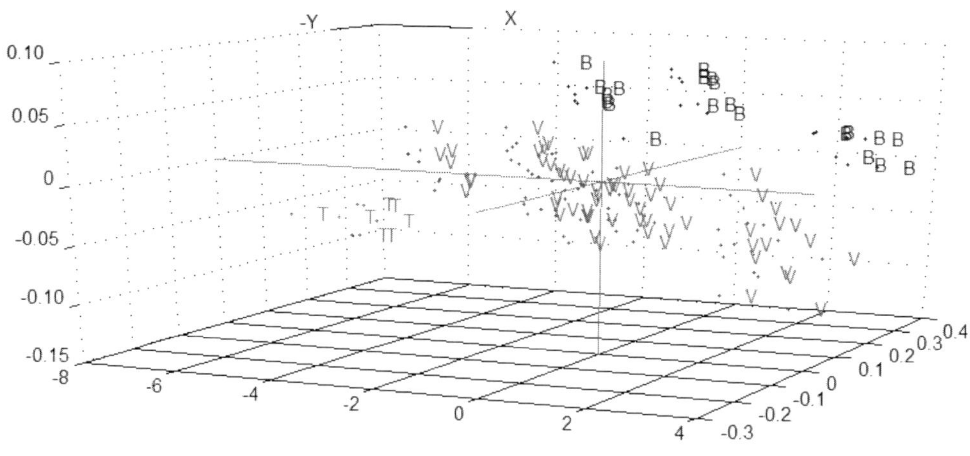

Figure 18. Principal Component Analysis scores plot of the measurements of 10 red wines (7 replicates) with phthalocyanine-based sensor array (Apetrei et al., 2012).

Finally, it has been demonstrated that the combination of electronic noses and electronic tongues can increase the amount of information extracted from a certain sample. A combination of an electronic nose (based on MOX sensors) and an electronic tongue (based on bisphthalocyanines) has been able to discriminate wines with different organoleptic characteristics (Apetrei et al., 2010; Apetrei et al., 2012).

Impedance Electronic Tongues

Impedance spectroscopy has also been used as transduction method in electronic tongues for the analysis of a variety of liquids. Arrays of impedimetric electrodes covered with various organic materials, including phthalocyanines, prepared by the Langmuir-Blodgett and the layer-by-layer techniques have been used to detect trace levels of phenolic compounds in water or to analyze complex solutions such as wines (Pioggia et al., 2007; Moreno i Codinachs et al., 2008; Labrador et al., 2010).

Bioelectronic Tongues

A bioelectronic tongue is a system formed by an array of biosensors with cross sensitivity coupled to appropriate software to interpret the data. These systems are advantageous because they combine the specificity of the enzyme-substrate reaction (or between complementary DNA strands or antigen-antibody) characteristic of biosensors with the ability of simultaneous detection of a large spectrum of compounds characteristic of the multisensing systems, providing a comprehensive information on the sample in a single step. The number of works in this field increases rapidly and includes potentiometric, amperometric or voltammetric bioelectronic tongues (Zeravik et al., 2009).

Combinations of biosensors containing enzymes (usually oxidases or dehydrogenases) have been used to analyze a variety of samples containing glucose, phenolic compounds etc. or wines (Cetó et al., 2012; Zeravik et al., 2009). Biosensors containing different phthalocyanines show distinct sensitivity and arrays can be constructed combining phthalocyanines and enzymes. A series of works in the field of beers have demonstrated that a bioelectronic tongue formed by sensors containing tyrosinase and different phthalocyanines can discriminate between beers with different characteristics. This e-tongue system proved capable of predicting alcohol and polyphenol content (Ghasemi-Vamankhasti et al., 2012).

Future Trends and Perspectives

Electrochemical techniques are methods suitable for determination of phenolic compounds in wines. Their relating with traditional analytical techniques (i.e., spectrometry, HPLC) allows identification of phenolic compounds in food matrices. On the other hand, the quantitative analyses of phenolic compounds in wines could be carried out. However, the development of novel sensors and biosensors based on nanotechnologies are expected to increase the performance characteristics of these devices. The multisensor system together with optimized multivariate data analysis methods could become a complementary method for wine analysis. The principal applicability could be the wine quality monitoring taking into account the high sensibility and selectivity of sensing devices and the possibilities for on-line or real time analysis.

ACKNOWLEDGMENTS

This work was supported by a grant of the Romanian National Authority for Scientific Research, CNCS-UEFISCDI, project number PN-II-ID-PCE-2011-3-0255.

REFERENCES

Amerine, M.A., Pangborn, R.M., Roessler, E.B., (1965). Principles of sensory evaluation of food. In: Food Science and Technology Monographs, 338-339, Academic Press.

Apetrei, C., Apetrei, I.M., Nevares, I., del Alamo, M., Parra, V., Rodríguez-Méndez, M.L., De Saja, J.A., (2007). Using an e-tongue based on voltammetric electrodes to discriminate among red wines aged in oak barrels or aged using alternative methods. Correlation between electrochemical signals and analytical parameters. *Electrochim. Acta* 52, 2588-2594.

Apetrei, C., Apetrei, I.M., Villanueva, S., de Saja, J.A., Gutierrez-Rosales, F., Rodriguez-Mendez, M.L., (2010). Combination of an e-nose, an e-tongue and an e-eye for the characterisation of olive oils with different degree of bitterness. *Anal. Chim. Acta* 663, 91-97.

Apetrei, C., Ghasemi-Varnamkhasti, M., (2013). Biosensors in food PDO authentication, in Comprehensive Analytical Chemistry, Gonzalvez, A., de la Guardia, M. (Eds.), 60, 279-297.

Apetrei, C., Rodríguez-Méndez, M.L., Parra, V., Gutierrez, F., De Saja, J.A., (2004). Array of voltammetric sensors for the discrimination of bitter solutions. *Sens. Actuat. B* 103, 145-152.

Apetrei, C., Alessio, P., Constantino, C.J.L., de Saja, J.A., Rodriguez-Mendez, M.L., Pavinatto, F.J., Fernandes, E.G., Zucolotto, V., Oliveira, O.N., (2011). Biomimetic biosensor based on lipidic layers containing tyrosinase and lutetium bisphthalocyanine for the detection of antioxidants. *Biosens. Bioelectron.* 26, 2513-2519.

Apetrei, I. M., Rodriguez-Mendez, M. L., Apetrei, C., de Saja, J. A., (2013). Fish freshness monitoring using an e-tongue based on polypyrrole modified screen-printed electrodes. *IEEE Sens. J.* 13, 2548-2554.

Apetrei, I.M., Rodríguez-Méndez M.L., Apetrei, C., Nevares, I., del Alamo, M., de Saja, J.A., (2012). Monitoring of evolution during red wine aging in oak barrels and alternative method by means of an electronic panel test. *Food Res. Int.* 45, 244-249.

Arrieta, A.A., Apetrei, C., Rodríguez-Méndez, M.L., De Saja, J.A., (2004). Voltammetric sensor array based on conducting polymer-modified electrodes for the discrimination of liquids. *Electrochim. Acta* 49, 4543-4551.

Aznar, O., Checa, A., Oliver, R., Hernández-Cassou, S., Saurina, J., (2011). Determination of polyphenols in wines by liquid chromatography with UV spectrophotometric detection. *J. Sep. Sci.* 34, 527-535.

Bakowska-Barczak, A., (2005). Acylated anthocyanins as stable, natural food colorants - A review. *Pol. J. Food Nutr. Sci.* 14, 107-116.

Bard, A. J., Faulkner, L. R., (2001). Electrochemical methods, John Wiley and Sons, New York.

Bănică, F. G. (2012). Chemical sensors and biosensors. Fundamentals and applications. John Wiley & Sons.

Borato, C. E., Leite, F. L., Oliveira, Jr., O. N., Mattoso, L. H. C., (2006). Efficient taste sensors made of bare metal electrodes. *Sensor Lett.* 4, 155-159.

Boulton, R., (2001). The copigmentation of anthocyanins and its role in the color of red wine: a critical review. *Am. J. Enol. Vitic.* 52, 67-87.

Bratov, A., Abramova, N., Ipatov, A., (2010). Recent trends in potentiometric sensor arrays-A review. *Anal. Chim. Acta* 678, 149-159.

Brouillard, R., Dubois, J.E., (1977). Mechanism of the structural transformations of anthocyanins in aqueous media. *J. Am. Chem. Soc.* 99, 1359-1364.

Buiarelli, F., Coccioli, F., Jasionowska, R., Merolle, M., Terracciano, A., (2007). Analysis of some stilbenes in Italian wines by liquid chromatography/tandem mass spectrometry. *Rapid Commun. Mass Sp.* 21, 2955-2964.

Buratti, S., Benedetti, S., Scampicchio, M., Pangerod, E.C., (2004). Characterisation and classification of Italian Barbera wines by using an electronic nose and amperometric electronic tongue. *Anal. Chim. Acta* 525, 133-139.

Carew, A. L., Smith P. A., Close, D. C., Curtin, C., Dambergs, R. G., (2013). Yeast effects on Pinot noir wine phenolics, colour and tannin composition. *J. Agric. Food Chem.* 61, 9892–9898.

Carralero Sanz, V., Mena, M. L., Gonzalez-Cortes, A., Yanez-Sedeno, P., Pingarron, J. M., (2005). Development of a tyrosinase biosensor based on gold nanoparticles-modified glassy carbon electrodes Application to the measurement of a bioelectrochemical polyphenols index in wines. *Anal. Chim. Acta* 528, 1-8.

Castellari, M., Matricardi, L., Arfelli, G., Galassi, S., Amati, A., (2000). Level of single bioactive phenolic in red wine as a function of the oxyen supplied during storage. *Food Chem.* 69, 61-67.

Castillo-Sánchez, J.J., Mejuto, J.C., Garrido, J., García-Falcón, S., (2006). Influence of wine-making protocol and fining agents on the evolution of the anthocyanin content, colour and general organoleptic quality of Vinhão wines. *Food Chem.* 97, 130-136.

Cetó, X., Gutiérrez, J. M., Gutiérrez, M., Céspedes, F., Capdevila, J., Mínguez, S., Jiménez-Jorquera, C., del Valle, M., (2012). Determination of total polyphenol index in wines employing a voltammetric electronic tongue. *Anal. Chim. Acta* 732, 172-179.

Chasagne, D., Ramirez, G., Lubbers, S., Charpentier, C., Feuillat M., Voilley, A., (2003). Sorption of aroma compounds to oak wood in wine ageing conditions. In: J. L. Le Quere, P. X. Etievant (Eds.), Flavour research at the dawn of the twenty - first century, 63-66, Cachan, Lavoisier.

Cheynier V., (2005). Polyphenols in foods are more complex than often thought 1,2,3. *Am. J. Clin. Nutr.* 81, 223S-229S.

Cheynier, V., Dueñas-Paton, M., Salas, E., Maury, C., Souquet, J. M., Sarni-Manchado, P., Fulcrand, H., (2006). Structure and properties of wine pigments and tannins. *Am. J. Enol. Vitic.* 57, 298-305.

Ciosek, P., Mamińska, R., Dybko, A., Wróblewski, W., (2007). Potentiometric electronic tongue based on integrated array of microelectrodes. *Sens. Actuat. B* 127, 8-14.

Ciosek, P., Wróblewski W., (2011). Potentiometric electronic tongues for foodstuff and biosample recognition-an overview. *Sensors* 11, 4688-4701.

Ciosek, P., Wroblewski, W. (2007). Sensor arrays for liquid sensing - electronic tongue systems. *Analyst* 132, 963-978.

Compton, R. G., Banks, C. E., (2007). Understanding voltammetry. World Scientific Publishing.

del Valle, M. (2010). Electronic tongues employing electrochemical sensors. *Electroanalysis* 22, 1539-1555.

Dhroso, A., Laschi, S., Marrazza, G., Mascini M. (2010). A fast electrochemical technique for characterization of phenolic content in wine. *Anal. Lett.* 43, 1190-1198.

Di Fusco, M., Tortolini, C., Deriu, D., Mazzei, F., (2010). Laccase-based biosensor for the determination of polyphenol index in wine. *Talanta* 81, 235-240.

Dobes, J., Zitka, O., Sochor, J., Ruttkay-Nedecky, B., Babula, P., Beklova, M., Kynicky, J., Hubalek, J., Klejdus, B., Kizek, R., Adam, V., (2013). Electrochemical tools for determination of phenolic compounds in plants. A review. *Int. J. Electrochem. Sci.* 8, 4520-4542.

Escuder-Gilabert, L., Peris, M., (2010). Review: Highlights in recent applications of electronic tongues in food analysis. *Anal. Chim. Acta* 665, 15–25.

Etievant P. X., (1981). Volatile phenol determination in wine. *J. Agric. Food Chem.* 29, 65-67.

Frankel, E.N., German, J.B., Kinsella, J.E., Parks, E., Kanner, J., (1993). Inhibition of oxidation of human low-density lipoprotein by phenolic substances in red wine. *Lancet* 341, 454-457.

Ghasemi-Varnamkhasti, M., Rodríguez-Méndez M.L., Mohtasebi, S.S., Apetrei, C., Lozano, J., Ahmadi, H., Razavi, S.H., de Saja, J.A., (2012). Monitoring the aging of beers using a bioelectronic tongue. *Food Control* 25, 216-224.

Gil, D. M. A., Rebelo, M. J. F., (2010). Evaluating the antioxidant capacity of wines: a laccase-based biosensor approach. *Eur. Food Res. Technol.* 231, 303-308.

Gomes, S.A.S.S., Nogueira, J.M.F., Rebelo, M.J.F., (2004). An amperometric biosensor for polyphenolic compounds in red wine. *Biosens. Bioelectron.* 20, 1211-1216.

Granero, A. M., Fernández, H., Agostini, E., Zón, M. A., (2008). An amperometric biosensor for trans resveratrol determination in aqueous solutions by means of carbon paste electrodes modified with peroxidase basic isoenzymes from *Brassica napus*. *Electroanalysis* 20, 858-864.

Gutierrez, M., Moo, V.M., Alegret, S., Leija, L., Hernandez, P. R., Munoz, R., del Valle, M., (2008). Electronic tongue for the determination of alkaline ions using a screen-printed potentiometric sensor array. *Microchim. Acta* 163, 81-88.

Harrington, R. J., (2008). Food and wine pairing: a sensory experience. John Wiley & Sons.

Holmin, S., Spångeus, P., Krantz-Rülcker, C., Winquist, F., (2001). Compression of electronic tongue data based on voltammetry - a comparative study. *Sens. Actuat. B* 76, 455-464.

Ignat, I., Volf, I., Popa, V.I., (2011). A critical review of methods for characterisation of polyphenolic compounds in fruits and vegetables. *Food Chem.* 126, 1821-1835.

Ishikawa, I., Noble, A.C., (1995). Temporal perception of astringency and sweetness in red wine. *Food Qual. Prefer.* 6, 27-33.

Ivarsson, P., Holmin, S., Höjer, N.-E., Krantz-Rülcker, C., Winquist, F., (2001). Discrimination of tea by means of a voltammetric electronic tongue and different applied waveforms. *Sens. Actuat. B* 76, 449-454.

Jakubec, P., Bancirova, M., Halouzka, V., Lojek, A., Ciz, M., Denev, P., Cibicek, N., Vacek, J., Vostalova, J., Ulrichova, J., Hrbac, J., (2012). Electrochemical sensing of total antioxidant capacity and polyphenol content in wine samples using amperometry online-coupled with microdialysis. *J. Agric. Food Chem.* 60, 7836-7843.

Kähkönen, M.P., Heinämäki, J., Ollilainen, V., Heinonen M., (2003). Berry anthocyanins: isolation, identification and antioxidant activities. *J. Sci. Food Agric.* 83, 1403-1411.

Kelebek, H., Selli, S., Canbas, A., Cabaroglu, T., (2009). HPLC determination of organic acids, sugars, phenolic compositions and antioxidant capacity of orange juice and orange wine made from a Turkish cv. Kozan. *Microchem. J.* 91, 187-192.

Kilmartin, P. A., Zou, H., Waterhouse, A. L., (2001). A cyclic voltammetry method suitable for characterizing antioxidant properties of wine and wine phenolics. *J. Agric. Food Chem.* 49, 1957–1965

Kilmartin, P. A., Zou, H., Waterhouse, A. L., (2002). Correlation of wine phenolic composition versus cyclic voltammetry response. *Am. J. Enol. Vitic.* 53, 294-302.

Kim, K. I., Lee, J. C., Robards, K., Choi, S. H., (2010). Immobilization of tyrosinase in carboxylic and carbonyl group-modified MWNT electrode and its application for sensing phenolics in red wines. *J. Nanosci. Nanotechnol.* 10, 3790-3798.

Ksenzhek, O., Petrova, S. Kolodyazhny, M. (2007). Redox Spectra of Wines. *Electroanalysis* 19, 389-392.

Labrador, R.H., Masot, R., Alcañiz, M., Baigts, D., Soto, J., Martínez-Mañez, R., García-Breijo, E., Gil, L., Barat, J.M., (2010). Prediction of NaCl, nitrate and nitrite contents in minced meat by using a voltammetric electronic tongue and an impedimetric sensor. *Food Chem.* 122, 864-870.

Larcher, R., Nicolini, G., Puecher, Chr., Bertoldi, D., Moser, S., Favaro G., (2007). Determination of volatile phenols in wine using high-performance liquid chromatography with a coulometric array detector. *Anal. Chim. Acta* 582, 55-60.

Latha, R. S., Lakshmi, P. K., (2012). Electronic tongue: An analytical gustatory tool. *J. Adv. Pharm. Technol. Res.* 3, 3-8.

Legin, A., Rudnitskaya, A., Lvova, L., Vlasov, Yu., Di Natale, C., D'Amico A., (2003). Evaluation of Italian wine by the electronic tongue: recognition, quantitative analysis and correlation with human sensory perception. *Anal. Chim. Acta* 484, 33-44.

Lesschaeve, I., Noble, A. C., (2005). Polyphenols: factors influencing their sensory properties and their effects on food and beverage preferences. *Am. J. Clin. Nutr.* 81, 330S-335S.

Li, H., Guo, A., Wang, H., (2008). Mechanisms of oxidative browning of wine. *Food Chem.* 108, 1-13.

Liao, H., Cai, Y., Haslam, E., (1992). Polyphenol interactions. Anthocyanins: co-pigmentation and color changes in red wines. *J. Sci. Food Agric.* 59, 299-305.

Lorrain, B., Ky, I., Pechamat, L., Teissedre, P. L., (2013). Evolution of analysis of polyhenols from grapes, wines, and extracts. *Molecules* 18, 1076-1100.

Lvova, L., Paolesse, R., Di Natale, C., D'Amico, A., (2006). Detection of alcohols in beverages: An application of porphyrin-based Electronic tongue. *Sens. Actuat. B* 1-2, 439-447.

Makhotkina, O., Kilmartin, P. A. (2013). Electrochemical oxidation of wine polyphenols in the presence of sulfur dioxide. *J. Agric. Food Chem.* 61, 5573-5581.

Makris, P.D., Kallithraka, S., Mamalos, A., (2006). Differentiation of young red wines based on cultivar and geographical origin with application of chemometrics of principal polyphenolic constituents. *Talanta* 70, 1143-1152.

Matemadombo, F., Apetrei, C., Nyokong, T., Rodríguez-Méndez, M.L., de Saja, J.A., (2012). Comparison of carbon screen printed and disk electrodes in the detection of antioxidants using CoPc derivatives. *Sens. Actuat. B* 166-167, 457-466.

Mateus, N., Oliveira, J., Santos-Buelga, C., Silva, A. M. S., de Freitas, V. (200). NMR structure characterization of a new vinylpyranoanthocyanin-catechin pigment (a portisin). *Tetrahedron Lett.* 45, 3455-3457.

Mattivi, F., Vrhovsek, U., Masuero, D., Trainotti D. (2009). Differences in the amount and structure of extractable skin and seed tannins amongst red grape varieties. *Aust. J. Grape Wine Res.* 15, 27-35.

Mazza, G., Miniati, E. (1993). Anthocyanins in fruits, vegetables and grains. CRC Press, Boca Raton.

McRae, J. M., Kennedy, J. A., (2011). Wine and grape tannin interactions with salivary proteins and their impact on astringency: a review of current research. *Molecules* 16, 2348-2364.

McRae, J.M., Falconer, R.J., Kennedy, J.A., (2010). Thermodynamics of grape and wine tannin interaction with polyproline: Implications for red wine astringency. *J. Agric. Food Chem.* 58, 12510-12518.

Moreno Arribas, M. V., Gómez-Cordovés, C., Martín Álvarez, P. J., (2008). Evolution of red wine anthocyanins during malolactic fermentation, postfermentative treatments and ageing with. *Food Chem.*, 109, 149-158.

Moreno i Codinachs, L., Kloock, J. P., Schöning, M. J., Baldi, A., Ipatov, A., Bratov, A., Jiménez-Jorquera, C., (2008). Electronic integrated multisensor tongue applied to grape juice and wine analysis. *Analyst* 133, 1440-1448.

Moreno-Arribas, M. V., Polo M. C. (Eds.), (2009). Wine chemistry and biochemistry, Springer Science.

Moreno-Barón, L., Cartas, R., Merkoçi, A., Alegret, S., del Valle, M., Leija, L., Hernandez, P.R., Muñoz, R., (2006). Application of the wavelet transform coupled with artificial neural networks for quantification purposes in a voltammetric electronic tongue. *Sens. Actuat. B* 113, 487-499.

Nicholson, R. S., (1965). Theory and application of cyclic voltammetry for measurement of electrode reaction kinetics. *Anal. Chem.* 37, 1351-1355.

Nicolaou, I.N., Kapnissi-Christodoulou, C.P., (2010). Analysis of polyphenols using capillary zone electrophoresis--determination of the most effective wine sample pre-treatment method. *Electrophoresis* 31, 3895-3902.

O'Byrne, P. (Ed.), (2009). Red wine and health, Nova Science Publishers.

Ortega-Regules, A., Romero-Cascales, I., Ros-García, J. M., Bautista-Ortín, A. B., López-Roca, J. M., Fernández-Fernández J. I., Gómez-Plaza E., (2008). Anthocyanins and tannins in four grape varieties (*Vitis vinifera L.*). Evolution of their content and extractability. *Int. J. Vine Wine Sci.* 42, 147-156.

Paolesse, R., Lvova, L., Nardis, S., Di Natale, C., D'Amico, A., Lo Castro, F., (2008). Chemical images by porphyrin arrays of sensors. *Microchim. Acta* 163, 103-112.

Parra, V., Arrieta, A.A., Fernández-Escudero, J.A., García, H., Apetrei, C., Rodríguez-Méndez, M.L., Saja, J.A., (2006). E-tongue based on a hybrid array of voltammetric sensors based on phthalocyanines, perylene derivatives and conducting polymers: Discrimination capability towards red wines elaborated with different varieties of grapes, *Sens. Actuat. B* 115, 54-61.

Rodriguez-Mendez, M. L., Parra, V., Apetrei, C., Villanueva, S., Gay, M., Prieto, N., Martinez, J., de Saja, J. A., (2008). Electronic tongue based on voltammetric electrodes modified with materials showing complementary electroactive properties. Applications. *Microchim. Acta* 163, 23-31.

Peynaud, E., (1984). Knowing and Making Wine, John Wiley & Sons Inc., New York.

Pianet, I., André, Y., Ducasse, M. A., Tarascou, I., Lartigue, J. C., Pinaud, N., Fouquet, E., Dufourc, E. J., Laguerre, M., (2008). Modeling procyanidin self-association processes and understanding their micellar organization: a study by diffusion NMR and molecular mechanics. *Langmuir* 24, 11027-11035.

Pigani, L., Culetu, A., Ulrici, A., Foca, G., Vignali, M., Seeber, R., (2011). Pedot modified electrodes in amperometric sensing for analysis of red wine samples. *Food Chem.* 129, 226-233.

Piñeiro, Z., Palma, M., Barroso, C. G. (2004). Determination of catechins by means of extraction with pressurized liquids. *J. Chromatogr. A* 1026, 19-23.

Pioggia, G., Di Francesco, F., Marchetti, A., Ferro, M., Ahluwalia, A., (2007). A composite sensor array impedentiometric electronic tongue: Part I. Characterization. *Biosens. Bioelectron.* 22, 2618-2623.

Pomar, F., Novo, M., Masa Vázquez, A., (2005). Varietal differences among the anthocyanin profiles of 50 red table grape cultivars studied by high performance liquid chromatography. *J. Chromatogr. A* 1094, 34-41.

Poncet-Legrand, C., Cartalade, D, Putaux, J.L., Cheynier, V., Vernhet, A., (2003). Flavan-3-ol aggregation in model ethanolic solutions: incidence of polyphenol structure, concentration, ethanol content, and ionic strength. *Langmuir* 19 10563-10572.

Prieto, N., Oliveri, P., Leardi, R., Gay, M., Apetrei, C., Rodriguez-Méndez, M.L., de Saja, J.A., (2013). Application of a GA-PLS strategy for variable reduction of electronic tongue signals, *Sens. Actuat. B* 183, 52- 57.

Ratola, N., Faria, J. L., Alves, A., (2004). Analysis and quantification of trans-resveratrol in wines from Alentejo region (Portugal). *Food Technol. Biotechnol.* 42, 125-130.

Rebelo, M. J., Rego, R., Ferreira, M., Oliveira, M. C., (2013). Comparative study of the antioxidant capacity and polyphenol content of Douro wines by chemical and electrochemical methods. *Food Chem.* 141, 566-573.

Ribéreau-Gayon, P., Glories, Y., Maujean, A., Dibourdieu, D., (Eds.), (2006). Handbook of Enology, The Chemistry of Wine: Stabilization and Treatments, Second Edition, John Wiley & Sons.

Riul Jr., A., Dantas, C. A. R., Miyazaki, C. M., Oliveira, O. N. (2010). Recent advances in electronic tongues. *Analyst* 135, 2481-2495.

Röck, F., Barsan, N., Weimar U., (2008). Electronic nose: current status and future trends. *Chem. Rev.* 108, 705-725.

Rudnitskaya, A., Rocha, S. M., Legin, A., Pereira, V., Marques J. C., (2010). Evaluation of the feasibility of the electronic tongue as a rapid analytical tool for wine age prediction and quantification of the organic acids and phenolic compounds. The case-study of Madeira wine. *Anal. Chim. Acta* 662, 82-89.

Sánchez Arribas, A., Martínez-Fernández, M., Chicharro, M., (2012). The role of electroanalytical techniques in analysis of polyphenols in wine. *TrAC-Trend. Anal. Chem.* 34, 78-96.

Sánchez-Rangel, J. C., Benavides, J., Basilio Heredia, J., Cisneros-Zevallos, L., Jacobo-Velázquez, D. A., (2013). The Folin-Ciocalteu assay revisited: improvement of its specificity for total phenolic content determination, *Anal. Methods* 5, 5990-5999.

Singleton, V.L., (1995). Maturation of wines and spirits: comparisons, facts and hypotheses. *Am. J. Enol. Vitic.* 46, 98-115.

Souza, L.P., Calegari, F., Zarbin, A. J., Marcolino-Junior, L. H., Bergamini, M. F., (2011). Voltammetric determination of the antioxidant capacity in wine samples using a carbon nanotube modified electrode. *J. Agric. Food. Chem.* 59, 7620-7625.

Stervbo, U., Vang, O., Bonnesen, C., (2007). A review of the content of the putativechemopreventive phytoalexin resveratrol in red wine. *Food Chem.* 101,449-457.

Sun, B., Leandro, M.C., de Freitas, V., Spranger, M. I., (2006). Fractionation of red wine polyphenols by solid-phase extraction and liquid chromatography. *J. Chromatogr. A* 1128, 27-38.

Vitaglione, P., Sforza, S., Galaverna, G., Ghidini, C., Caporaso, N., Vescovi, P.P., Fogliano, V., Marchelli, R., (2005). Bioavailability of trans-resveratrol from red wine in humans. *Mol. Nutr. Food Res.* 49, 495-504.

Vlasov, Yu., Legin, A., Rudnitskaya, A., Di Natale, C., D'Amico, A., (2005). Nonspecific sensor arrays ("electronic tongue") for chemical analysis of liquids. *Pure Appl. Chem.* 77, 1965-1983.

Wilson, A. D., Baietto, M., (2009). Applications and Advances in Electronic-Nose Technologies. *Sensors* 9, 5099-5148.

Winkel-Shirley, B., (2002). Biosynthesis of flavonoids and effects of stress. *Curr. Opin. Plant Biol.* 5, 218-223.

Winquist, F., Bjorklund, R., Krantz-Rülcker, C., Lundström, I., Östergren, K., Skoglund, T. (2005). An electronic tongue in the dairy industry. *Sens. Actuat. B* 111-112, 299-304.

Winter, M., Herrmann, K., (1986). Esters and glucosides of hydroxycinnamic acids in vegetables. *J. Agric. Food Chem.* 34, 616-620.

Zeravik, J., Hlavacek, A., Lacina, K., Skladal, P., (2009). State of the art in the field of electronic and bioelectronic tongues - towards the analysis of wines. *Electroanalysis* 21, 2509-2520.

Zhao, Q., Duan, C. Q., Wang J., (2010). Anthocyanins profile of grape berries of *Vitis amurensis*, its hybrids and their wines. *Int. J. Mol. Sci.* 11, 2212-2228.

In: Wine
Editor: Youssef El Rayess

ISBN: 978-1-63321-048-6

Chapter 3

Analytical Methods for Wine Polyphenols Analysis and for Their Antioxidant Activity Evaluation

***Youssef El Rayess*[*1,2,3], *Reine Barbar*[1], *Eli Adjélé Wilson*[4] and *Jalloul Bouajila*[4]**

[1]Faculty of Agricultural and Food Sciences,
Holy Spirit University of Kaslik, Jounieh, Lebanon
[2]Université de Toulouse, INPT, UPS,
Laboratoire de Génie Chimique, Toulouse, France
[3]Centre de Viticulture et d'Œnologie de Midi-Pyrénées,
Castanet-Tolosan, France
[4]Université de Toulouse, Faculté de pharmacie de Toulouse,
Laboratoire des IMRCP UMR CNRS 5623, Université Paul-Sabatier, Toulouse, France

Abstract

Wine phenolics are of considerable interest and received more and more attention due to their implication in the organoleptic appreciation of wines and to their bioactive functions. Due to their importance in enology sector, there is an increasing demand for highly sensitive and selective analytical methods for the study and assessment of polyphenols. In this chapter, we will review the analytical methods used for polyphenols determination in addition to the methods used for measuring the antioxidant activity.

Keywords: Polyphenol, spectrometry methods, liquid chromatography, cyclic voltammetry, antioxidant capacity, DPPH, ABTS

[*] Corresponding Author, address: Faculty of Agricultural and Food Sciences, Holy Spirit University of Kaslik, Jounieh, Lebanon. Email: youssefrayess@usek.edu.lb, youssef.elrayess@ensiacet.fr

Introduction

Polyphenols are a complex group of compounds divided into two main groups of compounds: non-flavonoids and flavonoids. The non-flavonoids compounds include the hydroxycinnamates, the hydroxybenzoates and the stilbenes while the flavonoids contain the flavan-3-ols, the flavonols, and the anthocyanins. Polyphenols constitute the most important quality parameters of wines since they contribute to their organoleptic characteristics, color, flavor, astringency, bitterness, enzymatic and non-enzymatic browning, haze formation due to their interaction with proteins, and aging behavior (Kennedy et al., 2006). In addition, these compounds have gained considerable interest due to research suggesting their many health benefits especially as antioxidants.

Many epidemiological studies have shown the positive effects of wine polyphenols on human health (German and Walzem, 2000). Therefore, determination for the chemical composition, the amount of polyphenols and the antioxidant activity of wine could be very useful for the interpretation of epidemiological studies. Given their importance in wines, the identification and quantification of wine polyphenols have received a great deal of attention and have been extensively studied. Polyphenol analysis involves an extremely wide variety of analytical techniques. Techniques used vary between relatively simple chemical methods and highly complex instrumental ones capable of detailed investigation of individual chemical constituents. The most common test is the total polyphenols index using absorbance at 280 nm or the Folin-Ciocalteu reagent. Identification and quantification of each phenolic compound are generally achieved by liquid chromatography. The cyclic voltammetry method has been used to characterize a range of wine phenolics, to estimate the phenolic content in wine and to evaluate the antioxidant activity of red wines. Nowadays, liquid chromatography coupled with mass spectrometry is the most effective tool for the structural characterization of low molecular weight polyphenols.

In the literature, many reviews concerning the chemistry of wine polyphenols can be found but few of these focus on the analytical methodology for the analysis of the compounds and for the assessment of their antioxidant capacity. Therefore, the chapter seeks to provide an overview of these analytical methods and to highlight the last developments in this sector.

Spectroscopic Analysis

A number of spectroscopic methods have been developed for wine phenolics measurement. These assays are based on different principles and are used to determine different structural groups in phenolic compounds. UV-Vis sepctophotometry is used to measure the color of grapes and intact red wines, but does not provide detailed information regarding individual pigments in a red wine.

Spectrophotometric Measurement of Red Wine Color

Color measurements are made on intact wine using a UV/Vis spectrophotometer. The color intensity or density is traditionally the sum of different absorbances ($A_{420} + A_{520} + A_{620}$)

taking into account the contributions of yellow, red and blue to overall color. This measurement gives an idea about the amount of color present in a wine. Sudraud (1958) was the first to propose the measure of color intensity by summing the absorbance at 420 nm and 520 nm. This method was modified lately by Glories (1984a; 1984b) by adding the absorbance measurement at 620 nm. Mazza et al. (1999) have proposed to include the absorbance measurement at 700 nm to account for problems due to turbidity. The wine turbidity can be eliminated by centrifuging the sample or by membrane filtration. The usefulness of color intensity measurement is very limited because it does not account for the color provided by anthocyanin-derived pigments such as Vitisin A and the ethyl-linked anthocyanin-flavanols whose λ_{max} values are at 512 and 540 nm respectively (Brise, 2007).

The color hue is calculated as the ration of A_{420}/A_{520} and is normally used to monitor wine aging. An increase in the hue value is expected for an aging wine, and describes a shift from purple red via brick red to brown tones of the wine color. The main drawback of the measurements is that the wine color is pH dependent and they can't be used to compare multiple wines but it can be used to monitor the evolution of a single wine.

Another method for wine chromatic characteristic determination is CIELab coordinates (Bakker et al., 1986). The CIELab color system is based on a sequential or continuous cartesian representation of 3 orthogonal axes: L^* representing Clarity or Luminance ($L^*= 0$ black and $L^* = 100$ colorless), a^* is green/red color component ($a^* > 0$ red and $a^* < 0$ green) and b^* corresponds to blue/yellow color component ($b^* > 0$ yellow and $b^* < 0$ blue). The main advantage of measurements based on CIELab color system is that they are made across 300 to 800 nm wavelength interval every 5 nm using distilled water as a reference. This will ensure that minor absorption values of pigments that contribute to red wine color at additional wavelenghts are included resulting in a more precise definition of color. The limitations of this method are that L^*, a^* and b^* provide limited information about a color and it is difficult to make conclusion when the wine color is described on a scale of "yellowness" and "redness". It is more meaningful to describe color by its L^* value, its intensity or brightness, its C^* value and the initial color observed or by its hue, H^* value. The Chroma C^* is given by the equation 1 and is derived from the a^* and b^* coordinates:

$$C^* = \sqrt{a^{*2} + b^{*2}} \quad [1]$$

The hue value (H^*) is used to describe the attribute of a color that is observed. It is given by equation 2

$$H^* = \tan^{-1}(a^*/b^*) \quad [2]$$

An additional parameter that can be added to CIELab color system is the saturation or s. It is determined as the proportion of chromaticity to lightness:

$$s = C^*/L^* \quad [3]$$

Somers and Evans (1974, 1977) established a set of spectroscopic color measurements which extend beyond the measure of wine color by giving an insight into the contributing elements such as anthocyanin equilibrium and phenolic composition. The original Somers assay is a four part assay where the wine is analyzed in its original state and is then treated with excess SO_2, excess acetaldehyde, and hydrochloric acid to investigate the anthocyanin equilibrium of the wine. First, the absorbance of the young red wine sample is read at 420 and 520 nm in its original state with regard to wine pH and SO_2 concentration, and from these values, the wine color parameters, wine color density and hue are calculated. The second reading is resulting from SO_2-resistant pigments present in the wine. Third, the original wine is treated with excess acetaldehyde, which allows the estimation of colored anthocyanins at wine pH. Finally, the wine is diluted with 1 M HCl, lowering the pH and converting all anthocyanins and many other pigments into their colored forms. The acidified solution is then monitored at A_{520} and A_{280} to give an indication of the concentration of total red pigments and total phenolics, respectively.

Spectrophotometric Measurement for Wine Polyphenols Quantification

The Folin-Ciocalteu assay is widely used for determining total phenolics. This is a colorimetric oxidation/reduction assay that measures all phenolic molecules with no differentiation between gallic acid, monomers, dimers and larger phenolic compounds. This assay relies on the transfer of electrons in alkaline medium from phenolic compounds to phosphomolybdic/phosphotungstic acid complexes. The oxidation by Folin-Ciocalteu reagent causes a color change from yellow to blue that is determined spectroscopically at 760 nm. The problem of this assay is that the Folin-Ciocalteu reagent is non-specific and can oxidize other compounds than the phenolics one like bisulfite, aminoacids, fructose and ascorbic acids. So the total phenolic content is overestimated with this method and needs correction. A gallic acid standard curve is used and results are typically expressed as gallic acid equivalents (GAE).

As a faster alternative to this method, the use of what is known as total polyphenol index (I_{280}) is now arising. In this case, wine absorbance is measured directly after dilution with water at 280 nm and straightly correlated with phenolic content because this method is based on the fact that compounds with aromatic rings absorb UV light. This method has two main drawbacks: i) it has interferences from non-phenolic compounds that also contain aromatic rings, ii) it offers no information about the type of phenolics measured while the vanillin and Bate Smith assays have been used to estimate total proanthocyanidins.

More specific assays such as vanillin assay, dimethylaminocinnamaldehyde (DMACA) assay, Bate Smith assay have been established to determin the content of proanthocyanidins (condensed tannins).

The vanillin assay is specific for flavan-3-ols, dihydrochalcones and proanthocyanidins which have a single bond at the 2,3-position and possess free meta-oriented hydroxy groups on the B ring. The principle of the Vanillin-HCl assay is as follows: vanillin is protonated in an acid solution (generally HCl), giving a weak electrophilic carbocation that reacts with the flavonoid ring at the 6 or 8 position. This intermediate compound is dehydrated to give a red colored compound. The red colored compound has a specific absorption at a range between 500 and 550 nm. Catechin, a monomeric flavan-3-ol, is often used as a standard in the vanillin assay. According to the literature (Price et al., 1978; Gupta and Haslam, 1980), this may lead to overestimate the contents of proanthocyanidin. The vanillin assay in methanol is more sensitive towards polymeric proanthocyanidins than monomeric flavan-3-ols. This assay is generally recognized as a useful method for the detection and quantification of proanthocyanidins in wine due to its simplicity, sensitivity and specificity. The method can be used for quantifying proanthocyanidins in the range of 5–500 μg with precision and accuracy of greater than 1 μg when optimum concentrations of reactants and solvents are selected.

The DMACA assay is analogous to the vanillin assay in its chemistry. The DMACA is yellow-pigmented but in the presence of alcohol and tannins as in wine, it binds the tannin giving a vivid blue color measured at 640 nm. This last wavelength is outside of the spectra of anthocyanins thus eliminating interference. DMACA recognizes the meta-orientated hydroxyls on the A ring of flavonoid which imparts specificity.

The specific reaction of condensed tannins identification is the Bate-Smith reaction (figure 1). The principle of this reaction is as follows: under heating in acidic medium, the condensed tannins release anthocyanins with red color characteristic of this reaction. The hydrolysis of interflavanic bond forms a highly oxidizable benzylic cation which quickly transforms into flavilium cation of red color at the reaction pH.

Both categories of tannins (condensed and hydrolysable) are known to interact with proteins forming soluble and insoluble complexes. This approach of tannins precipitation has been developed to propose several assays for tannins quantification. Hagerman and Butler (1978) proposed the first tannin-protein precipitation assay by the addition of the bovine serum albumin (BSA) as the precipitant to tannins. This method was lately adapted to must and wine analysis by Haberston et al. (2002). After 15 minutes of incubation, the tannin-protein precipitate is formed and the solution is centrifuged. The pellet is washed and re-suspended in an alkaline solution resulting in a colored complex. The spectrophotometric measurement is done at 510 nm before and after adding ferric or iron chloride to the solution. The iron chloride reagent does not react with the monohydroxylated phenolics and anthocyanins. So, this method may underestimate the amount of tannins in the wine. This method can be combined with bisulfite bleaching. This allows the separation of polymeric pigments into small fractions which are not bleachable or precipitable and large fractions which are not bleachable but precipitable.

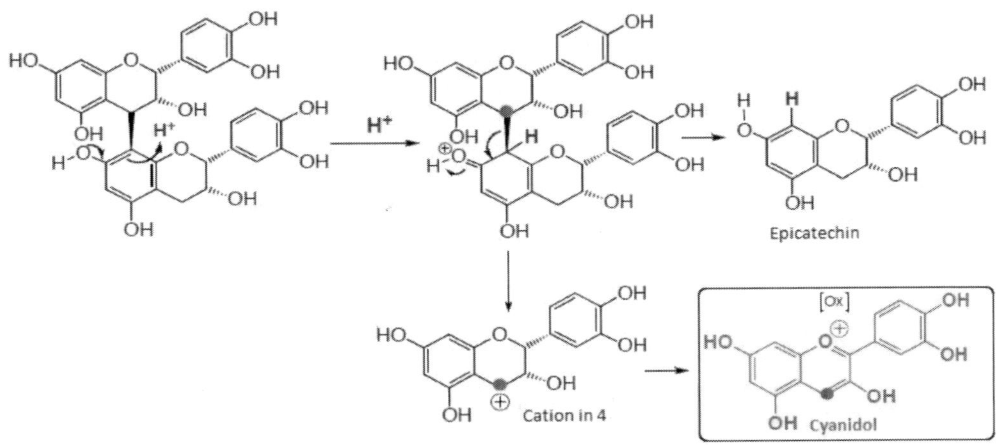

Figure 1. Bate-Smith reaction for condensed tannins identification.

The Gelatin index was also one of the first protein precipitation assays and is known also as Glories method. Briefly, tannins are measured by absorbance at 550 nm after acid hydrolysis. Then, all the tannins are precipitated for 3 days with an excess of gelatin, and measured again. The index is a ratio between the two tannin measurements. This assay is tedious because it relies upon a long precipitation step.

Llaudy et al. (2004) proposed the purified chicken egg protein, ovalbumin, as the precipitant of tannins. This method is the same as Gelatin index with two exceptions: the method to measure the total phenolics is the absorbance at 280 nm and the incubation time is 10 minutes. After precipitation with ovalbumin, the remained phenolics in the supernatant are assumed to be non-tannins phenolics. Llaudy et al. (2004) suggested that ovalbumin method is less sensitive than the gelatin method because the absolute tannin concentrations given by the ovalbumin methodology are lower than those obtained by gelatin index. It must be noticed that the incubation is not the same and the ovalbumin method is much quicker and not tedious.

Another method to quantify proanthocyanidin based on precipitation using methyl cellulose has been published. The proanthocyanidins form an insoluble complex with methyl cellulose. The absorbance of phenolics at 280 nm before and after proanthocyanidin precipitation (subtractive approach) can be obtained, thus enabling selective measurement of tannin only. This methyl cellulose precipitable (MCP) tannin assay allows complete precipitation of tannins from red wine and from grape homogenate extracts.

Chromatographic Techniques for Grape and Wine Polyphenols Separation and Quantification

Several chromatographic techniques such as liquid chromatography (LC), gas chromatography (GC) and thin-layer chromatography (TLC) have been applied in analysis of grape and wine phenolics.

High performance liquid chromatography (HPLC) is the widely used technique for both separation and quantification of wine polyphenols. HPLC is a commonly used analytical separation technique that combines high resolution and easy automation with modest simple requirements. Reversed-phase (RP) LC on C_{18} or normal-phase (NP) LC on C_{18} stationary phases are the most common methods used for phenolic analysis. The RPLC on C_{18} are columns packed with spherical particles of silica bonded with octadecyl chains. Monolitic columns have been proposed as new stationary phase for polyphenols analysis. Replacing conventional columns with monolithic columns can allow shorter assay times, faster column equilibration and larger void volume between the packed particles. Monolithic columns have also the advantages over conventional particle-columns such as higher flow rates at lower backpressures. Castellari et al. (2002) were the first to test monolithic column for wine phenolics analysis. They concluded that monolithic column could operate at a higher flowrate than a conventional RP column with a shorter washing and re-equilibration times. This produces a faster separation of the phenolic compounds and contributes to improve S/N ratio (Signal/Noise). Du Toit et al. (2006) used the monolithic column in order to ascertain the effect of micro-oxygenation on the levels of 21 non-colored phenolics and anthocyanins. A RP-monolithic column was also used by Liazid et al. (2010) for the analysis of 13 common wines phenolic compounds. They developed a rapid (8 minutes), reliable and reproducible method for the quantification of polyphenols in real samples. Recently, core-shell columns have been tested for wine polyphenols analysis. Manns and Mansfield (2012) showed that core-shell columns resulted in a dramatically improved selectivity, resolution and throughput in each tested sample. The core-shell method offers a rapid throughput design, lower solvent consumption, low detection levels and high reproducibility.

Acetonitrile or methanol and their aqueous forms are the dominant mobile phases used in HPLC quantification of phenolics. The pH of the mobile phase is maintained in the range pH 2-4 to avoid ionization of phenolics during identification. A gradient elution system is more commonly applied than an isocratic elution system.

Some phenolic compounds found in wine show characteristic absorbance in the UV-vis region, and their chromatograms at 280 nm are widely used to study phenolic compounds because their absorption at this wavelength is suitable for the detection of a large number of such compounds, which can be easily detected by HPLC because of its versatility and precision. Flavanols show two bands absorption maxima at 210 and 278 nm while anthocyanins show three regions of absorption maxima around the 265-275 nm and 465 and 560 nm (Merken and Beecher; 2000). Therefore, wine polyphenols are frequently analyzed using the UV-Vis and photodiode array (PDA) detectors. Analysis of condensed tannins by HPLC is difficult because the separation of large molecules (degree of polymerization > 4) is not possible. Therefore, the analysis of condensed tannins is achieved by acid-catalyzed cleavage of the interflavan bond in the presence of a nucleophilic reagent like benzyl mercaptan (thiolysis) or phloroglucinol (phloroglucinolysis). The cleavage of the interflavan bond in acidic medium yields a carbocation from the upper and extension units of the molecule whereas the lower part is released as such. The carbocation reacts with the nucleophiles to give a stable adduct. HPLC analysis of the reaction mixture allows the separation of lower units from derivatized units. This analysis gives access to the average degree of polymerization (DP) of the fractions and to its composition in monomeric flavan-3-ols.

Fluorescence spectroscopy detection has been proposed to detect phenolics in wine. Several researches have been conducted on the fluorescence detection of wine polyphenols (Vinas et al., 2000; Rodriguez-Delgado et al., 2001; Vitrac et al., 2002, Gurbuz et al., 2007; Gomez-alonso et al., 2007; Porgali and Buyuktuncel, 2011; Silva et al., 2011). Some of these researches have proposed a combination of absorption and fluorescence detection in order to discriminate the overlapping peaks. Silva et al. (2011) showed that flavanol monomer, dimer and trimer can be detected and quantified by fluorescence spectroscopy (excitation wavelength = 280 nm and emission wavelength = 320 nm). They showed that it was a rapid and simple method with good performances in terms of selectivity and sensitivity.

The UV spectra of some phenolics are very similar making their identification ambiguous. Therefore, liquid chromatography (LC) coupled to mass spectrometry (MS) was developed to characterize polyphenols structures. LC/MS coupled with multiple mass spectrometry (MS/MS and MS^n) is the most useful tool for structure elucidation of low molecular weight polyphenols in grapes and wines as well as being used for characterization of high molecular weight compounds such as tannins and proanthocyanidins.

Mass spectrometric measurements are carried out in the gas phase on ionized analytes. The principle of mass spectrometers consists of using the difference in mass-to-charge ratio (m/z) of ionized molecules to separate one from another. This requires first that the various molecular species of interest have been charged (often by loss or addition of protons) and transferred into the gas phase, and that they are then separated as a function of their *m/z* values. These two steps are achieved by the mass spectrometer source and analyzer, respectively. Electrospray ionization (ESI), matrix assisted laser desorption ionization (MALDI) and atmospheric pressure ionization (API) including atmospheric pressure chemical ionization (APCI) and atmospheric pressure photo-ionization (APCI) are the most used techniques for volatilization and ionization of polyphenolic compounds. Three different principles may be applied to achieve mass separation: separation on the basis of time of flight (TOF), separation by selective ejection of ions from a three-dimensional trapping field (ion trap and Fourier transform ion cyclotron).

Samples are injected in solution into the ESI, APCI and TSP sources, which can thus be coupled to HPLC systems, whereas the use of MALDI and FAB (fast atom bombardment) require mixing of the sample with a matrix solution. ESI, APCI, FAB, and MALDI are operated in both the negative and positive ion modes, generating anions such as $[M-H]^-$ and $[M+Cl]^-$ or cations such as $[M+H]^+$, $[M+Na]^+$ and $[M+K]^+$, respectively; ESI yields multiply charged ions ($z = 1, 2, 3 \ldots$), while MALDI and FAB yield mostly singly charged ions ($z = 1$) (Fulcrand et al., 2008).

The use of mass spectrometry for the identification and quantification of wine polyphenols goes back to the end of 80's and the beginning of 90's (Lee and Jaworski, 1988; Cilliers and Singleton, 1990). Anthocyanins are detected as flavylium cations when operating in the positive ion mode and in acidic medium, providing signals of high intensity. Baldi et al. (1995) used the HPLC/MS for anthocyanic compounds of *Vitis vinifera* L. This technique led to the identification of the 3-glucosides, the 3-acetylglucosides and the 3-p-coumaroylglucosides of delphinidin, cyanidin, petunidin, peonidin and malvidin. de Villiers et al. (2004) reported a method for anthocyanins analysis in five South African red wine cultivars using LC-UV-MS method. They were able to identify forty four anthocyanins and derived products. These data were used to differentiate between cultivars. In 2009, pelargonidin-3-glucosides and its camaroyl and acetyl derivatives were been detected in *Vitis vinifera* grapes by using LC/MS (Castillo-Munoz et al., 2009). He el al. (2010) confirmed the existence of these pelargonidin derivatives in non-teinturier *V. vinifera* grapes. For the purpose of a varietal assessment, Picariello et al. (2014) used the LC/MALDI-TOF for the detection of the 3,5-O-diglucoside anthocyanins and their acylated derivatives. These compounds are specific of non-*V. vinifera* varieties and furnish valuable clues for the varietal assessment of red grapes.

For wine anthocyanins, Heier et al. (2002) identified sixty six different mono and diglucosides anthocyanins from a single red wine using LC/MS methods. In addition, they also identified the acetic acid ester and coumaric acid ester of the diglucosides. de Villiers et al. (2009) proposed an optimized method of LC for wine anthocyanins separation. In 2011, the same authors used these optimized conditions of LC and coupled the chromatography separation with MS detection. In this study, they used also the kinetic plot method in order to evaluate the most favorable experimental configuration for high-efficiency anthocyanin determination. With all these ameliorated conditions, a total of 101 anthocyanins and derived pigments (among them 17 grape derived anthocyanins, 6 anthocyanidin-di-glucosides and 7 oligomeric anthocyanins) were tentatively identified in a single analysis of this red wine sample. Gordillo et al. (2010) studied the evolution of wine anthocyanins of Syrah Spanish wine during first year of barrel aging. A total of 32 anthocyanic pigments were identified, including 14 major compounds from grape and 18 minor derivatives formed during the vinification. The anthocyanin profile changed towards more chemical complexity, giving that vitisin-like pyranoanthocyanins were the predominant minor pigments during the first month of aging. As wine became older, a progressive increase on the content of 4-vinylcatechin, 4-vinylphenol and 4-vinylcatechol compounds took place.

Anthocyanin-derived pigments are called pyranoanthocyanins which are known to be formed from the reaction of anthocyanins with small molecules such as acetaldehyde, acetoacetic acid, pyruvic acid, vinylphenol, vinylguaiacol, vinylcatechol and vinylcatechin. Pyranoanthocyanins (mainly carboxypyranoanthocyanins) are thought to contribute to the orange hues observed during wine maturation. Potrisins are new family of anthocyanin derived compounds presenting a bluish color in acidic conditions. Oliveira et al. (2010) investigated the used of LC-DAD-MS for the elucidation of pyranoanthocyanin in Porto wine. They found new compound with an attractive and rare turquoise blue color at acidic conditions which is formed by two pyranoanthocyanin moieties linked together through a methyne bridge.

Mass spectrometry (MS) is a very effective and highly sensitive method for characterizing proanthocyanidins and procyanidins from complex matrices as wines (Flamini, 2003 and Reed et al., 2005). LC/ESI-MS analysis of proanthocyanidins is usually performed by reversed phase chromatography even if normal phase chromatography using silica columns provided a satisfactory separation of oligomers based on their molecular weight (Flamini, 2013). Cheynier et al. (1997) studied grape seed extracts using LC/ESI-MS operating in negative-ion mode. They found procyanidins from trimers to hexadecamers and their acylated derivatives. Wu et al. (2005) used LC/ESI-MS positive-ion mode for polyphenols determination from grape seed extract. The chromatogram showed the signals of protonated catechin and epicatechin at *m/z* 291, protonated catechin/epicatechin gallate at *m/z* 443, protonated catechin/epicatechin dimer at *m/z* 579 and protonated catechin/epicatechin gallate dimers and trimers at *m/z* 731, *m/z* 883 and *m/z* 867 respectively. Operating in positive ionization mode provides at the same time the signals of procyanidins and proanthocyanidins protonated molecules. Nunez et al. (2006) showed the presence of non-galloylated and monogalloylated flavan-3-ols up to octamers, and di- and trigalloylated flavan-3-ols up to heptamers in seed extracts from grapes of Cabernet-Sauvignon, Tempranillo and Graciano using NP-HPLC/ESI-MS. The presence of galloylated A-type procyanidins in grape seeds was evidenced by Passos et al. (2007) using a combination of ESI-MS and ESI-tandem (MS/MS) in positive-ion mode. When studying the real influence of the heterogeneity in the degree of grape ripeness on wine composition and quality, Kontoudakis et al. (2011) used HPLC/ESI-MS in negative ion mode to determine the monomeric flavan-3-ols ((+)-catechin, (-)-epicatechin and (-)-epicatechin-3-O-gallate) and the flavan-3-ols dimers (procyanidin B1, B2, B3 and B4). Delcambre and Saucier (2012) reported the identification of 14 flavan-3-ol glycosylated compounds from Merlot red wine and grape seed extracts. The analyses were performed using UHPLC coupled to an ESI-quadrupole time of flight mass spectrometry system. The identification is based on four aglycons ((+)-catechin, (-)-epicatechin, (-)-epigallocatechin and (-)-epicatechin-3-O-gallate). More recently, Di Lecce et al. (2014) used the hydride quadrupole time of flight and triple quadrupole mass spectrometry for the determination of the phenolic compounds in skin, pulp and seed of the Albarino grape. Other than monomers, dimers and trimers, they found 2 new hybrid polyphenols, (+)-catechin hexoside and (-) epicatechin hexoside, never been reported in the grape berry.

Electroanalytical Techniques for Wine Polyphenols Analysis

Polyphenols are electroactive compounds that are easily oxidized at inert electrodes. The electrochemical profile of polyphenols, close to their redox behavior, is mainly driven by the stability of electro-generated phenoxyl radicals, which as a consequence determines the overall electrode reactions. The redox behavior is characterized by proton-electron transfer mechanisms in wide range of pH, whose extension is defined by chemico-structural properties as pKa and Ks. Thus, the electrochemical oxidation of these compounds at moderate potentials has been exploited for their detection. Therefore, the electrochemical characterization of polyphenols is a useful approach to evaluate the mechanisms involved in the stability, antioxidant activity, pro-oxidant effect and other chemical structural aspects of their redox behavior (Hendrickson et al., 1994; Zoulis and Efstathiou, 1996; Simic et al., 2007).

Among the electrochemical methods, the cyclic voltammetry is the most representative example of electroanalytical approaches, but as coupled techniques, in which electrochemical detectors are employed in chromatographic as well as electrochemical biosensors. These techniques have advantages over other analytical methods. They have rapid response, higher sensitivity, low detection limits, as well as the possibility to improve the selectivity by using suitable electrode conditions. Also, these techniques allow selective detection of polyphenols with good sensitivity, even in very complex samples as wine, and responses are independent of the optical path length or their turbidity.

Cyclic Voltammetry (CV) is an electrochemical technique which measures the current that develops in an electrochemical cell under conditions where voltage is in excess of that predicted by the Nernst equation. CV is performed by cycling the potential of a working electrode, and measuring the resulting current. This technique was the first electroanalytical method used for characterization of polyphenols and determination of total polyphenols content in wine. Figure 2 shows a cyclic voltammogram resulting from a single electron reduction and oxidation. The reduction process occurs from (a) the initial potential to (d) the switching potential. In this region, the potential is scanned negatively to cause a reduction. The resulting current is called cathodic current (i_{pc}). The corresponding peak potential occurs at (c), and is called the cathodic peak potential (E_{pc}). The E_{pc} is reached when all of the substrate at the surface of the electrode has been reduced. After the switching potential has been reached (d), the potential scans positively from (d) to (g). This results in anodic current (I_{pa}) and oxidation to occur. The peak potential at (f) is called the anodic peak potential (E_{pa}), and is reached when all of the substrate at the surface of the electrode has been oxidized.

Cyclic voltammetry has been used to discriminate red wines; to estimate the total phenolic content in wine, to study the antioxidant activity of red wines and provide a measure of the "redox spectra of wines"; to determine the levels of total polyphenols and sulfur dioxide, to evaluate the oxidation status of white wine and perceived astringency of red wines (Makhotkina and Kilmartin, 2010).

Kilmartin and co-workers are the most active researchers in the field since 2001. In order to estimate the antioxidant capacity of wine, they developed a method based on CV with glassy-carbon electrodes (GCEs). This type of electrode was the best one for this purpose because it minimizes interferences from ethanol. It was shown that low potential peaks were correlated to the presence of the phenolic compounds with high antioxidant activity while the high potential peaks correspond to harder-to-oxidize functional groups.

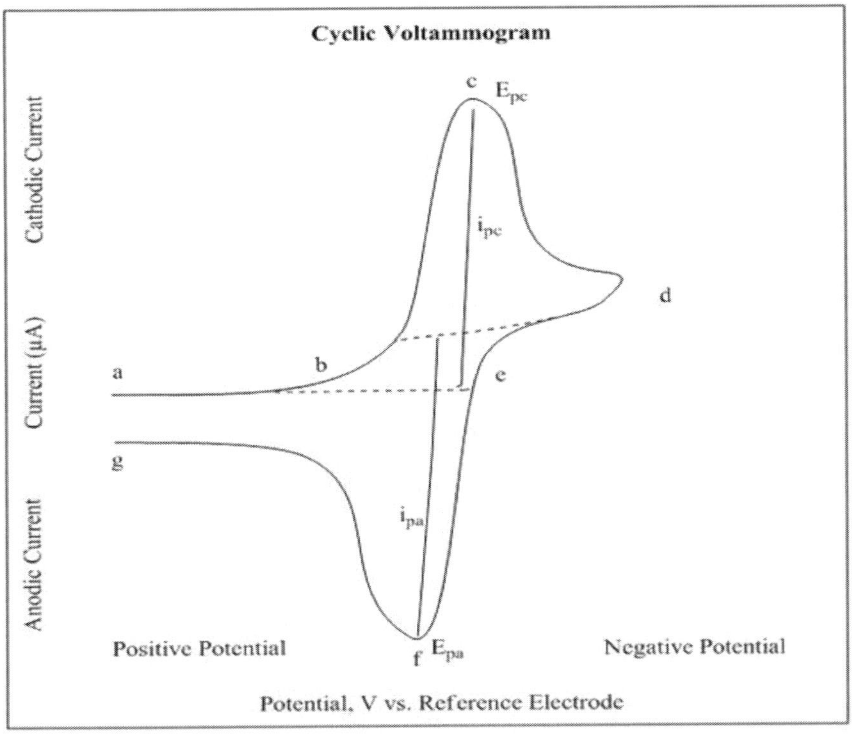

Figure 2. Cyclic voltammogram of a single electron oxidation-reduction.

Other investigations have been realized to evaluate the total phenolic content of different wines. They were calculated from the area under voltammetric recordings up to +500 mV in model wine solutions and were expressed as (+)-catechin equivalents. These results correlated with those obtained by Folin-Ciocalteu colorimetric assay for red wines and they corresponded to electrochemically easily-oxidizable polyphenols (Kilmartin et al., 2002; De Beer et al., 2004). This technique present some disadvantages because it measures the content of phenolic compounds containing only gallate and catechol groups in monomers, oligomers and polymers such as flavanols proanthocyanidins, flavanols and phenolic acids. It doesn't include the majority of anthocyanins and the phenolics of white wine.

Makhotkina and Kilmartin (2012) studied the polyphenol content of Sauvignon blanc grape juice by CV. From an examination of the cyclic voltammograms of these six fractions obtained by semi-preparative HPLC, the first anodic peak at 0.4 V for the cyclic voltammograms of grape juice appeared to be due mainly to the oxidation of caffeic-type hydroxycinammic acids (caftaric acid and the grape reaction product) and catechin. The second peak at 0.5 V was due to oxidation of quercetin-type flavonols, and HPLC–MS analyses confirmed that the main flavonols present in the Sauvignon blanc juices were quercetin 3-O-glucoside and quercetin 3-O-glucuronide. The third peak at 0.9 V had contributions from the oxidation of non-phenolic compounds. A good correlation was obtained between the total phenolic content in juices, measured by the Folin-Ciocalteu procedure, and the area under anodic current for scans taken to 700 mV.

In 2013, while studying the antioxidant capacity and polyphenol content of Douro wines, Rebelo et al. found a non-linear correlation between CV results and ABTS or Folin-Ciocalteau data for all types of wines (white, muscatel, ruby, tawny and red wines).

Overview of the Assay Methods Used to Estimate Wine Antioxidant Properties

Wine contains antioxidants that have protective and preventative effects against free-radical oxidations (Lugemwa et al., 2013). Many studies have shown that the lower incidence of cardiovascular disease mortality in southern France than in other European countries, despite a high consumption of saturated fats, was correlated with an increased intake of red wine (Renaud and de Lorgeril, 1992; Rimm et al., 1996 and Tunstall-pedoe et al., 1999), this is the French paradox. In fact, wine contains many polyphenolic compounds ((+)-catechin, quercetin, resveratrol and others…), which are powerful antioxidants that prevent lipoprotein free radical-mediated oxidation (Soleas et al., 1997; Frankel et al., 1995 and Teissedre et al., 1996). The reported evidences of beneficial health effects of phenolic compounds also include reducing plasma oxidation stress, slowing aging and reduction of hydroperoxide formation (Meyer et al., 1997; Sato et al., 1996). Moreover, phenolic compounds are regarded as preservatives against microbes and oxidation for food (Rodriguez-vaquero et al., 2007; Rhodes et al., 2006).

Several methods were employed to evaluate the antioxidant capacities of phenolic compounds extracted from various grapes, different parts of grapes and from wine. On one hand, there are chemical assay used such as 1,1-diphenyl-2-picrylhydrazine (DPPH) radical scavenging assays, 2,2-azinobis(3-ethylbenzthiazoline-6-sulphonic acid) (ABTS) radical scavenging method, oxygen radical absorbance capacity (ORAC) assay, N,N-Dimethyl-p-phenylene diamine (DMPD) assay, ferric reducing antioxidant power (FRAP) assay and total radical trapping antioxidant potential (TRAP) assay. On the other hand, *in vivo* methods are used like the low density Lipoprotein (LDL) oxidation. Others assays are based on the capacity of wine compounds to inhibit or to act on defense mechanisms of the body including enzymes such as xanthine oxidase, superoxide dismutase, catalase, and glutathione peroxidase, etc…

Chemical Assay

Chemical assays have been applied in order to evaluate possible differences between different types of wine (red, white, sherry wines and other …). For example, Fernández-Pachón et al. (2004) have explored the relation between phenolic composition and antioxidant values by searching for correlations with the total polyphenolic contains as well as examining the contribution of the different polyphenolic families to the total antioxidant values. They have selected methods including the measurement of scavenging capacity against a given radical (ABTS, DPPH and DMPD assays) and the protection of a protein from suffering oxidation the oxygen radical absorbance capacity (ORAC) for the evaluation of antioxidant features of wines. Their feasibility, advantages and disadvantages, practical limitations and usefulness are considered.

Others scientist such as Alonso et al. (2002) have used these methods for studying the antioxidant activity of wine byproducts like grape marcs, stalks, and lees of both wines, white and red varieties.

DPPH Method

The DPPH radical scavenging assay was first described by Blois in 1958. Later, it was slightly modified by numerous researchers. It is one of the most extensively used antioxidant assays for plant samples and for wine compounds. In fact, during these last 20 years, we note an increase of publications which items are based on Wine and DPPH (Figure 3A). Between 2009 and 2012 years, there were more than 50 publications about these subjects. Furthermore, Figure 3B representing the number of publications about Wine & DPPH cited in each year, shows that there is an increase of citations about these subjects during these same years, and with more than 1400 citations between 2010 and 2013.

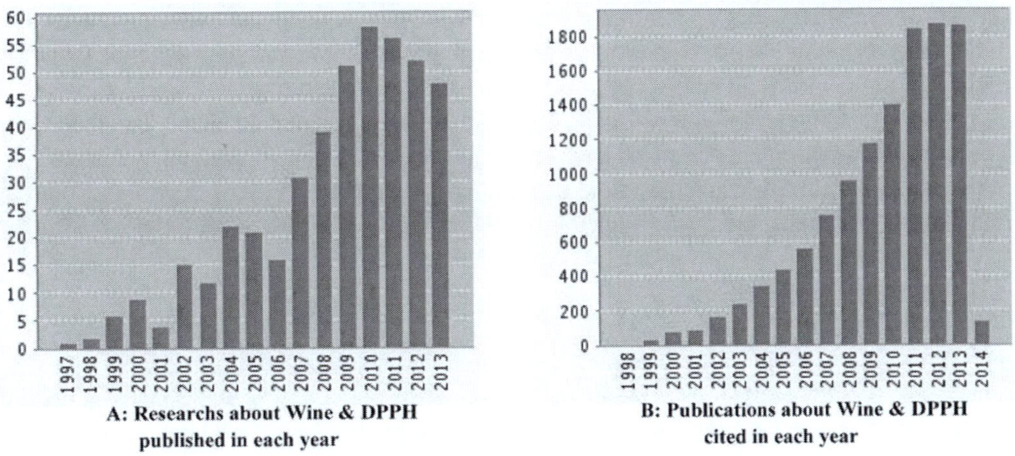

Figure 3. Citation report about Wine and DPPH method indexed within the database (Source: Web of knowledge, 02/05/2014".

DPPH$^{\cdot}$ is a stable free radical that reacts with compounds that can give a hydrogen atom. This method is based on the scavenging of DPPH through the addition of an antioxidant for example R-OH that decolorizes the DPPH$^{\cdot}$ solution. The reduction of DPPH$^{\cdot}$ as indicated below (Figure 4) is followed by monitoring the decrease in its absorbance at a characteristic wavelength during the reaction.

In its radical form, DPPH$^{\cdot}$ absorbs at 515 nm, but upon reduction by an antioxidant, DPPH-H form is obtained and the absorption disappears. The reaction mixture is composed of a solution of DPPH in methanol at known concentration in which a known volume is added to the antioxidant samples solutions at various concentrations of methanol. Some minutes later (between 15 to 30 min), the absorbance is measured. A large decrease in the absorbance of the reaction mixture indicates significant free radical scavenging capacity of the compounds present in sample and thus antioxidant activity. In the same way, a high absorbance of the reaction mixture indicates low free radical scavenging capacity.

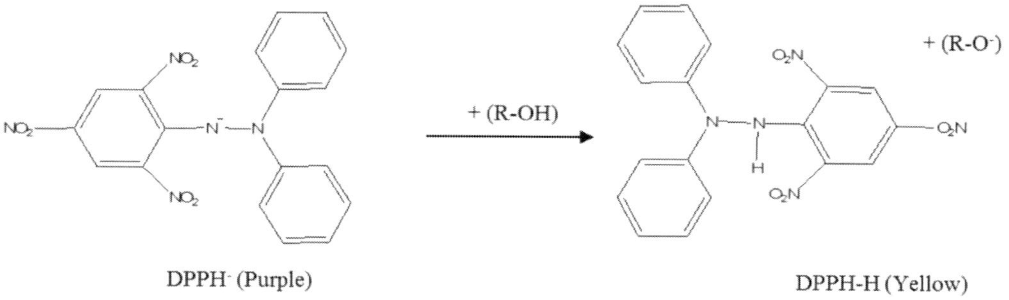

Figure 4. Reaction of DPPH scavenging by an antioxidant.

According to the literature, the antiradical activity could be defined as various manners. For example, Brand-Williams et al. (1995) determined a percentage of stable $DPPH^-$ and transferred the values onto a graph showing the percentage of residual stable $DPPH^-$ as a function of the molar ratio of antioxidant to $DPPH^-$. They defined the antiradical activity as the quantity of antioxidant (AO) necessary to decrease the concentration of initial $DPPH^-$ by 50% (Efficient Concentration (EC_{50}) = (AO (mol/L)/$DPPH^-$(mol/L)).

In the case of wine, Lugemwa et al. (2013) defined the antiradical activity as the volume of wine in the solutions needed to decrease the initial DPPH concentration by 50% together with the quantity of phenolic compound in mg/L to obtain the LC_{50} values in mg of phenolic compound/L. The capacity to scavenge the DPPH radical is calculated using the equation [4]:

$$I = [((ODo - ODs)/ODo) \times 100] \quad [4]$$

where I is the inhibition percentage; ODo is the absorbance of the negative control (containing 100 µL of MeOH instead of the sample); and ODs is the absorbance of the samples. The percent inhibition was plotted against volumes of wine and the volume needed to decrease DPPH concentration by 50% was calculated from the graph ($R^2 \geq 0.95$).

As for Fernandez-Pachon et al. (2004) they analyzed and compared the antioxidant activity of 41 wines by using trolox, a water-soluble analogue of Vitamin E, as reference. Results are expressed as trolox equivalent antioxidant capacity (TEAC).

ABTS Method

Rice-Evans and Miller (1994) were the first to develop the ABTS radical scavenging method and then they used it for measurement of the antioxidant ability of wine samples.

Along with the DPPH method, the ABTS assay is one of the most extensively used antioxidant methods for plant samples. During the last 20 years, there is an increase of researches concerning the study about wine antioxidant using ABTS assay (Figure 5A). However, we note that this increase isn't constant and contrary to the DPPH assay, there were only 24 publications about these subjects in 2011 and 2012. As far as the publications about Wine and ABTS items cited, there were around 1200 citations in the same years (Figure 5B).

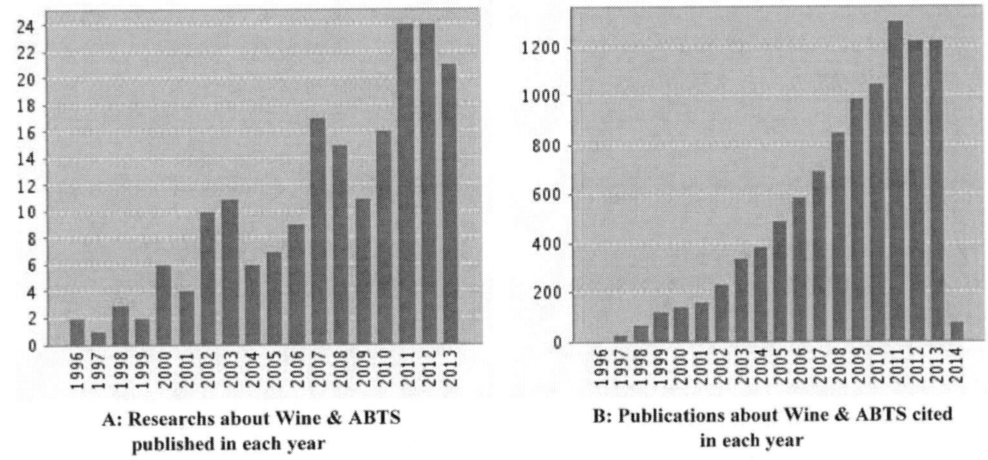

Figure 5. Citation report about Wine and ABTS method indexed within the database (Source: Web of knowledge, 02/05/2014).

The original ABTS method was based on the activation of metmyoglobin with hydrogen peroxide in the presence of ABTS to produce a radical cation in the presence or absence of antioxidant according the following reaction (Figure 6).

$$\text{Metmyoglobine} + H_2O_2 \longrightarrow \text{Ferrylmyoglobine} + H_2O$$

$$\text{ABTS} + \text{Ferrylmyoglobine} \longrightarrow \text{ABTS}^{\bullet+} + \text{Metmyoglobine}$$

Figure 6. The chemical method of ABTS radical scavenging method.

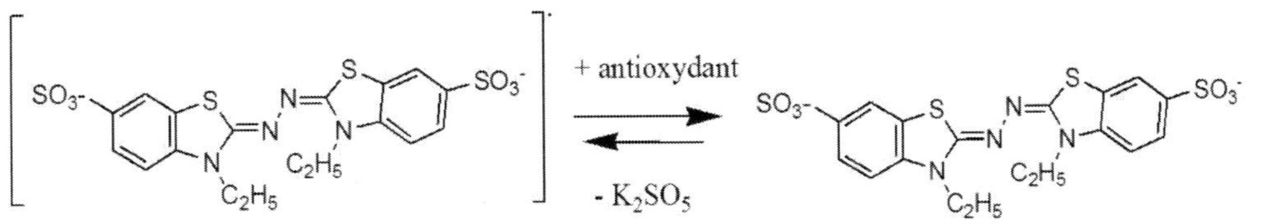

Figure 7. Reaction of the ABTS radical in the presence of the antioxidant compound during the ABTS assay.

However, this was criticized because the faster reacting antioxidants might also contribute to the reduction of the ferrylmyoglobin radical. In this case, the assay was modified on the basis of a decolorization technique allowing a direct generation of ABTS radical in a stable form (Re et al., 1999). Thus, the assay consisted of an enzymatic system containing a peroxidase enzyme, its substrate of oxidation (hydrogen peroxide) and the chromophore (ABTS). A blue/green radical cation $ABTS^{\bullet+}$ is generated by the oxidation of ABTS with potassium persulfate. The absorbance of the radical solution is inhibited when it is exposed to an antioxidant (Figure 7).

The reduction in the presence of hydrogen-donating antioxidants is measured with a spectrophotometer at 734 nm. Re et al. (1999), have shown that the effect of the antioxidant concentration and the duration of the inhibition of the radical cation absorption have to be taken into account when the antioxidant activity is determined.

Usually, trolox is used as a positive control. Thus, the antioxidant activity is expressed in terms of the trolox-equivalent antioxidant capacity of the extract (TEAC/mg or TEAC/mM). According to the literature, the reaction mixture and conditions could slightly vary.

ORAC Method

Cao et al. (1993) were the first who developed the ORAC method. Thanks to the Figure 8, we can note that since in 2001, there was an increase of the use of ORAC assay for the measurement of wine antioxidant activity. In fact, during these last decades, the assay has been widely used in many recent studies of wine. However, the number of researches published and the publications cited about wine and ORAC were lower than those about wine and DDPH or wine and ABTS. The ORAC method is less used than the two methods quoted above.

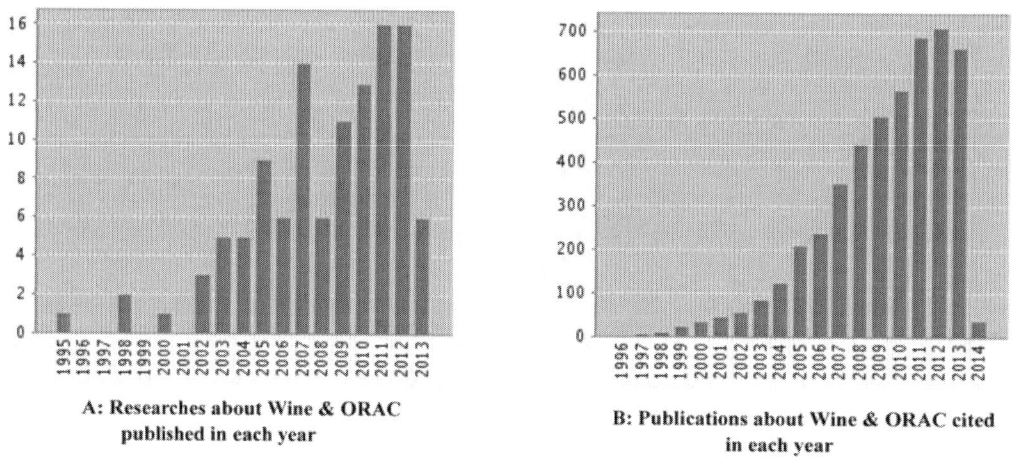

Figure 8. Citation report about Wine and ORAC indexed within the database (Source: Web of knowledge, 02/05/2014).

The original method consisted of the measurement of the decrease in the fluorescence of a protein as a result of the loss of its conformation when it suffers oxidative damage caused by a source of peroxyl or free radicals and oxidant sources. The method allowed the measure of the antioxidants ability to protect the protein from oxidative damage. The b-phycoerythrin (b-PE) was the protein used in the original method, but it had a series of disadvantages, such as inconsistency between batches, photosensitivity and interaction with phenolic compounds owing to nonspecific protein binding. In order to solve this problem, Ou et al. (2001) proposed using the fluorescein (30,60-dihydroxyspiro[isobenzofuran-1[3H],9'[9H]-xanthen]-3-one) or the dichlorofluorescein (H_2DCF-dA; 2',7'-dichlorodihydrofluorescein diacetate) as the target protein because they are more stable and less reactive. Then, probe reaction with radicals is followed by loss of fluorescence over time.

ORAC assay is based on the inhibition of radical induced oxidations and reflects radical chain breaking antioxidation activity by H atom transfer. In fact, the presence of an oxidant leads to the formation of peroxyl radicals that directly attack the protein molecule and reduce its fluorescence. The fluorescence decay curve was measured for extend period, for example 80 min, using emission and excitation wavelengths of 584 and 544 nm, respectively.

The results were calculated by measuring the difference of net areas under the curve (AUCnet) taken from the spectrofluorometer between the white and the sample colors, expressed as trolox equivalent. Nowadays, it is the sole method that takes the free radical reaction to completion and uses an AUC technique for quantification. The ORAC assay provides a controllable source of peroxyl radicals that model reactions of antioxidants with lipids in both food and physiological systems, and it can be adapted to detect both hydrophilic and hydrophobic antioxidants by altering the radical source and solvent (Ou et al., 2001; Huang et al., 2002; Prior et al.,2003).

DMPD Assay

DMPD assay's reaction is rapid and the stable end point is taken as a measure of the antioxidant efficiency. This method was used by Flogliano et al. (1999) and Rivero-Pérez et al., (2007) in order to evaluate "antioxidant profile" of wine samples collected in Campania, Italy, and red wines from Spain respectively.

However, contrary to the methods quoted above, the DMPD assay is very little used for measuring the antioxidant activity of wine. In fact, during these last years, the number of publications about wine and DMPD assay doesn't exceed three (Figure 9) in spite of an increase of publications cited about these items.

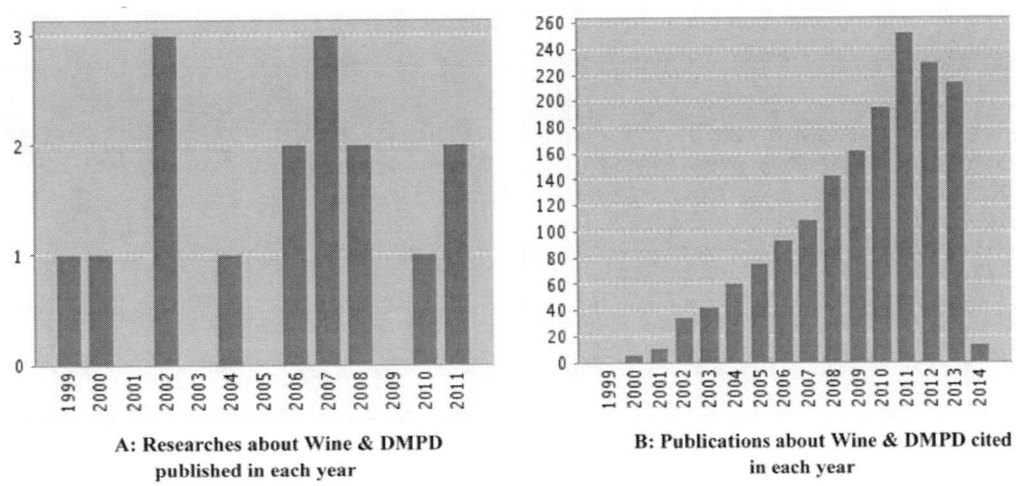

Figure 9. Citation report about Wine and DMPD method indexed within the database (Source: Web of knowledge. 02/05/2014).

The principle of this radical-scavenging method is that in the presence of a suitable oxidant and at an acidic pH, the DMPD can form a stable and colored radical cation ($DMPD^{\bullet+}$) which presents a maximum of absorbance at 505 nm. (Figure 10, step 1). The presence of an antioxidant compounds (AO) which is able to transfer a hydrogen atom to $DMPD^{\bullet+}$, causes a decolorization of the solution that is proportional to their amounts (figure 10, step 2).

$$\text{DMPD (uncolored)} + \text{oxidant } (Fe^{3+}) + H^{+} \longrightarrow DMPD^{\bullet+} \text{ (purple)}$$
$$DMPD^{\bullet+} \text{ (purple)} + \text{AOH} \longrightarrow DMPD^{+} \text{ (uncolored)} + \text{AO}$$

Figure 10. The chemical reaction of DMPD assay.

The absorbance at 505 nm was measured and was transformed in antioxidant activity using trolox as reference. Results are expressed as trolox equivalent antioxidant capacity (TEAC).

FRAP Assay

The FRAP method was developed in 1996 by Benzie et al. in order to measure the ferric reducing power in plasma at low pH. Then, the method was adapted and used for the assay of antioxidants in botanicals (Prior et al., 2005). The reaction allows the measurement the reduction of ferric 2,4,6-tripyridyl-s-triazine (Fe^{3+}-TPTZ) to a colored product (Fe^{2+}-TPTZ) (Figure 11). The absorption of Fe^{2+} - TPTZ is recorded at 593 nm.

$$Fe^{3+}\text{ - TPTZ} + \text{Reducing antioxidant} \longrightarrow \underset{\text{(blue)}}{Fe^{2+}\text{ - TPTZ}} + H_2O$$

Figure 11. The chemical reaction of Frap assay.

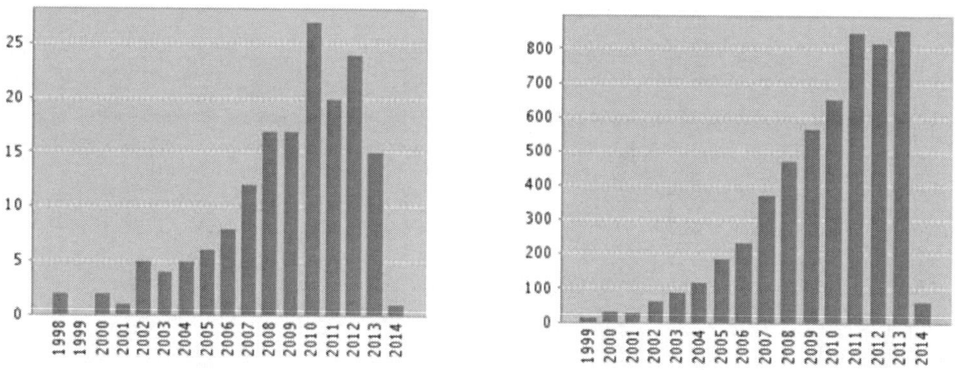

A: Researches about Wine & FRAP published in each year

B: Publications about Wine & FRAP cited in each year

Figure 12. Citation report about Wine and FRAP method indexed within the database (Source: Web of knowledge, 02/05/2014).

The reaction detects compounds with redox potentials of <0.7 V. The reducing power appears to be related to the degree of hydroxylation and extent of conjugation of reductive agent like phenolics (Rivero-Pérez et al., 2007). The results were expressed as moles of Fe-(II), using linear calibration obtained with different concentrations of $FeSO_4$. Sometimes, the results could be expressed as Trolox equivalent antioxidant capacity (TEAC) or ascorbic acid equivalent antioxidant capacity (AEAC).

According to figure 12, it is obvious that the FRAP method was used by several scientists these last years. There was an increase of papers published about wine and FRAP's items until 2010 and an increase of publications cited about these items until 2011.

For example, the FRAP assay was used in order to determine the antioxidant capacity of red wine produced in California or in Spain (Gil et al., 2000; Rivero-Pèrez et al., 2007). Other scientists as Katalinic et al. (2004) have applied this assay for investigation of antioxidant capacity of selected wines and their correlation with wine phenolic composition and antioxidant capacity. The FRAP assay is simple, speedy, inexpensive, and robust and does not require specialized equipment. It can be performed using automated, semiautomatic, or manual methods. However, this assay does not measure thiol antioxidants, such as glutathione.

TRAP Assay

The TRAP method is based on measuring oxygen consumption during a controlled lipid oxidation reaction induced by thermal decomposition of 2,2′-Azobis(2-amidinopropane)hydrochloride (AAPH) (Ghiselli et al., 1995). The antioxidants could interfere in the reaction between peroxyl radicals generated by AAPH and a target probe. Various methods have used fluorescence of R-phycoerythrin, or absorbance of ABTS as the reaction probe (De lange et al. 1989; Bartosz et al., 1998). For the assay, the requirements are that the probe must be reactive with radicals at low concentrations allowing a spectroscopic change between the native and oxidized probe. Thus, any radical chain reaction beyond probe oxidation should occur. The oxidation of the probe could be followed optically or by fluorescence (Prior et al., 2005).

During the induction period (lag phase), in the presence of wine's antioxidant, the oxidation is inhibited. The length of induction period is compared to that of trolox (as an internal standard) and then quantitatively related to the antioxidant capacity of the wine. TRAP values could be calculated by percent reduction of a reaction and expressed as a lag time or reaction time of the sample compared to corresponding times for trolox or expressed as mmol of trolox per kg (solid foods) or per L (beverages) of sample.

The TRAP assay was most often used for measurements of antioxidant activity in wine. As shown in Figure 13, during the last decades, the number of researches about wine antioxidants using TRAP method remained constant. Moreover, the publications cited about these items is in a constant increasing, that demonstrates a growing interest for the subject.

In the literature, several applications of this technique for the assessment of wine antioxidant capacity can be found. For example, Pellegrini et Al., (2003) have applied TRAP assay to assess the total antioxidant capacity of beverages like red and white wines commonly consumed in Italy to obtain robust data useful for determining the potential intake of antioxidants in Italian population studies. TRAP assay was also applied for the in vivo measurements of antioxidant capacity in plasma. In fact, Serafini et al. (1998) have determined the effect of the non-alcoholic fraction of red and white wines on the TRAP and on the total plasma phenolic levels in vivo in humans.

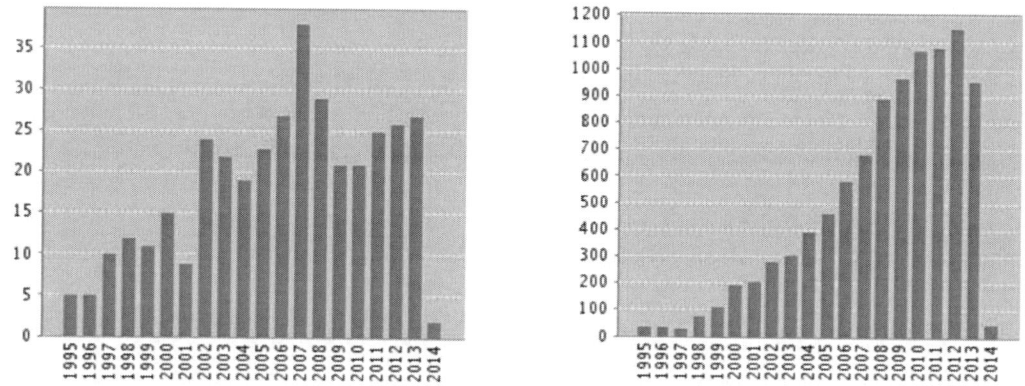

A: Researches about Wine & TRAP published in each year

B: Publications about Wine & TRAP cited in each year

Figure 13. Citation report about Wine and TRAP method indexed within the database (Source: Web of knowledge 02/05/2014).

Comparison of Chemical Methods

According to some scientists, on methodological point of view, the DPPH methods and FRAP methods are recommended because they are easy and accurate methods for measuring the antioxidant activity (Gil et al., 2000). Moreover, it is important to note that DPPH has been described as more specific for lipophilic antioxidants, DMPD and FRAP for hydrophilic antioxidants, and ABTS and ORAC for both classes. The DMPD method should be used with caution in those extracts rich in organic acids.

From quantitative point of view, it must be considered that each method works with a different mechanism and sensibility, that explain the quantitative differences detected among methods. Although, each method has its own merits and drawbacks, most of publications dealing with the subject about wine antioxidant activity, have found that the most common and reliable methods are the DPPH and ABTS methods. In recent years, these two assays have been modified and improved for the determination of wine phenolic antioxidant activity (Table 1).

Thanks to table 1, we can note that the number of OH group and its position on the structure of compound could influence the antioxidant capacity of phenolic molecules. In fact, when the OH added onto the flavonoid nucleus, the activity is enhanced. However, when the OH is substituted by the OCH_3 groups, the activity is diminished. Several scientists have confirmed theses remarks (Arora et al., 1998; Majo et al., 2005). For example, the *o*-diphenoxyl groups in resveratol were shown to exhibit higher antioxidant activity than other compositions (Qian et al., 2009).

Table 1. Comparison of IC_{50} values of wine phenolic compounds obtained with DDPH and ABTS assays

Wine phenolic compound	Structure	IC_{50} DPPH (mg/L)	IC_{50} ABTS (mg/L)	References
Butylhydroxyanisol (BHA)	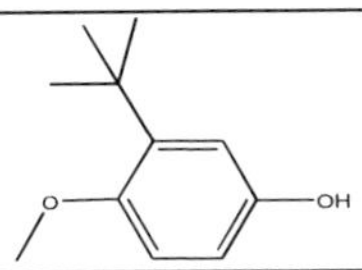	10.97 3.08	4.95 -	Scherer et al., 2009
Caffeic acid	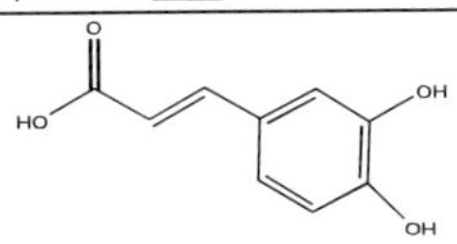	1.80±0.05 - 12.4±0.7	- 5.92 -	Harish Nayaka et al., 2010 Chuan-Ling et al., 2009 Erkan et al., 2011
Catechin	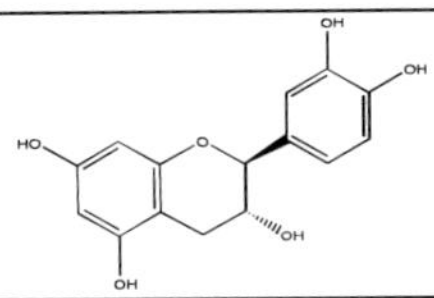	- 1.96±0.4 16.75	1.88 - 2.26	Snijman et al., 2009 Iacopini et al., 2008 Li et al., 2005
Chlorogenic acid	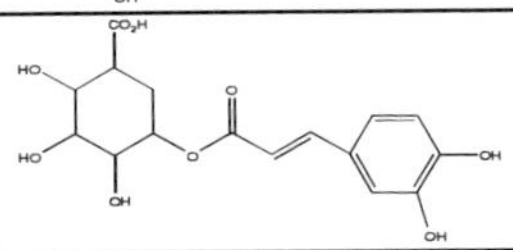	2.75 35.6±2.1	- -	Scherer et al., 2009 Erkan et al., 2011
Epicatechin	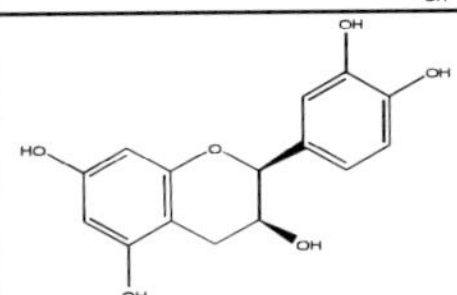	1.99±0.2		Iacopini et al., 2008
Ferulic acid	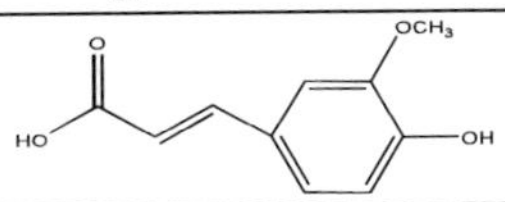	>19.418 49.6±2.3 5.84	2.45 - -	Li et al., 2005 Erkan et al., 2011 Scherer et al., 2009
Gallic acid	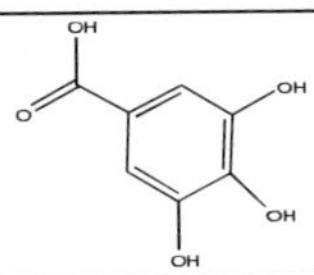	1.1±0.04 2.23	- -	Harish Nayaka et al., 2010 Erkan et al., 2011
Isoquercitrin	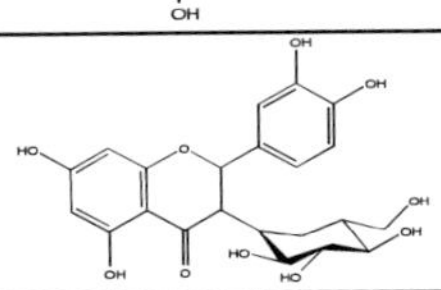	-	5.99	Snijman et al., 2009
Kaempferol	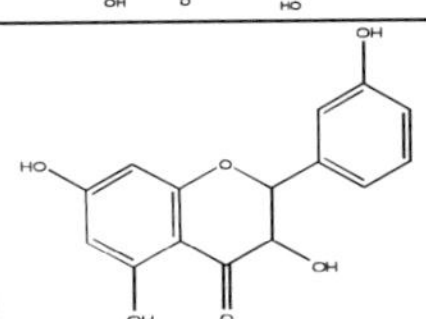	26.6±2.5	-	Erkan et al., 2011
Myricetin	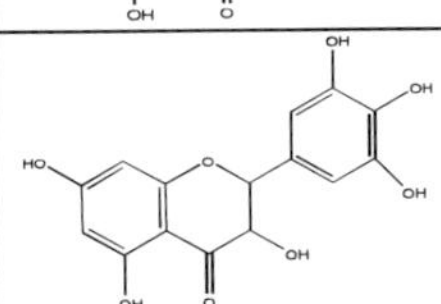	438.60	-	Erkan et al., 2011

Table 1. (Continued)

Wine phenolic compound	Structure	IC_{50} DPPH (mg/L)	IC_{50} ABTS (mg/L)	References
Naringenin	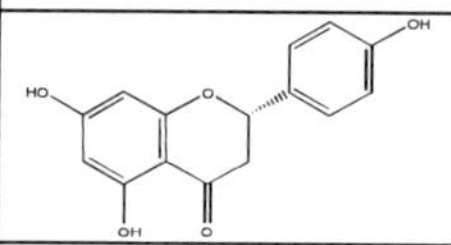	>81.68	-	Lee et al.2006
P-coumaric acid	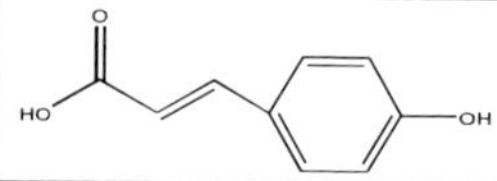	1.90±0.04 1105.3±4.3	- -	Harish Nayaka et al., 2010 Erkan et al., 2011
P-hydroxybenzoïc acid	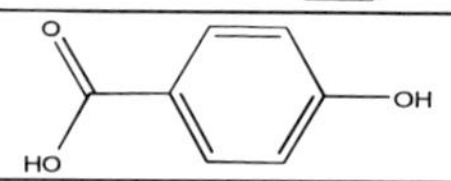	33.75±1,72	-	Harish Nayaka. et al., 2010
Protocatechuic acid	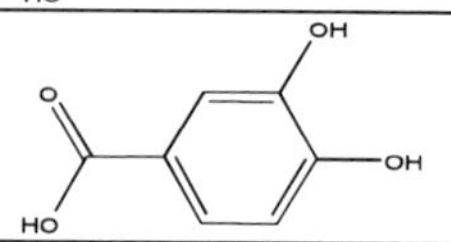	1.35±0.08 1.45	- -	HarishNayaka et al., 2010 Scherer et al., 2009
Quercetin	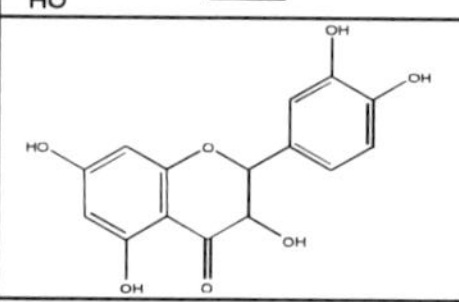	0.09±0.01 2.266±0.9 10.58±0.07	0.14±0.03 1.2±0.048 6.16±0.27	Oliveira et al., 2011 Lee et al., 2006 Wua et al., 2007
Resveratrol	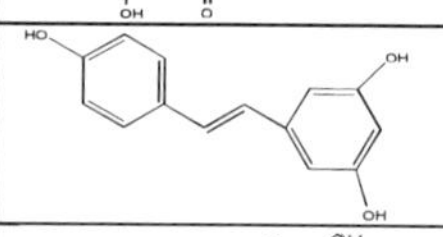	13.21±0.1		Iacopini et al., 2008
Rutin	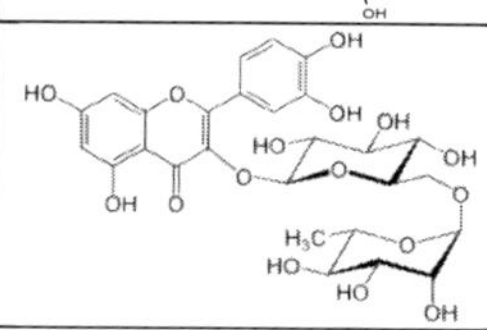	- 4.56±0.2 4.97±0.16 -	0.51±0.01 6.39 - -	Vijayabaskaran et al., 2010 Snijman et al., 2009 Iacopini et al., 2008 Badami et al., 2005
trans-cinnamic acid	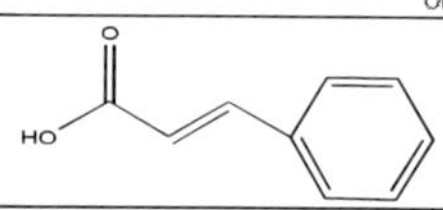	4.60±0.34 -	- -	Harish Nayaka et al., 2010 Scherer et al., 2009
Vanillic Acid	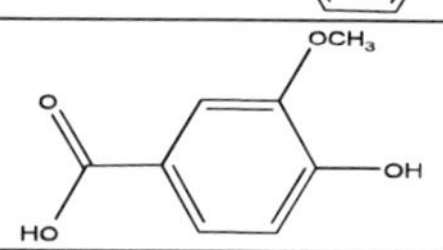	49.5±2.4	-	Harish Nayaka et al., 2010

Biological Assay

In the literature, the biological assays are most focused on the antioxidant ability of naturals polyphenols derived from wine, such as resveratrol, quercetin, catechin and others.

LDL Oxidation

The LDL method was developed primarily as a measure of antioxidant status and then adapted to assess antioxidant capacity in a more physiologically relevant system (Prior et al.,

2005). LDL is isolated from blood samples, and its oxidation is initiated by AAPH. During the formation of hydroperoxides from unsaturated fatty acids, conjugated dienes are generated due to the rearrangement of the double bonds. The resulting conjugated dienes exhibit an important absorption at 234 nm (Handelman et al., 1999; Neudörffer et al., 2004). According to Handelman et al. (1999), there is a good relationship between LDL oxidation using AAPH and the ORAC value on a limited group of samples.

Thanks to this method, Rifici et al. (2002), have studied the impacts of several phenolic compounds found in red wine, individually and in combination, on the cell-mediated oxidation of lipoproteins. The effects of these compounds on LDL oxidation were measured by the production of conjugated dienes, lipid peroxides, and trinitrobenzene sulfonic acid. Thus, their results indicated that catechin and epicatechin were the major contributors to the antioxidant activity of red wine.

Other examples, *in vivo* experiments have shown that the resveratrol increases plasma antioxidant capacity and decreases lipid peroxidation which is strongly associated with the risk of coronary heart disease (Holvoet et al., 2004) and myocardial infarction (Fernández-Mar et al., 2012). Studies in rats and humans indicated that phenolic compounds can suppress pathological increases in the peroxidation of lipids in vivo, but nowadays the mechanism direct, indirect or both, is not clear yet (Rifici et al., 2002, Gresele et al., 2011).

Enzymatic Systems

The antioxidant capacity of a compound could be depend upon its capacity to interact with the enzymatic pathways involved in reactive oxidant species (ROS) formation (Gresele et al., 2011). Phenolic compounds may affect indirectly the redox status by increasing the ability of endogenous antioxidants, activation of key regulating enzymes or by inhibiting enzymatic systems involved in ROS formation such as xanthine oxidase (XOD), glutathione peroxidase or reductase (GPx or GR), superoxide dismutase (SOD) and catalase (CAT) etc…

The XOD pathway has been implicated in the oxidative injury to tissues, especially after ischemia-reperfusion where xanthine dehydrogenase and xanthine oxidase are involved in the metabolism of xanthine to uric acid (Van Hoorn et al., 2003).

Xanthine dehydrogenase is the form of the enzyme present under physiologic conditions, but its configuration is changed to XOD during ischemic conditions. In the reperfusion phase (i.e., reoxygenation), XOD reacts with molecular oxygen releasing superoxide free radicals. Two flavonoids, quercetin and silibin, can inhibit XOD activity resulting in decreased oxidative injury (Shoskes 1998).

Kasdallah-Grissa et al. (2007) have evaluated the capacity of resveratrol to prevent hepatic oxidative status as an expression of the cellular damage caused by free radicals, and to act on antioxidant defense mechanism during chronic ethanol treatment. In this aim, they have performed analyses of antioxidant enzymes activity by using various and specific kits with spectrophotometer for the determination of SOD, GPx, GR and CAT levels. Thus, their results showed that liver activity of antioxidant enzymes, such as SOD, GPx and CAT were ameliorated by resveratrol supplemented diet during ethanol administration.

Conclusion

Wine polyphenols are a much complex group. Nowadays, there is a growing interest in substances exhibiting antioxidant properties which can provide health benefits to human. This chapter provides information on the simple and advanced analytical methods used for the qualitative and quantitative assessment of wine polyphenols. Nowadays, the necessary technology is available to achieve the desired analytical goals concerning the separation and quantification of wine polyphenols. Improvement is still needed because many unidentified compounds still remain especially the polymeric fractions.

Hundreds of publications on the antioxidant of effects of wine polyphenols have been appeared over the past two decades. Several methods were presented and discussed in this chapter. These methods were divided into two groups: chemical assays and biological methods. The obtained results with these tests differ from those obtained in real biological systems which depend on the bioavailability of the antioxidants and their metabolism.

References

Alonso, A., Guillean, D., Barroso, C., Puertas, B., Garcia, A., (2002). Determination of antioxidant activity of wine byproducts and its correlation with polyphenolic content. *J. Agric. Food Chem*, 50, 5832– 5836.

Arora, A., Nair, M.G., Strasburg, G.M., (1998). Structure-activity relationships for antioxidant activities of a series of flavonoids in a liposomal system. *Free Radic. Biol. Med.* 24, 1355–1363.

Badami S., Prakash O.M., Dongre., S.H., Suresh., B., (2005). In Vitro antioxidant properties of *Solanum pseudocapsicim* leaf extracts. *Indian J. Pharmacol.* 37, 251-252.

Bakker, J., Bridle, P., Timberlake, C.F., (1986). Tristimulus measurements (CIELAB 76) of port wine color. *Vitis* 25, 67-78.

Baldi, A., Romani, A., Mulinacci, N., Vincieri, F.F., Casetta, B., (1995). HPLC/MS application to anthocyanins of *Vitis vinifera* L. *J. Agric. Food Chem.* 43, 2104-2109.

Bartosz, G., Janaszewska, A., Ertel, D., Bartosz, M., (1998). Simple determination of peroxyl radical-trapping capacity. *Biochem. Mol. Biol. Int.* 46, 519-528.

Bendaoud, H., Romdhane, M., Souchard, J. P., Cazaux, S., Bouajila, J., (2010). Chemical composition and anticancer and antioxidant activities of *Schinus Molle* L. and *Schinus Terebinthifolius* Raddi Berries Essential Oils. *J. Food Sci.* 75, 466-472.

Benzie, I. F. F., Strain, J. J., (1996). The Ferric reducing ability of plasma (FRAP) as a measure of "antioxidant power": the FRAP assay. *Anal. Biochem.* 239, 70-76.

Blois, M.S., (1958). Antioxidant determinations by the use of a stable free radical. *Nature* 181, 1199–1200.

Brand-Williams, W., Cuvelier, M.E., Berset, C., (1995). Use of free radical method to evaluate antioxidant activity. *LWT - Food Sc. Technol.* 28, 25–30

Cao, G., Alessio, H. M., Cutler, R. G., (1993). Oxygen-radical absorbance capacity assay for antioxidants. *Free Radic. Biol. Med.* 14, 303–311.

Castellari, M., Sartini, E., Fabiani, A., Arfelli, G., Amati, A., (2002). Analysis of wine phenolics by high-performance liquid chromatography using a monolithic type column. *J. Chromatogr A.* 973, 221-227.

Castillo-Munoz, N., Fernandez-Gonzalez, M., Gomez-Alonso, S., Garcia-Romero, E., Hermosin-Gutierrez, I., (2009). Red-color related phenolic composition of Garnacha Tintorera (*Vitis vinifera* L.) grapes and red wines. *J. Agric. Food Chem.* 57, 7883-7891.

Cheynier, V., Doco, T., Fulcrand, H., Guyot, S., Le Roux, E., Souquet, J. M., Rigaud, J., Moutounet, M., (1997). ESI-MS analysis of polyphenolic oligomers and polymers: New methods for analysing old plant polyphenols. *Analusis* 25, M32-M37.

Chuan-Ling, S., Jin-Kyu, K., Young-Soo, B., Shu-Ming, L., (2009). Phenolic compounds in the leaves of populusussuriensis and their antioxidant activities. *Planta Med.* 75, 1165-1167.

Cilliers, J., Singleton, V., (1990). Autoxidative phenolic ring opening under alkaline conditions as a model for natural polyphenols in food. *J. Agric. Food Chem.* 38, 1797-1798.

De Beer, D., Harbertson, J. F., Kilmartin, P. A., Roginsky, V., Barsukova, T., Adams, D. O., Waterhouse, A. L., (2004). Phenolics: A comparison of diverse analytical methods. *Am. J. Enol. Vitic.* 55, 389-400.

De Lange, R. J., Glazer, A. N. (1989). Phycoerythrin fluorescence-based assay for peroxy radicals: a screen for biologically relevant protective agents. *Anal. Biochem.* 28, 300-306.

De Villiers, A., Cabooter, D., Lynen, F., Desmet, G., Sandra, P., (2009). High performance liquid chromatography analysis of wine anthocyanins revisited: effect of particle size and temperature. *J. Chromatogr A.* 1216, 3270-3279.

De Villiers, A., Vanhoenacker, G., Majek, P., Sandra, P., (2004). Determination of anthocyanins in wine by direct injection liquid chromatography-diode array detection mass spectrometry and classification of wines using discriminant analysis. *J. Chromatogr A.*1054, 195-204.

Delcambre, A., Saucier, C., (2012). Identification of new flavan-3-ol monoglycosides by UHPLC-ESI-Q-TOF in grapes and wine. *J. Mass Spectrom.* 47, 727-736.

Di Lecce, G., Arranz, S., Jáuregui, O., Tresserra-Rimbau, A., Quifer-Rada, P., Lamuela-Raventós, R. M., (2014). Phenolic profiling of the skin, pulp and seeds of Albariño grapes using hybrid quadrupole time-of-flight and triple-quadrupole mass spectrometry. *Food Chem.*145, 874-882.

Du Toit, W.J., Lisjak, K., Marais, J., du Toit, M., (2006). The effect of micro-oxygenation on the phenolic composition, quality and aerobic wine-spoilage microorganisms of different south african red wines. *S. Afr. J. Enol. Vitic.* 27, 57-67.

Erkan, N., Cetin, H., Ayranci, E., (2011). Antioxidant activities of *Sideritiscongesta Davis* et Huber-Morath and *Sideritisarguta Boiss* et Heldr: Identification of free flavonoids and cinnamic acid derivatives. *Food Res. Int.* 44, 297–303.

Fernández-Mar, M.I., Mateos, R., García-Parrilla, M.C., Puertas, B., Cantos-Villar, E., (2012). Bioactive compounds in wine: Resveratrol, hydroxytyrosol and melatonin: A review. *Food Chem.* 130, 797–813.

Fernández-Pachón, M. S., Villano, D., Garcıa-Parrilla, M. C., Troncoso, A. M., (2004). Antioxidant activity of wines and relation with their polyphenolic composition. *Anal. Chimi. Acta*, 513, 113-118.

Flamini, R., (2003). Mass spectrometry in grape and wine chemistry. Part I: Polyphenols. *Mass Spectrom. Rev.* 22, 218-250.

Flamini, R., (2013). Recent applications of mass spectrometry in the study of grape and wine polyphenols. *ISRN Spectroscopy*, 2013.

Fogliano, V., Verde, V., Randazzo, G., Ritieni, A., (1999). Method for measuring antioxidant activity and its application to monitoring the antioxidant capacity of wines. *J. Agric. Food Chem.* 47, 1035-1040.

Frankel, E.N., Waterhouse, A.L., Teissedre, P.-L., (1995). Principal phenolic phytochemicals in selected Californian wines and their antioxidant activity in inhibiting oxidation of human low-density lipoproteins. *J. Agric. Food Chem.* 43, 890–894.

Fulcrand, H., Mané, C., Preys, S., Mazerolles, G., Bouchut, C., Mazauric, J.-P., Souquet, J.-M., Meudec, E., Li, Y., Cole, R., Cheynier, V., (2008). Direct mass spectrometry approaches to characterize polyphenol composition of complex samples. *Phytochemistry* 69, 3131-3138.

German, B., Walzem, R., (2000). The health benefits of wine. *Annu. Rev. Nutr.* 20, 561-593.

Ghiselli, A., Serafini, M., Maiani, G., Azzini, E. Ferro-Luzzi, A., (1995). A fluorescence-based method for measuring total plasma antioxidant capability. *Free Radic. Biol. Med.* 18, 29–36.

Gil, M. I. Tomas-Barberan F., A, Hess-Pierce, B., Holcroft, D., M., Kader, A., A., (2000). Antioxidant activity of pomegranate juice and its relationship with phenolic composition and processing. *J. Agric. Food Chem.* 48, 4581-4589.

Glories, Y., (1984 a). La couleur des vins rouges. I- Les équilibres des anthocyanes et des tannins. *Conn. Vigne Vin.* 18, 195-217.

Glories, Y., (1984 b). La couleur des vins rouges. II- Mesure, origine et interprétation. *Conn. Vigne Vin.* 18, 253-271.

Gomez-Alonso, S., Garcia-Romero, E., Hermosin-Gutierrez, I., (2007). HPLC analysis of diverse grape and wine phenolics using direct injection and multidetection by DAD and fluorescence. *J. Food Comp. Anal.* 20, 618-626.

Gordillo, B., López-Infante, M. I., Ramírez-Pérez, P., González-Miret, M. L., Heredia, F. J., (2010). Influence of prefermentative cold maceration on the color and anthocyanic copigmentation of organic tempranillo wines elaborated in a warm climate. *J. Agric. Food Chem. 58*, 6797-6803.

Gresele, P., Cerletti, C., Guglielmini, G., Pignatelli, P., de Gaetano, G., Violi, F., (2011). Effects of resveratrol and other wine polyphenols on vascular function: an update. *J. Nutr. Biochem.* 22, 201-211.

Gupta, R.K., Haslam, E., (1980). Vegetable tannins: structure and biosynthesis. J.H. Hulse (Ed.), Polyphenols in Cereals and Legumes, International Development Research Centre, Ottawa, 15–24.

Gurbuz, O., Göçmen, D., Dag delen, F., Gursoy, M., Aydin, S., Sahin, I., Buyukuysal, L., Usta, M., (2007). Determination of flavan-3-ols and *trans*-resveratrol in grapes and wine using HPLC with fluorescence detection. *Food Chem.* 100, 518-525.

Hagerman, A.E., Butler, L.G., (1978). Protein precipitation method for the quantitative determination of tannins. *J. Agric. Food Chem.* 26, 809-812.

Handelman, G. J., Cao, G., Walter, M. F., Nightingale, Z. D., Paul, G. L., Prior, R. L., Blumberg, J. B, (1999). Antioxidant capacity of oat extracts. Inhibition of low-density lipoprotein oxidation and oxygen radical absorbance capacity. *J. Agric. Food Chem.* 47, 4888-4893.

Harbertson, J.F., Picciotto, E.A., Adams, D.O., (2003). Measurement of polymeric pigment in grape berry extracts and wines using a protein precipitation assay combined with bisulfite bleaching. *Am. J. Enol. Vitic.* 54, 301-306

Harish Nayaka, M.A., Sathisha, U.V., Dharmesh, M., (2010). Cytoprotective and antioxidant activity of free, conjugated and insoluble-bound phenolic acids from swallow root (Decalepishamiltonii). *Food Chem.* 119, 1307–1312.

He, J.-J., Liu, Y.-X., Pan, Q.-H., Cui, X.-Y., Duan, C.-Q., (2010). Different Anthocyanin Profiles of the Skin and the Pulp of Yan73 (Muscat Hamburg × Alicante Bouschet) Grape Berries. *Molecules* 15, 1141-1153.

Heier, A., Blaas, W., Dross, A., Wittkowski, R., (2002). Anthocyanin analysis by HPLC/ESI-MS. *Am. J. Enol. Vitic.* 53, 78-86.

Hendrickson, H. P., Kaufman, A. D., Lunte, C. E., (1994). Electrochemistry of catechol-containing flavonoids. *J. Pharm. Biomed. Anal.12*, 325-334.

Holvoet P., (2004) Oxidized LDL and coronary heart disease. *Acta Cardiol* (59) 479-484.

Huang, D., Ou, B., Hampsch-Woodill, M., Flanagan, J., Prior, R. L., (2002). High-throughput assay of oxygen radical absorbance capacity (ORAC) using a multichannel liquid handling system coupled with a microplate fluorescence reader in 96-well format. *J. Agric. Food Chem.* 50, 4437-4444.

Iacopini, P., Baldi, M., Storchi, P., Sebastiani, L., (2008). Catechin, epicatechin, quercetin, rutin and resveratrol in red grape: Content, in vitro antioxidant activity and interactions. *J. Food Compos. Anal.* 21, 589-598.

Kasdallah-Grissa A., Mornagui B., Aouani E., Hammami M., El May M., Gharbi N., Kamoun A., El-Fazaâ S. (2007). Resveratrol, a red wine polyphenol, attenuates ethanol-induced oxidative stress in rat liver. *Life Sci.* 80, 1033–1039

Katalinic, V., Milos, M., Modun, D., Music, I., Boban, M., (2004). Antioxidant effectiveness of selected wines in comparison with (+)-catechin. *Food Chem.* 86, 593–600.

Kennedy, J., Saucier, C., Glories, Y., (2006). Grape and wine phenolics: history and perspective. *Am. J. Enol. Vitic.* 57, 239-248.

Kilmartin, P. A., Zou, H., Waterhouse, A. L., (2001). A cyclic voltammetry method suitable for characterizing antioxidant properties of wine and wine phenolics. *J. Agric. Food chem.* 49, 1957-1965.

Kilmartin, P. A., Zou, H., Waterhouse, A. L., (2002). Correlation of wine phenolic composition versus cyclic voltammetry response. *Am. J. Enol. Vitic.* 53, 294-302.

Kontoudakis, N., Esteruelas, M., Fort, F., Canals, J. M., De Freitas, V., Zamora, F., (2011). Influence of the heterogeneity of grape phenolic maturity on wine composition and quality. *Food Chem.* 124, 767-774.

Lee, B., Lee, J., Gal, S., Moon, Y., Park, K., (2006). Selective ABTS radical-scavenging activity of prenylated flavonoids from *Cudrania tricuspidata. Biosci. Biotechnol. Biochem.* 70, 427-432.

Lee, C.Y., Jaworski, A.W., (1988). Phenolics and browning potential of white grapes grown in New York. *Am. J. Enol. Vitic.* 39, 337-340.

Li, G., Min, B.S., Zheng, C., Lee, J., Oh, S.R., Ahn, K.S., Lee, K.H., (2005). Neuroprotective and free radical scavenging activities of phenolic compounds from Hoveniadulcis. *Arch. Pharm. Res.* 28, 804-809.

Liazid, A., Barbero, G., Palma, M., Brigui, J., Barrosso, C., (2010). Rapid determination of simple polyphenols in grapes by LC using a monolithic column. *Chromatographia* 72, 417-424.

Llaudy, M.C., Canals, R., Canals, J.M., Rozea, N., Arola, S.L., Zamora, F., (2004). Nez method for evaluating astringency in red wine. *J. Agric. Food Chem.* 52, 742-746.

Lugemwa, F. N., Snyder, A. L., Shaikh, K., (2013). Determination of radical scavenging activity and total phenols of wine and spices: A randomized study. *Antioxidants*, 2, 110-121.

Majo, D.D., Giammanco, M., Guardia, M.L., Tripoli, E., Giammanco, S., Finotti, E., (2005). Flavanones in Citrus fruit: Structure–antioxidant activity relationships. *Food Res. Int.* 38, 1161–1166.

Makhotkina, O., Kilmartin, P. A., (2010). The use of cyclic voltammetry for wine analysis: Determination of polyphenols and free sulfur dioxide. *Anal. Chim. Acta 668*, 155-165.

Makhotkina, O., Kilmartin, P. A., (2012). The phenolic composition of Sauvignon blanc juice profiled by cyclic voltammetry. *Electrochim. Acta* 83, 188-195.

Manns, D., Mansfield, A.K., (2012). A core-shell column approach to a comprehensive high-performance liquid chromatography phenolic analysis of *Vitis vinifera* L. and interspecific hybrid grape juices, wines, and other matrices following either solid phase extraction or direct injection. *J. Chromatogr A.* 1251, 111-121.

Mazza, G., Fukumoto, L., Delaquis, P., Girard, B., Ewert, B., (1999). Anthocyanins, phenolics, and color of Cabernet Franc, Merlot and Pinot Noir wines from British Columbia. *J. Agric. Food Chem.* 47, 4009-4017.

Merken, H.M., Beecher, G.R., (2000). Liquid chromatographic method for the separation and quantification of prominent flavonoid aglycones. *J. Chromatogr A.* 897, 177-184.

Meyer, A.S., Yi, O.S., Pearson, D.A., Waterhouse, A.L., Frankel, E.N., (1997). Inhibition of human lowdensity lipoprotein oxidation in relation to composition of phenolic antioxidants in grapes (*Vitis vinifera*). *J. Agric. Food Chem.* 45, 1638–1643.

Neudörffer, A., Bonnefont-Rousselot, D., Legrand, A., Fleury, M.-B., Largeron, M., (2004). 4-Hydroxycinnamic ethyl ester derivatives and related dehydrodimers: relationship between oxidation potential and protective effects against oxidation of low-density lipoproteins. *J. Agric. Food Chem.* 52, 2984-2091.

Núñez, V., Gómez-Cordovés, C., Bartolomé, B., Hong, Y. J., Mitchell, A. E., (2006). Non-galloylated and galloylated proanthocyanidin oligomers in grape seeds from *Vitus vinifera* L. cv. Graciano, Tempranillo and Cabernet Sauvignon. *J. Sci. Food Agric.* 86, 915-921.

Oliveira, J., Azevedo, J., Silva, A. M., Teixeira, N., Cruz, L., Mateus, N., de Freitas, V., (2010). Pyranoanthocyanin dimers: A new family of turquoise blue anthocyanin-derived pigments found in port wine. *J. Agric. Food Chem. 58*, 5154-5159.

Oliveira, M.S.C., Morais, S., Magalhães, D., Batista, W., Vieira, Í., Craveiro A., de Manezes, J.E.S., Carvalho, A. F., de Lima, G.P.G., (2011). Antioxidant, larvicidal and antiacetylcholinesterase activities of cashew nut shell liquid constituents. *Acta Tropica* 117, 165–170.

Ou, B., Hampsch-Woodill, M., Prior, R.L., (2001). Development and validation of an improved oxygen radical absorbance capacity assay using fluorescein as the fluorescent probe. *J. Agric. Food Chem.* 49, 4619–462.

Passos, C. P., Cardoso, S. M., Domingues, M. R. M., Domingues, P., Silva, C. M., Coimbra, M. A., (2007). Evidence for galloylated type-A procyanidins in grape seeds. *Food chem.*105, 1457-1467.

Pellegrini, P.; Serafini, M.; Colombi, B.; Del Rio, D.; Salvatore, S.; Bianchi, M.; Brighenti, F., (2003). Total antioxidant capacity of plant foods, beverages and oils consumed in Italy assessed by three different *in vitro* assays. *J. Nutr.* 133, 2812-2819.

Picariello, G., Ferranti, P., Garro, G., Manganiello, G., Chianese, L., Coppola, R., Addeo, F., (2014). Profiling of anthocyanins for the taxonomic assessment of ancient purebred *V. vinifera* red grape varieties. *Food chem.* 146, 15-22.

Porgali, E., Buyuktuncel, E., (2012). Determination of phenolic composition and antioxidant capacity of native red wines by high performance liquid chromatography and spectrophotometric methods. *Food Res. Int.* 45, 145-154.

Price, M.L., Van Scoyoc, S., Butler, L.G., (1978). A critical evaluation of the vanillin reaction as an assay for tannin in sorghum grain. *J. Agric. Food Chem.* 26, 1214–1217.

Prior, R. L., Hoang, H., Gu, L., Wu, X., Bacchiocca, M., Howard, L., Hampsch-Woodill, M., Huang, D., Ou, B., Jacob, R., (2003). Assays for hydrophilic and lipophilic antioxidant capacity (oxygen radical absorbance capacity (ORACFL)) of plasma and other biological and food samples. *J. Agric. Food Chem.* 51, 3273-3279.

Prior, R.L., Wu, X., Schaich, K., (2005). Standardized methods for the determination of antioxidant capacity and phenolics in foods and dietary supplements. *J. Agric. Food Chem.* 53, 4290-4302.

Qian, Y.P., Cai, Y.J., Fan, G.J., Wei, Q.Y., Yang, J., Zheng, L.F., Li, X.Z., Fang, J.G., Zhou, B., (2009). Antioxidant-based lead discovery for cancer chemoprevention: the case of resveratrol. *J. Med. Chem.* 52, 1963–1974.

Re, R., Pellegrini, N., Proteggente, A., Pannala, A., Yang, M., Rice-Evans, C.A., (1999). Antioxidant activity applying an improved ABTS radical cation decolorization assay. *Free Radical Biol. Med.* 26, 1231–1237.

Rebelo, M. J., Rego, R., Ferreira, M., Oliveira, M. C., (2013). Comparative study of the antioxidant capacity and polyphenol content of Douro wines by chemical and electrochemical methods. *Food Chem.* 141, 566-573.

Reed, J. D., Krueger, C. G., Vestling, M. M., (2005). MALDI-TOF mass spectrometry of oligomeric food polyphenols. *Phytochemistry* 66, 2248-2263.

Renaud, S., de Lorgeril, M., (1992). Wine, alcohol, platelets, and the French paradox for coronary heart disease. *Lancet* 339, 1523–1526.

Rhodes, P.L., Mitchell, J.W., Wilson, M.W., Melton, L.D., (2006). Antilisterial activity of grape juice and grape extracts derived from Vitis vinifera variety Ribier. *Int. J. Food Microbiol.* 107, 281–286.

Rice-Evans, C., Miller, N.J., (1994). Total antioxidant status in plasma and body fluids. *Methods Enzymol,* 234, 279–293.

Rifici, V., Schneider, S., Khachadurian, A., (2002). Lipoprotein oxidation mediated by J774 murine macrophages is inhibited by individual red wine polyphenols but not by ethanol. *J. Nutr.* 132, 2532-2537.

Rimm, E.B., Klatsky, A., Grobbe, D., Stampfer, M.J., (1996). Review of moderate alcohol consumption and reduced risk of coronary heart disease: Is the effect due to beer, wine, or spirits. *Br. Med. J.* 312, 731–736.

Rivero-Pérez, M.D., Muñiz, P., Gonzalez-Sanjosé, M.L., (2007). Antioxidant profile of red wines evaluated by total antioxidant capacity, scavenger activity, and biomarkers of oxidative stress methodologies. *J. Agric. Food Chem.* 55, 5476-5483.

Rodriguez-Delgado, M.A., Malovana, S., Perez, J.P., Borges, T., Garcia Montelongo, F.J., (2001). Separation of phenolic compounds by high-performance liquid chromatography with absorbance and fluorimetric detection. *J. Chromatogr A.* 912, 249-257.

Rodriguez-Vaquero, M.J., Alberto, M.R., Manca-de-Nadra, M.C., (2007). Antibacterial effect of phenolic compounds from different wines. *Food Control* 18, 93–101.

Sato, M., Ramarathnam, N., Suzuki, Y., Ohkubo, T., Takeuchi, M., Ochi, H., (1996). Varietal differences in the phenolic content and superoxide radical scavenging potential of wines from different sources. *J. Agric. Food Chem.* 44, 37–41.

Scherer, R., Godoy, H. T., (2009). Antioxidant activity index (AAI) by the 2,2-diphenyl-1-picrylhydrazyl method. *Food Chem.* 112, 654–658.

Serafini, M., Maiani, G., and Ferro-Luzzi, A., (1998). Alcohol-free red wine enhances plasma antioxidant capacity in humans. *J. Nutr* 128, 1003-1007.

Shoskes, D.A., (1998). Effect of bioflavonoids quercetin and curcumin on ischemic renal injury: a new class of renoprotective agents. *Transplantation* 66, 147-152.

Silva, C.L., Pereira, J., Wouter, V.G., Giro, C., Camara, J.S., (2011). A fast method using a new hyfrophilic-hydrophobic balanced sorbent in combination with ultra-high performance liquid chromatography for quantification of significant bioactive metabolites in wines. *Talanta,* 86, 82-90.

Simić, A., Manojlović, D., Šegan, D., Todorović, M., (2007). Electrochemical behavior and antioxidant and prooxidant activity of natural phenolics. *Molecules* 12, 2327-2340.

Snijman, P. W., Joubert, E., Ferreira, D., Li, X. C., Ding, Y., Green, I. R., Gelderbow, W. C. A., (2009). Antioxidant activity of the dihydrochalcones aspalathin and nothofagin and their corresponding flavones in relation to other rooibos (Aspalathuslinearis) flavonoids, epigallocatechinGallate, and Trolox. *J. Agric. Food Chem.* 57, 6678–6684.

Soleas, G.J., Diamandis, E.P., Goldberg, D.M., (1997). Wine as biological fluid: History, production, and role in disease prevention. *J. Clin. Lab. Anal.* 11, 287–313.

Somers, T.C., Evans, M.E., (1974). Wine quality: Correlations with color density and anthocyanin equilibria in a group of young red wines. *J. Sci. Food Agric.* 25, 1369 – 1379.

Somers, T.C., Evans, M.E., (1977). Spectral evaluation of young red wines: anthocyanin equilibria, total phenolics, free and molecular SO_2, "chemical age". *J. Sci. Food Agric.* 28, 279-287.

Sudraud, P., (1958). Interprétation des courbes d'absorption des vins rouges. *Ann. Technol. Agric.* 7, 203-208.

Teissedre, P.-L., Frankel, E.N., Waterhouse, A.L., Peleg, H., German, J.B. (1996). Inhibition of in vitro human LDL oxidation by phenolic antioxidants from grapes and wine. *J. Sci. Food Agric.* 70, 55–61.

Tunstall-Pedoe, H., Kuulasmaa, K., Mahonen, N., Tolonen, H., Ruokokoski, E., Amouyel, P., (1999). Contribution of trends in survival and coronary-event rates to changes in coranary heart disease mortality: 10-Year results from 37 WHO MONICA project populations. Monitoring trends and determinants in cardiovascular disease. *Lancet* 353, 1547–1557.

Van Hoorn, D. E., Van Norren, K., Boelens, P. G., Nijveldt, R. J., Van Leeuwen, P. A., (2003). Biological activities of flavonoids. *Science & Medicine 9*, 152-161.

Vijayabaskarna, M., Venkatswarmurthy, N., Babu, G., Permural, P., Jayakar, B., (2010). *In vitro* antioxidant evaluation of *Pseudartha Viscida* Linn. *Int. J. Current Pharm. Res.* 2, 21-23.

Vinas, P., Lopez-Erroz, C., Marin-Hernandez, J.J., Hernandez-Cordoba, M., (2000). Determination of phenols in wines by liquid chromatography with photodiode array and fluorescence detection. *J. Chromatogr A.* 871, 85-93.

Vitrac, X., Monti, J.-P., Vercauteren, J., Deffieux, G., Mérillon J.-M., (2002). Direct liquid chromatographic analysis of resveratrol derivatives and flavanonols in wines with absorbance and fluorescence detection. *Anal. Chim. Acta* 458, 103-110.

Wu, Q., Wang, M., Simon, J. E., (2005). Determination of proanthocyanidins in fresh grapes and grape products using liquid chromatography with mass spectrometric detection. *Rapid Commun. Mass Spectrom.* 19, 2062-2068.

Wua, C.R., Huang, M.Y., Lin, Y.T., Ju, H.Y., Ching, H., (2007). Antioxidant properties of Cortex fraxiniand its simple coumarins. *Food Chem.* 104, 1464-1471.

Zoulis, N. E., Efstathiou, C. E., (1996). Preconcentration at a carbon-paste electrode and determination by adsorptive-stripping voltammetry of rutin and other flavonoids. *Anal. Chim. Acta* 320, 255-261.

In: Wine
Editor: Youssef El Rayess

ISBN: 978-1-63321-048-6

Chapter 4

IMPACT OF WINEMAKING TECHNIQUES ON PHENOLIC COMPOUNDS COMPOSITION AND CONTENT OF WINE: A REVIEW

Chantal Ghanem[1,2], Lara Hanna-Wakim[1], Nancy Nehmé[3,4], Jean-Pierre Souchard[5,7], Patricia Taillandier[6,7] and Youssef El Rayess[1,6,7]*

[1]Faculty of Agricultural and Food Sciences, Holy Spirit University of Kaslik, Jounieh, Lebanon
[2]Lebanese Agronomy Research Institute, Fanar Station, Fanar, Lebanon.
[3]Faculty of Sciences, Holy Spirit University of Kaslik, Jounieh, Lebanon.
[4]Lebanese University Faculty of Agriculture and Veterinary Sciences, Dekwaneh, Lebanon
[5]Laboratoire des Interactions Moléculaires et Réactivité Chimique et Photochimique UMR CNRS 5623, Université de Toulouse, Toulouse Cedex 9, France
[6]Université de Toulouse, INPT, UPS, Laboratoire de Génie Chimique, Toulouse, France
[7]Centre de Viticulture et d'Œnologie de Midi-Pyrénées, Castanet-Tolosan, France

ABSTRACT

Phenolic compounds play an important role in the quality of wines. The composition of these compounds is not stable and evolves during winemaking. Winemaking techniques play an important role in the extraction of polyphenols from the grapes and consequently in the further stability of wine properties. In the course of winemaking and aging, changes in polyphenolic composition occur due to the participation of these

* Corresponding author: Faculty of Agricultural and Food Sciences, Holy Spirit University of Kaslik, Jounieh, Lebanon. Email: youssefrayess@usek.edu.lb, youssef.elrayess@ensiacet.fr.

compounds in numerous reactions such as copigmentation, cycloaddition, polymerization and oxidation. In this chapter, we will present the impact of traditional and new processes of winemaking on wine phenolic composition. It will be focused on maceration type and time, fermentation process, Aging step, finning and clarification methods, membrane processes and filtration techniques as well as the microoxygenation step on wine polyphenols.

Keywords: Phenolic compounds, maceration, fermentation, membrane processes, aging

INTRODUCTION

Phenolic compounds or polyphenols represent a large group of molecules which are present in grapes and wines. These compounds constitute a decisive factor in red wine quality and contribute to wine organoleptic characteristic such as color, taste, astringency and bitterness. They also confer to the wine the capacity of aging. The chemical composition of these compounds is discussed in other chapters. The antioxidant properties of phenolic compounds have been associated with health-promoting effects. Nowadays, the anticarcenogenic ability and the neuroprotective effect of these compounds are slightly proven and still under investigation.

Scientific papers showed clearly that several factors affect polyphenols biosynthesis and accumulation through berry ripening (Koundardas et al., 2009; Spayd et al., 2002; Zoecklein et al., 2008; Poni et al., 2009). Among these factors, the cultivars varieties (clones, and rootstock), the environmental factors (agro-pedological, topographical and climatic factors) and the cultural practices (training system, row vine spacing, pruning, bunch thinning, bud and leaf removal, water, fertilizers and pesticides management) play a crucial role in the determination of the quantitative as well as the qualitative phenolic composition.

After grape harvest, the winemaking process begins. Regardless the geographical zone, the winemaking process scheme is almost the same with some steps modification. The general scheme of winemaking is presented in figure 1. During this process, the diffusion and extraction of the grape polyphenols take place and a perpetual evolution of the phenolic composition of the must at the beginning and of the wine later, occurs with the participation of biochemical and chemical phenomena. New technologies and processes (membrane processes, flash release, etc.) have been introduced to wine industry in response to various challenges as climate change, wine with low alcohol content, better quality, higher production, new products etc. It is obvious that the traditional and new processes hugely impact the qualitative and quantitative composition of phenolic compounds.

Therefore, in this chapter, we will review the impact of traditional and new processes of winemaking on wine phenolic composition. It will be focused on the incidence of maceration type and time, fermentation process, aging step, finning and clarification methods, membrane processes and filtration techniques as well as the microoxygenation step on wine polyphenols. The factors influencing the grape polyphenols content will not be discussed despite their importance.

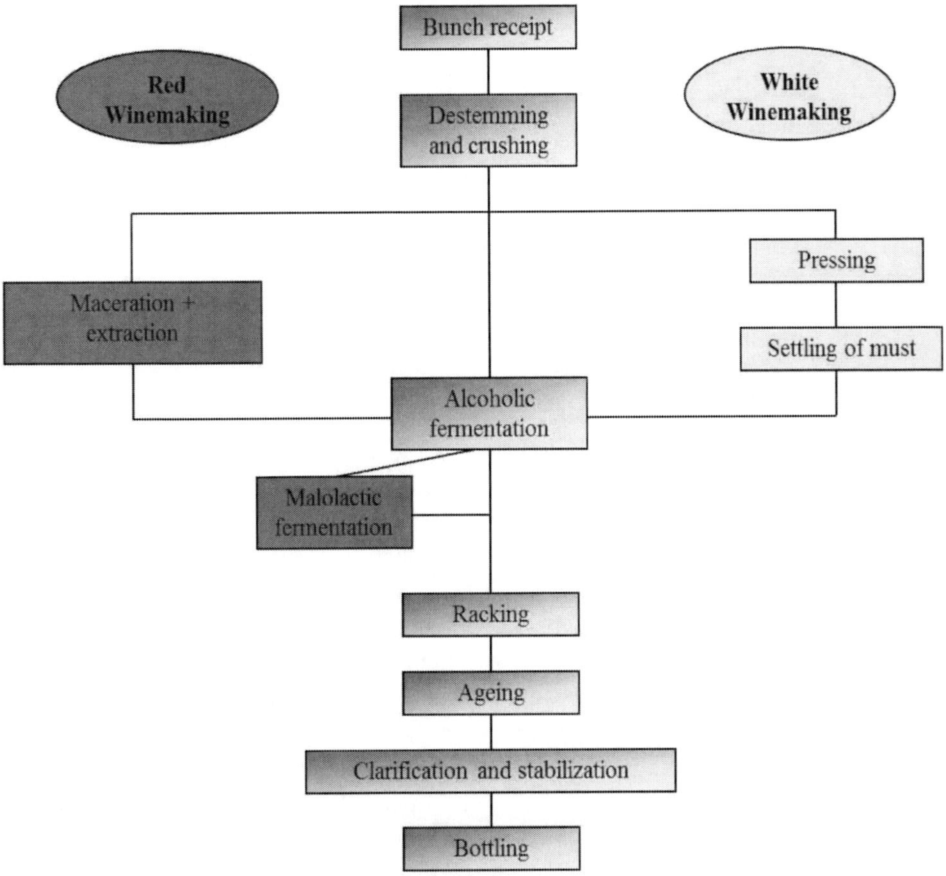

Figure 1. The winemaking process of red and white wines.

IMPACT OF EXTRACTION PROCESSES AND PROCEDURES

Most grape phenolics are localized in the skins and seeds. During winemaking, phenolic compounds and other compounds contained in the grape are transferred to the wine by diffusion while the contact between the juice and the solid part of grapes is established.

Diffusion is the process by which a compound moves from a region of high concentration toward a region of lower concentration. The diffusion period in winemaking is called maceration and it is affected by several factors as grape variety and maturity, temperature of must or wine, duration of juice and grape skin and seed contact, concentration of alcohol and sulfur dioxide and use of enzymes.

In order to extend the extraction that occur during conventional maceration, and to achieve organoleptic properties beyond those offered by conventional maceration during fermentation, extended contact with skins may occur before (pre-fermentation extended maceration) or after fermentation (post-fermentation extended maceration). Depending on the temperature levels, the pre-fermentation extended maceration could be divided into two categories: i) cold maceration or cold soak for low levels of temperature, ii) heating maceration.

The effect of cold soak technique as regards with control vinification on the concentration of anthocyanins and proanthocyanidins in Monastrell wine, was studied by Busse-Valverde et al. (2010) and the results have shown an increase in the anthocyanin extraction, mainly the extraction of malvidin-3-glucoside with higher grade of co-pigmentation and polymerization, therefore better stabilization than traditional wines. Gómez-Míguez et al. (2006) and Gordillo et al. (2010) reported a similar anthocyanin concentration after seven days of low temperature pre-fermentation maceration of Syrah and Tempranillo wine. Gil-Muñoz et al. (2009) found that cold maceration technique led to the highest anthocyanin content at the end of alcoholic fermentation in Cabernet Sauvignon wines. Consequently cold macerated wines tended to show higher chromatic stability than traditional maceration (Gordillo et al., 2010). When the results of the different practices were compared in Monastrell wines (Busse-Valverde et al., 2010), the proanthocyanidins concentration was greatest when cold soak was used (an increase of 33% in the proanthocyanidins concentration). In Cabernet Sauvignon wines, the proanthocyanidins content was higher when cold soak was used, with a total increase of 13.2%. Alvarez et al. (2006) also found a positive effect of low temperature pre-fermentative maceration on the concentration and polymerization of proanthocyanidins and on the stability of Monastrell wine color. Cold soak increased the ratio of anthocyanin to proanthocyanidins (67%) in the wine after maceration, suggesting a possible increase in the proportion of the anthocyanin-proanthocyanidin adducts against total polymers, which must affect the quality of the resultant wine after long-term storage (Cheynier et al., 1999). These authors also stated that the phenolic concentration was not related to the duration of the treatments since the results did not improve when pre-fermentation maceration time was increased but the effect was more evident when grapes were not completely mature. Alvarez et al. (2006) found an important decrease of the proanthocyanidins, of the mean degree of polymerization (mDP) and of the percentage of epigallocatechin (EGC) and an increase of the percentage of galloylation were reported when cold soak maceration was used. It might be expected that, with the application of these low temperature techniques, which are supposed to help the physical degradation of skin cell walls, the concentration of skin proanthocyanidins would increase in the wines, but these results indicated that the increase in proanthocyanidins is mainly due to an increase in seed proanthocyanidins. These results also agree with those obtained by the study of Busse-Valverde et al. (2010) who showed that proanthocyanidins concentrations in Monsatrell and Cabernet Sauvignon wines are increased with cold treatment and this increase seems to be related to the extraction of seeds proanthocyanidins. The cold Treatment is more effective when it is realized with less mature grapes (Álvarez et al. 2006). This result is according to previous studies for wines of other grape varieties (Couasnon, 1999).

Pre-Fermentation Heating Maceration

Historically, this type of maceration is coupled to the fermentation in liquid phase. Practically, it is used to quickly handle the entry of large volumes of grape harvest.

Technically, this practice is used to extract phenolic compounds, denature alteration enzymes and destruct vegetal aromas of grapes. Many variations of pre-fermentation heating maceration exist:

i) The pre-fermentation heating maceration followed by direct pressing: grapes are heated to 70-75°C. The maceration must last between 6 to15 hours to obtain the same amount of polyphenols as a classical vinification.
ii) The pre-fermentation heating maceration followed by maceration during fermentation: its principle is the same as the first pre-fermentation heating maceration but heating maceration lasts 2 hours and then the bunch is cooled down.
iii) The thermo-vinification process: this technique consists in bunch heating to 70-75°C for a short duration (30-40 minutes). The bunch is then pressed and cooled. Wines are generally less rich in phenolic compounds comparing to a classical vinification.
iv) The flash release (FR) process: it consists in heating the grapes quickly at high temperature (>95 °C) with biological vapor (i.e., steam produced from the water present in the grape, without dilution) at atmospheric pressure and then placing them under a strong vacuum (pressure closed to 60 hPa) which causes instant vaporization. The vaporization induces weakness in the cells wall and cooling of the treated grapes which favorites the polyphenol extraction. It is generally coupled to fermentation in liquid phase.

In 2000, Berger and Cottereau studied the winemaking of fruity red wines by pre-fermentation maceration under heat. The trials were conducted in the Beaujolais using a pre-fermentation heating to 70°C lasting from 8 to 16 hours. They found that this technique increase significantly the color (+40%) and tannins content (+55%) comparing to traditional Beaujolais vinification. It was also judged that the pre-fermentation technique influenced the wine aromas with red fruits notes.

The influence of bunch heating technique on the phenolic composition of red wines (Pinot noir, Lemberger and Cabernet Franc), regarding those of the control wines, was studied by Netzel et al. (2003), and the results have shown an efficient extraction of anthocyanins (located in the skin of red grapes), flavonols (especially quercetin glycosides accumulate in the skin), resveratrol (stored within the grape cells in the form of glucosides) and total flavan-3-ols (highest concentration in the seeds), while the level of individual monomeric (catechin and epicatechin) and dimeric (proanthocyanidins B1 and B2) flavan-3-ols were similar to or less than the control wine. These results were in accordance with those obtained by Borazan and Bozan (2013). In contrast, the phenolic acids (found in the skin, juice, solid pulp, and seeds) and tyrosol (produced from tyrosine by yeast during fermentation which is the only phenolic compound produced in significant amounts from non-phenolic precursors) did not show these effects.

A preliminary study carried out on flash-release (FR) by Moutounet et al. (2000) showed an increase of 50% in the total phenolic compounds than that observed in the control wines. In 2006, Morel-Salmi et al. applied FR treatment on three grape varieties in different vintage (Grenache, Mourvedre, Carigan), and the results, presented in table 1, showed that FR wines contained larger amounts of flavonols, anthocyanins, catechins and proanthocyanidins (tannins) than the control wines. The average chain length of proanthocyanidins (mean degree of polymerization, mDP) in control and FR wines were almost identical. The FR-treated wines contained higher percentages of galloylated units and lower proportions of epigallocatechin (EGC) units than the control wines (table 1). Again, this presumably reflects the fact that the extraction of tannins from seeds is greater than that of the skins (Morel-Salmi et al., 2006). This study also showed that FR increased total anthocyanins, total polyphenol index (TPI), color intensity (CI) and it lowered sulfite bleaching resistance than in the corresponding control wines. FR increased the tannin-to-anthocyanin ratio. This increase in the ratio T/A allows the conversion of anthocyanins to T-A dimers adducts that show the same color properties as anthocyanin. Formation of T-A adducts increased with the oxygenation, tannin-to-anthocyanin ratio and with FR and heating (Fulcrand et al., 2004).

Carbonic Maceration

Carbonic maceration consists of placing whole grapes in a closed tank under CO_2 atmosphere. The tank is kept at a moderate temperature (20-30°C) for 1-2 weeks. The carbon dioxide gas permeates through the grape skins and begins to stimulate fermentation at an intracellular level. The entire process takes place inside each single, intact berry. Then the juice is run off, the pomace pressed, and the free- run and press wines are usually assembled prior to normal alcoholic and malolactic fermentation (Ribéreau et al., 2006).

The influence of fermentation with carbonic maceration, on the contents of catechins, proanthocyanidins, and anthocyanins in Tinta Miúda red wines, was studied by Sun et al. (2001). They reported that the carbonic maceration wine contained the highest amounts of catechins, oligomeric and polymeric proanthocyanidins comparing to traditional process of winemaking, which might be explained by the fact that the phenolic compounds released from the solid parts of the grape cluster, using the carbonic maceration technique, are well-protected against oxidation or other physicochemical reactions during intracellular fermentation/maceration (Sun et al., 2001). On the other hand, analysis of individual and total anthocyanins by Sun et al. (2001) has shown that the concentrations of total anthocyanins and nearly all individual anthocyanins in the carbonic maceration wine were lower than traditional wine. Moreover, the carbonic maceration wine had less colored density and higher hue than the control wine.

Sun and Spranger (2005) also reported highest procyanidin levels in carbonic maceration Tinta Miúda red wines. It was shown also that carbonic maceration afforded wines with most stability in color density for 26 months' storage. On the opposite, Spranger et al. (2004) detected higher catechins and procyanidin levels in Castelão red wines made by classical techniques. Castillo-Sanchez et al. (2008) also found that procyanidin and catechin levels in traditional wines were much higher than in carbonic maceration wines

While studying the influence of winemaking protocol on the evolution of the anthocyanin content, Castillo-Sanchez et al. (2006) showed also that carbonic maceration led to lower anthocyanin levels and less intense coloration than the conventional pumping over and the rotary vats. They claimed also that during storage, the carbonic maceration wines underwent less color degradation than the others.

Table 1. Effect of Flash Release on the Wine Polyphenol and Proanthocyanidin Composition (mg/L) (Morel-Salmi et al., 2006)

		Anthocyanins	Flavonols	Hydroxycinnamic acids	Catechins	Proanthocyanidins	DPm	% gall	% EGC
Grenache 2003	control	106.1 ± 8.6	20.8 ± 1.4	460.8 ± 12.8	85.2 ± 12.8	751.2 ± 46.9	4.00 ± 0.16	3.42 ± 0.16	10.51 ± 0.4
	flash release	110.9 ± 2.1	30.8 ± 1.1	355.7 ± 7.3	143.3 ± 1.9	997.2 ± 59.1	3.3 ± 0.13	4.8 ± 0.3	7.52 ± 0.3
Mourvèdre 2003	control	173.2 ± 4.7	36.1 ± 2.0	47.9 ± 5.8	33.9 ± 1.4	819.5 ± 52.1	4.78 ± 0.18	3.1 ± 0.17	13.9 ± 0.6
	flash release	198.7 ± 1.7	67.9 ± 0.3	26.8 ± 1.4	46.8 ± 3.1	1281.7 ± 308.5	4.5 ± 0.2	4.9 ± 0.3	9.5 ± 0.3
Grenache 2004	control	83.5 ± 0.5	5.4 ± 0.6	316.1 ± 5.4	55.4 ± 2.6	564.1 ± 35.2	3.95 ± 0.13	4.04 ± 0.28	12.1 ± 0.85
	flash release	87.6 ± 0.7	9.4 ± 0.1	270.8 ± 5.7	113.8 ± 3.4	851.5 ± 36.0	3.11 ± 0.06	5.3 ± 0.16	7.5 ± 0.07
Carignan 2004	control	161.4 ± 1.9	13.4 ± 1.2	94.1 ± 3.8	26.0 ± 0.7	308.7 ± 9.2	3.97 ± 0.07	2.08 ± 0.1	19.0 ± 0.66
	flash release	210.3 ± 5.6	27.5 ± 0.7	103.2 ± 2.8	32.3 ± 0.8	356.2 ± 11.2	4.02 ± 0.12	3.0 ± 0.12	17.6 ± 0.2

Castillo-Sanchez et al. (2008) found that carbonic maceration produced wines with less color density and higher hue than the conventional process of winemaking. These results were in agreement with those obtained by Timberlake and Bridle (1976). Although, the carbonic maceration protocol might have been expected to increase the release of anthocyanins from the grape skin due to the longer overall time spent macerating and fermenting and to the higher temperature used (Lorincz et al., 1998), these effects seem to have been overweighed by the effect of reducing post-crushing fermentation time to 2–3 days, which reduced the duration of intimate contact between skin and must. Similar results were obtained by Spranger et al. (2004) where they detected higher anthocyanin levels in classical fermentation Castelão red wines obtained than in carbonic maceration Castelão red wines.

Post-Fermentation Re-Heating

It is a method consisting in prolonging the fermentative maceration by post-fermentation re-heating to approximately 45°C for 42 hours, in order to complete the liberation of grape skin constituents fulfilled by pre-fermentative and fermentative maceration. Discordant results are reported in literature on the effect of post-fermentation re-heating on red wine quality. A study by Koyama et al. (2007) on the influence of heating at the end of maceration during red winemaking from Cabernet Sauvignon showed that, contrary to expectations, the anthocyanin concentration was not increased. Flavonols showed an extraction similar to anthocyanins, while heat treatment decrease the level of proanthocyanidins, their mDP and the galloylation rate (%G). Barra et al. (2005) obtained higher increase in anthocyanin content (+10.7% of malvidin 3-glucoside), in color intensity (+13%) and total sensory score, in Pinot noir wine by heating at the end of maceration than by control vinification. Similar results were reported by Gerbaux et al. (1993). Potential variables, e.g., grape variety, berry maturity, heat conditions and fermentation scale might have affected the results.

MACERATION ENZYMES

The grape skin cell walls formed mainly by polysaccharides (pectins, hemicellulose and cellulose) are limiting barrier that prevent the release of polyphenols into the must during fermentation. Maceration enzymes may help in phenolic extraction and, at the same time, may modify the stability, taste and structure of red wines, because it is not only anthocyanins that are released from skins, but also tannins bound to the cell walls. These may be extracted due to the action of pectinases (polygalacturonase, pectin lyase and pectin esterase activities), hemicellulases and cellulases and the extracted compounds help to stabilize wine color and increase mouth feel sensations (Canal-Llaubères and Pouns, 2002). In the literature, the effect of the addition of enzymes on the phenolic content remains unclear because of some contradictory results.

The effect of enzymes treatment on phenolic composition was tested on a Monastrell wine by Bautista-Ortín et al. (2007) and it was compared to two other enological practices (running-off a part of must and tannins addition). The authors noticed that the addition of enzymes (pectinase + mannanase + glucanase activities) promoted higher values of total phenols (OD280) than in the control wine as also observed by Parley (1997) and Pardo et al. (1999). It seems that the action of the enzyme facilitates a higher extraction of proanthocyanidins from both skin and seeds but without changing their proportion or composition, as compared to control wines (Busse-Valverde et al., 2011). On the opposite, Ducasse et al. (2010) found a higher proanthocyanidin content in Merlot wines treated with enzymes but, surprisingly, not in the percentage of skin-derived proanthocyanidins, but with an increase of seed proanthocyanidins. Regarding anthocyanin concentrations, some authors have reported an increase in the anthocyanin levels (Bautista-Ortin et al., 2005; Kammerer et al., 2005; Romero-Cascales et al., 2012), whereas others have reported a decrease in the anthocyanin levels (Kelebek et al. 2007; Parley et al., 2001; Revilla and Gonzales-Sanjose, 2003;). Borazan and Bozan (2013) studied also the effect of pectolytic enzymes on the phenolic composition of Okozguzo wines. They found that the wines treated by the pectolytic enzyme addition had a lower monomeric flavan-3-ol content than the untreated wines, and that the amount of monomeric anthocyanins extracted did not increase with the addition of enzymes.

This different observations could be due to a different activities present in enzyme commercial preparations.

Effect of Yeasts and Bacteria

Yeasts and bacterial metabolism during fermentations produce a large array of metabolites which contribute to the aroma and flavor of wine.

The influence of yeast used for winemaking on phenolic compounds is still poorly understood; but it is known that yeasts interact with polyphenols by 3 mechanisms:

- Adsorption of phenolic compounds on yeast cell wall
- Extraction of phenolic compounds
- Excretion of parietal polysaccharides (mannoproteins) which can interact with tannins for better stabilization and sensorial perception of the wine.

Hayasaka et al. (2007) studied the impact of two different yeasts, *Saccharomyces cerevisiae* (SC) and *Saccharomyces bayanus* (SB) on the phenolic composition of red wine made from the same batch of Cabernet Sauvignon grapes. The color properties and pigment profiles of SC and SB wines were compared at 8 days and 387 days after yeast inoculation. Anthocyanin concentration was found to be lower in SB wines than in SC wines at day 8 and 387, but SB wine exhibited greater wine color density. The anthocyanin concentration did not correlate with wine color density. The levels of pigmented polymers and SO_2 non-bleachable pigments were found to be higher in SB wine at day 387, demonstrating that the formation of stable pyranoanthocyanins and pigmented polymers was enhanced by SB yeast. It was demonstrated that the formation of acetaldehyde-mediated pigments was enhanced by the use of the SB yeast. The compositional analysis suggested that the differences in color properties and pigment profiles of SC and SB wines were largely due to the greater production of acetaldehyde-mediated pigments by the use of SB yeast.

Caridi et al. (2004) studied the effect of two yeasts strains on the phenolic profile of red wine. They reported that the Strain Sc2659, compared to strain Sc1483, produced a wine with significantly higher values of color, color intensity, total polyphenols and monomeric anthocyanins. Also, the content of flavonoids, total anthocyanins, flavans and proanthocyanidins was higher in the wine produced by strain Sc2659, but the differences from the strain Sc1483 were not significant. The levels of non- anthocyanic flavonoids were significantly lower. Therefore, strain Sc2659 protects during winemaking the phenolics and the anthocyanins of the must better than strain Sc1483.

Two commercial yeast strains (Fermirouge and Rhône 2323) were tested during the winemaking process of Monastrell grapes to determine their influence on color and phenolic composition of the resulting wines during alcoholic fermentation and maturation. The results showed that in 2002, the wines did not present great differences but in 2003 higher color intensity and phenolic compounds content were detected when one of the commercial strains was used. The maximum values of monomeric anthocyanins were found when Rhône 2323 (L2) was used. In 2003, differences in hydroxybenzoic acids, flavan-3-ols and total anthocyanins were also found. Rhône 2323 (L2) wines presented the largest concentration of these compounds (Bautista-Ortin et al., 2007).

Yang Sun et al., (2011) studied the effect of six commercial wine yeast strains (BM4x4, RA17, RC212, D254, D21 and GRE) on the profiles of polyphenols in cherry wines. They showed that BM4x4 fermented wine had the highest total phenolics and tannins among the six wines tested, whereas RC212 fermented wine had the highest content of total anthocyanins. Therefore a wide range of concentrations of total anthocyanins, total phenolics and tannins were revealed depending on yeast strains.

Regarding low molecular weight phenolic compounds, it is known that some phenolic acids can inhibit the growth of lactic acid bacteria while others can stimulate malolactic fermentation carried out by *Oenococus oeni*. During this process, hydroxycinnamic acids and their derivatives are the main compounds modified. The decrease in the concentration of *trans*-caftaric and *trans-p*-coutaric acids until disappearance, along with an increase in the corresponding free forms, *trans*-caffeic and *trans-p*-coumaric acids could be linked to lactic acid bacteria metabolism.

It has been described that *Lactobacillus hilgardii* can degrade gallic acid and catechin (Alberto et al., 2004). *Pediococcus pentosaceus* can also reduce the quercetin levels (Locascio et al., 2006). *Oenococcus oeni* was found to be able to metabolize anthocyanins and other phenolics by a glycosidase action producing important wine aroma compounds (De Revel et al., 2005; Bloem et al., 2008).

Bloem et al. (2006) studied the production of vanillin from simple phenols by wine-associated lactic acid bacteria. They found that bacteria were not able to form vanillin from eugenol or vanillic acid. However, they showed that *Oenococcus oeni* could convert ferulic acid to vanillin.

Cabrita et al. (2008) reported that hydroxycinnamic acids and their derivatives were the main compounds modified by malolactic fermentation, independently of the use or not of commercial lactic bacteria. In fact, it seems clear that the decrease in the concentrations of caftaric, coutaric and fertaric acids, and the increase in the concentrations of caffeic, *p*-coumaric and ferulic acids are linked to lactic acid bacteria metabolism.

REACTION BETWEEN ANTHOCYANINS AND TANNINS: IMPACT OF MICRO-OXYGENATION

Anthocyanins are the most significant components, responsible for the purple-red color of young red wines. They are unstable and participate in reactions during fermentation and maturation to form more complex pigments, which mainly arise from the interaction between anthocyanins and other phenolic compounds, especially flavan-3-ols. Several mechanisms have been proposed and confirmed for the formation of these new pigments:

a) Direct anthocyanin-tannin condensation reactions (A^+-T product). The products are colorless flavenes, which can be oxidized to the corresponding flavylium ions, finally developing into yellow xanthylium salts. These reactions take place during fermentation, and O_2 is required (Liao et al. 1992; Santos-Buelga et al. 1999; Ribéreau-Gayon et al. 2000b).
b) Direct tannin-anthocyanin condensation reactions (T^+-A). The products are colorless, but are rapidly dehydrated into a reddish-orange form. This reaction is stimulated by higher temperatures, and O_2 is not required. It occurs predominantly during bottle aging (Remy et al. 2000; Ribéreau-Gayon et al. 2000b; Hayaska and Kennedy, 2003).
c) Reactions between anthocyanins and flavanols mediated by acetaldehyde to give a resulting product, with an ethyl bond, that can be protonated to form a colored compound. (Timberlake and Bridle, 1976; Francia-Aricha et al., 1997). Acetaldehyde can be derived from ethanol oxidation or from yeast metabolites.
d) Cycloaddition reactions to form pyranoanthocyanin compounds. Anthocyanins react with yeast metabolites or wine oxidation products (e.g., vinyl phenols, acetaldehyde and pyruvic acid). Vitisin-B is the specific compound resulting from ethenol (aldo-enol transformation of acetaldehyde) and malvidin-3-glucoside. Phenylpyrano-anthocyanins, carboxypyranoanthocyanins and pyranoanthocyanins are respectively the results of the reaction between anthocyanins and vinylphenols, pyruvic acid and acetaldehyde (Atanasova et al., 2002; Mateus et al., 2003; Fulcrand et al., 2006; Rentzsch et al., 2007).
e) Addition reactions between anthocyanins and oxidized phenolic compounds (i.e., ortho-quinones) (Cheynier, 2006; Guyot et al. 1996).
f) Depolymerization and repolymerization reactions of tannins during wine aging. These transformations can occur in the presence or absence of oxygen; however, the resulting structures will differ, depending on the pathways taken (Vidal and Aagaard, 2008). Oxygen brings about the production of different aldehydes, with acetaldehyde being the most abundant. Subsequently, acetaldehyde can react rapidly with tannin molecules. The resulting products are not as important as are direct C4–C8 and C4–C6 polymerization reactions between procyanidin molecules and are hence less astringent (Ribéreau-Gayon et al., 1983; Tanaka et al., 1994).
g) Copigmentation of anthocyanins. The phenomenon of copigmentation is due to molecular association between anthocyanins (intramolecular copigmentation) or between anthocyanins and other non-colored organic molecules (intermolecular

copigmentation). Copigmentation is important in color modification in young red wines, promoting an increase in the maximum absorption wavelength

All of these reactions result in the formation of more stable compounds that stabilize wine color since they partly resist discoloration by SO_2 and provide better color stability at wine pH.

Micro-oxygenation (MOX) is a technique that consists in introducing small and measured amounts of oxygen into wines with the objective of improving wine color, aroma and texture and involves the use of specialized equipment to regulate the oxygen doses applied (Parish et al. 2000; Paul, 2002). The term does not usually include the passive oxygen exposure that occurs during barrel aging nor the range of winemaking practices (such as pumping over and racking) where oxygen exposure may be intentional but is not well measured (Rieger, 2000). An important stipulation of micro-oxygenation is to introduce O_2 into the wine at a rate equal to or slightly less than the wine's ability to consume that, avoiding accumulation of dissolved oxygen (Du Toit et al. 2006). It is for this reason that the success of MOX depends strongly on controlling the rate of oxygen exposure. Typical dosage rates are relatively small, ranging from 2 to 90 mg O_2/L of wine/month (Dykes, 2007). Studies on MOX applications indicate that it can be performed at any time during the winemaking process. However, the best results are achieved when oxygen is added at the end of alcohol fermentation and before beginning malolactic fermentation (Parish et al., 2000; Castellari et al., 1998, González-Sanjosé et al., 2008).

A study conducted by Sánchez-Iglesias et al. (2009) on the effect of MOX on the phenolic fraction of Tempranillo wines during two consecutive vintages, showed significant higher contents of total anthocyanins, pyruvic derivates and polymerization pigments than the control wines, in which most of the pigments belonged to the group of flavanol-anthocyanin (direct and ethyl-bridged) derivatives (Arapitsas et al. 2012). Similar results were observed by (Atanasova et al. 2002) in blended red wine (var. Cabernet Sauvignon and Tannat) with a decrease in the percentage of copigmentation. As regards of chromatic parameters all of the micro-oxygenation wines showed significantly higher values of color intensity and percentage of blue, with a lower percentage of red and yellow than the control wines (Sánchez-Iglesias et al. 2009). These data agree with those already described for the greater anthocyanin drop, together with higher percentages of polymeric anthocyanins and greater contents of pyruvic derivatives (Revilla et al. 2001; Revilla et al. 2002). On the other hand a study carried out by Bastante et al. (2011) on the effects of micro-oxygenation before malolactic fermentation on Cencibel red wines, showed a decrease of the content of flavan-3-ols versus non micro-oxygenation wines. The micro-oxygenation treatment, together with the aforementioned lower content of flavan-3-ols, suggests that the oxygen addition activated the reactions between free anthocyanins and flavan-3-ols. As a consequence, new anthocyanin-derived pigments more stable to pH changes and bisulphite bleaching were formed (Escribano-Bailón et al. 2001). The latter was supported by the increase of percentage of polymerization and the lower value of copigmented anthocyanin (Hermosín-Gutiérrez et al. 2005).

The formation of polyphenolic compounds and pyranoanthocyanins during MOX could be enhanced by the presence of oak. This latter contains high amounts of hydrolyzable tannins such as ellagitannins and gallotannins. These compounds have high gallolated content that is more efficiently oxidized than the majority of the grape-derived phenolic compounds which are non-gallolated (Schmidtke et al., 2011).

Barrel Aging

Aging in wooden barrels is a process used to stabilize the color and enrich the sensorial characteristics of wine. Many compounds are released from wood into the wine; oxygen permeation through the wood favors formation of new anthocyanin and tannins derivates (De Rosso et al., 2009). During barrel aging, the total anthocyanins monoglucosides (the monoglucosides of delphinidin, cyanidin, peonidin, petunidin and malvidin, together with their acetyl and coumaryl derivates) decreased, but the percentage of pigments in the red form increased from 15 to 45%. This transformation of colorless anthocyanins (free anthocyanins) into the colored form (polymerized compounds) compensates for their loss and leads to the increase in color density (Cano- Lopez et al., 2010). In other hand, a drop in free and total anthocyanins was thus observed, with the concentration of anthocyanins dropping from about 850 mg/L to 400 mg/L within six months (Atanosova et al., 2002). The concentration of direct adducts (T-A^+ or T^+-A) increased after six months in new barrels (Cáno-López et al., 2010). As regards their chromatic characteristics the color intensity (the sum of the yellow, red and blue colors) increased from 8 to 10 and 12 to 16 between 3-6 months after barreling. The percentages of red color was lower than that of the control wine but the percentages of yellow and blue were higher due to pigments resistant to SO_2 discoloration. Such a difference in color density can be observed visually. In South African Pinotage and Shiraz wines, it was found that the origin of the barrel (American, French or Russian) did not affect the difference in color intensity, color hue or total red pigments. The Total phenol content (expressed as optical density at 280nm) increased in barrel aged wines as regards of the control wine due to the extraction of phenolic compounds from oak (phenolic acids, ellagitannins, wood aldehydes) (Gómez-Cordoves et al., 1995).

Several of these positive modifications in wine phenolics occurring during wood aging are due to: (1) the release of ellagitannins from wood to wine. These compounds have 15 OH groups per molecule and are highly reactive toward oxygen penetrating through wood. In the presence of oxygen, the ellagitannins will be more easily oxidized than the majority of grape constituents such as anthocyanins to produce hydrogen peroxide. When hydrogen peroxide reacts with ferrous iron to yield the hydroxyl radical, this highly unstable radical reacts almost immediately. It does not react selectively with antioxidants such as phenolics, but instead reacts with all substances present in solution, almost in proportion to their concentration (Gómez-plaza and Cano-López, 2011). Expected products in wine would be the oxidation of alcohol to acetaldehyde (Wildenradt et al., 1974), molecule that is incorporated into red wine phenolic polymers (Drinkine et al., 2007). As a consequence, a modification of red wine color (Timberlake et al., 1976) occurs. The phenolic compounds released from wood may also directly interact with colorant matter of wine giving condensation products bringing to a bathochromic shift of color absorbance (Quideau et al., 2005). (2) Condensation reactions occur between wine phenolics and aldehydes released from oak barrels (Es-Safi et al., 2000; Sousa et al., 2005). In this regard, it has been recently shown that the vanillin, one of the main aldehydes released from oak wood, leads to the formation of an anthocyanin–catechin purple pigment (Sousa et al., 2007). Due to the fact that some acetaldehyde-derived flavanol–anthocyanin polymers are insoluble (Escribano-Bailon et al., 2001), a precipitation of phenolics also occurs. According to several authors, this might explain the losses of astringent material observed as a result of wood aging (Haslam et al., 1980; Vivas et al., 1996). Another mechanism that has to be considered is the adsorption of wine phenolics on wood. In a study performed using a model solution, it has been observed that at least 50% of the resveratrol content can be sorbed by the wood (Barrera-Garcia et al., 2007), indicating that the wood sorption process was selective for the most hydrophobic compound. Different phenolic molecules are involved with the bitterness, astringency and fullness of red wine, but it is mainly the flavanols that are responsible for these tastes and flavors. A very young red wine might be harsh, course, very astringent and even bitter. During aging of red wine in barrels the wine becomes softer and less astringent. It is mainly the acetaldehyde-induced polymerization that contributes to the polymerization of flavanols. The resulting products are not as reactive towards proteins as their constituents. However, direct C4-C8 and C4-C6 polymerization reactions between procyanidin molecules produce products that are more reactive towards proteins and are hence more astringent than those formed from acetaldehyde-induced condensation reactions (Cheynier et al., 1997). In the case of flavanols, where the C6 and C8 positions can be occupied, polymers larger than trimers have been isolated. Both types of reactions produce procyanidins with a limit of 8 or 10 flavan units. The interaction of anthocyanin molecules with procyanidins can also influence the taste of wine because they can form the terminal subunits, thus preventing further polymerization (Ribéreau-Gayon et al., 2000; Monagas et al., 2005).

AGING ON LEES

The definition of wine lees given by EEC regulation No. 337/79 states that ‘‘wine lees is the residue that forms at the bottom of recipients containing wine, after fermentation, during storage or after authorized treatments, as well as the residue obtained following the filtration or centrifugation of this product” (Pérez-Serradilla and Luque de Castro, 2008). When wine is kept in contact with lees, the yeast covering is naturally and slowly degraded and most nutrient supplies are depleted. This microbiological phenomenon, known as autolysis, is mainly induced through different enzymatic activities of the yeast itself. This degradation in wine enriches this latter with products (polysaccharides, peptides and fatty acids) from different cell parts (Mazauric and Salmon, 2005). The importance of wine lees in the aging technique impact on phenolic compound composition comes from the fact that they can adsorb phenolic compounds and release to wine some compounds, among them enzymes and mannoproteins.

The compounds released can influence the structural integration of the wine in terms of phenols, body, aroma and wine stability (Palomero et al., 2009). Results showed that mannoproteins released during yeast lees autolysis can interact with phenolic compounds, improving the color stability and reducing the wine astringency by decreasing tannin aggregation and precipitation (Feuillat et al., 2000; Poncet-Legrand et al., 2000; Fornairon-Bonnefond et al., 2002).

It has been generally reported that anthocyanins content in wines decreases after contact with lees (Mazauric and Salmon, 2005; Mazauric and Salmon, 2006). This decrease is due to the adsorption of anthocyanins on wine lees. Mazauric and Salmon (2005) showed that this adsorption follows biphasic kinetics: an initial and rapid fixation is followed by a slow, constant and saturating fixation that reaches its maximum after about 1 week. Other authors

(Delcroix et al., 1994; Cunier, 1997) explained that anthocyanins decrease during wine aging is due to the degradation of anthocyanins by β-glucosidase enzymes released by yeast lees.

Evolution of red wine anthocyanins with or without aging on lees was studied by Moreno-Arribas et al. (2008), and the results showed that wines aged in the presence of lees, had the highest values of anthocyanins-glucosides (delphinidin-3-glucoside, petunidin-3-glucoside, peonidin-3-glucoside, malvidin-3-glucoside) and anthocyanins-cinnamoyl-glucosides (delphinidin-3-(6''-P-coumarylglucoside), malvidin-3-(6''caffeoylglucoside), malvidin-3-(6''coumaroylglucoside), than the wines aged without lees. On the other hand, the results showed that the formation of anthocyanin-vinylphenol adducts seems to be favored by yeast and lactic acid bacteria from lees. Similar results were reported by Pozo-Bayón et al. (2004).

One of the disadvantages of aging on lees is that they consume oxygen. Oxygen plays an important role in the stabilization of wine color by enhancing condensation reactions between flavonoids mediated by acetaldehyde and in the cycloaddition reactions between pyruvic acid and anthocyanins. Therefore, the consumption of oxygen can reduce the condensation and polymerization reactions between phenolic compounds and can result in decreasing of wine color. The aging on lees also favors the formation of reduction aromas. It seems that this problem can be resolved by combining the lees and oak aging because oak aging favors the transfer of oxygen into wine.

Hernandez et al. (2006) studied the impact of aging on lees on 38 non-anthocyanin phenolic compounds. They didn't observe any significant difference in the content of these compounds after 14 months of Aging. They reported an increase in hydroxycinnamic acids during aging on lees in oak barrels and they explained this result by the enzymatic activity of yeast and lactic acid bacteria and the hydro-alcoholysis of oak wood. Same results were obtained by Del Barrio-Galan et al. (2012) when studying the effect of the aging on lees on the low molecular weight phenols of Tempranillo red wine aged in oak barrels.

FILTRATION AND MEMBRANE TECHNIQUES

Wines, after alcoholic and malolactic fermentations, are a complex medium and need to be clarified and stabilized. Stabilization could be divided into physico-chemical and microbiological stabilization. The microbiological stabilization is ensured by filtration techniques while the physico-chemical stabilization is achieved through fining process, cold stabilization and addition of stabilizers agents. The filtration methods used in wine industry could be divided into 2 groups:

i) Precoat filtrations using exogenic additives as diatomaceous earth, perlite and cellulose
ii) Filtrations using filtering media as membranes and pads

All filtrations methods have an incidence on the chemical and organoleptic composition of wines (Serrano, 1998). For precoat filtrations, the impact depends on the permeability of the diatomaceous earth. Polyphenols loss in precoat filtrations is noticed through adsorption of the compounds in the exogenic additives.

Polyphenols losses were more studied in cross-flow microfiltration because it was shown that these compounds with polysaccharides are the main responsible of membrane fouling. At the beginning, several authors (Poirier et al., 1985, Belleville et al., 1990 and 1992) have reported that the colloidal deposit on ceramic membranes has an intense red color and therefore they pointed out the implication of polyphenols in membrane fouling. The involvement of wine polyphenols in the membrane fouling was been identified by washing the fouled membrane with acidified methanol. Significant increases in permeability were obtained. This fact can be attributed to the elimination of the layers of phenolic compounds because the other wine constituents are insoluble in this solvent (Cameira Dos Santos, 1995). According to Czekaj et al. (2000) and El Rayess et al. (2011; 2012), an increase in polyphenol concentration in wine leads to a decrease of membrane permeability and thus to an increase of membrane fouling.

Researchers have also demonstrated that membrane materials exhibit different fouling behaviors with wine compounds. Ulbricht et al. (2009) showed that polysaccharides and polyphenols adsorption occurs more on polar (hydrophilic) polyethersulfone (PES) membranes that on non-polar (hydrophobic) polypropylene membranes. In fact, Polyphenols are amphipathic molecules with hydrophobic aromatic rings and hydrophilic phenolic hydroxyl groups. So their adsorption involves both hydrophobic effects and the formation of hydrogen bonds. El Rayess et al. (2012) have reported that the most plausible mechanism for membrane fouling by tannins is a fast interaction between tannins and the ceramic membrane (adsorption) quickly followed by tannins-tannins interaction leading to aggregates that could block the pores and form a deposit at the top surface of the membrane. Recently, it was shown that cross-flow microfiltration significantly decreased the mean degree of polymerization (mDP) of tannins by 25% and it selectively removed the high polymerized proanthocyanidin. It was also reported that this technique lowered the levels of catechin, dimers and anthocyanins comparing to the control (Oberholster et al., 2013).

The effect of other membrane processes (Reverse osmosis, nanofiltration, electrodialysis,...) on polyphenol content in wine was also studied. Membrane processes including nanofiltration, reverse osmosis, pervaporation and membrane contactor can be used to reduce alcohol in wines. These techniques form alternatives to traditional techniques. There are two methods to reduce alcohol content in wine: i) reduction of sugar concentration of musts; ii) de-alcoholization of wine (Mietton-Peuchot, 2010). The electrodialysis is used for tartaric stabilization while the bipolar membrane electrodialysis serves to acidify or de-acidify the wine.

Gomez-Benitez et al. (2003) showed a negligible impact of electrodialysis on color intensity. Granès et al. (2009) also demonstrated that bipolar membrane electrodialysis had no effect on polyphenol contents in wine. Cottereau et al. (2007) reported that The REDUX® process (association of ultrafiltration and nanofiltration to reduce the sugar concentration of musts) allows the concentration of polyphenols in wines due to volume reduction. In 2011, Bogianchini et al. evaluated the phenolic profile and the antioxidant activity of commercial dealcoholized wines by reverse osmosis. They found that the reverse osmosis process didn't significantly affect any phenolic acids regardless to their chemical structure and alcoholic degree but the antioxidant activity decreased in average 40% compared to untreated wine. The antioxidant activities and phenolic compounds of these products were monitored for seven months. No significant changes were observed. In 2012, Liguori et al. tested the osmotic distillation for wine de-alcoholization and they tested its effect on wine phenolics. No significant differences in chemical analyses between crude and dealcoholized wine were found. The last observation is in agreement with the results obtained by Gambuti et al. (2011) while studying the influence of partial de-alcoholization by membrane contactor on red wines quality.

FINING AGENTS

Fining is used to clarify and stabilize wines. The purpose of fining is to cause haze-forming particles to combine with additional agents, leading to flocculation, clarity, and stabilization. Fining agents are used to eliminate or reduce undesirable substances in wine. Table 2 summarized the common fining agents, their sources and their applications in enology. Three major mechanisms of action of fining agents include charge-charge (electrical) interaction, bond formation, or absorption/adsorption. Wine components and the type of fining agent determine the mechanism of action. When compounds of opposite charges interact, larger particles form and settle. In the case of bond-formation, chemical bonds (i.e., hydrogen bonds) form between fining agents and wine components. Absorption occurs when compounds are engulfed by the fining agent. Alternatively, when the substance is bound to the agent's surface, the substance is adsorbed.

Winemakers use several chemical substances (table 2), the choice of which depends on the nature of the wine and the compounds that are going to be eliminated (Gómez-Plaza et al. 2000).

Table 2. Common fining agents used in winemaking

Fining agent	Source	Purpose of application
Gelatin	Animal Tissue	Removal of tannin and brown polymeric pigments
Isinglass	Fish bladder	Reduce phenolic compounds; add fruitiness to wine
Casein	Milk	Reduce wine haze and tannin content
Egg Albumen	Egg whites	Reduce wine haze and tannin content
Bentonite	Clay, volcanic deposits	Protein removal
Tannin	Wood and grapes seeds	Targets phenolic and proteins compounds
Sparkalloid	Alginate	Clarification and settling aid
Polyvinyl-polypyrrolidone	Synthetic polymer	Reduce polyphenols

Bentonite is negatively charged clay. The clay consists of complex hydrated aluminum silicate with exchangeable cationic components. Calcium and sodium bentonite are two forms that are commercially available for wine use. The mode of action of bentonite is electrostatic. The flat surface of a hydrated bentonite platelet is negatively charged. Positively charged particles adsorb onto the surface of the bentonite. Bentonite is principally used to remove proteins (protein stabilizer) from white wine and juice. It also attracts other positively charged compounds such as anthocyanins, other phenolics and nitrogen. Bentonite is not reactive towards small phenolic compounds but binds only large phenolic compounds such as anthocyanins and may also bind phenolic compound complexes with protein (Kalkan Yildirim, 2011)

Egg albumin and Gelatin are positively charged proteins used to remove excess negatively charged tannins from wine (Kalkan Yildirim, 2011). They are most commonly used to reduce the level of astringency and bitterness in the press fraction of wines, with reference to soften red wines (Stankovic et al. 2012). Egg albumen is colloidal in nature and has a positively charged surface that attracts negatively charged tannins in red wines. It is unsuitable for white wines treatment. Whereas, gelatin is primarily used to soften red wines but can also be used to reduce the phenol level and brown color in white juice before fermentation. Gelatin reduces astringency in red wines by lowering tannin levels and tends to remove more, higher molecular weight galloylated proanthocyanidols than lower molecular weight tannins (Sarni-Manchado et al., 1999). After the formation of gelatin-tannin complex, this complex may interact with anthocyanins, causing their removal.

Casein fining preparations are used in particular for the treatment of astringency and for the clarification of white and rosé wines, but are also sometimes used with red wines. Casein is a positively charged protein that flocculates in acidic media such as wine. When added to wine, casein adsorbs and mechanically removes suspended material as it settles. Casein is difficult to mix into the juice/wine as it is relatively insoluble in acidic solutions and should be mixed in water with a pH value above 8 or made alkaline prior to mixing.

Isinglass is a positively charged fining agent derived from the air bladder of a sturgeon. It is available as sheet or flocculated isinglass. Isinglass is used principally in white still and sparkling wines and to clean up the aroma, improve clarity and modify the finish without significantly modifying tannin levels.

Polyvinyl polypyrrolidone (PVPP) is a high molecular weight fining agent made of cross-linked monomer of polyvinlypyrrolidone. It complexes with phenolic and polyphenolic components in wine by adsorption and attracts low molecular weight tannins. It removes bitter compounds and browning precursors in both red and white wines. PVPP is quick acting with no preparation required.

Several studies in the literature treated the impact of fining agents on the phenolic composition of wines. A study done by Stankovic et al. (2012), on the effect of fining agents, on red Pinot noir variety of different ages, showed that fined wines lead to significant reduction of color intensity, ionized anthocyanins, and a low reduction of colorless anthocyanins, relative to unfined wines. Castillo-Sanchez et al. (2006) investigated the impact of PVPP, casein, egg albumin and gelatin on the evolution of anthocyanins and color of Vinhão wines. They found that all fining agents induced loss of color density and anthocyanin content but surprisingly, they noticed that PVPP caused more loss of color than the other fining agents. Several authors also found a decrease in anthocyanin content with fining (Castillo-Sanchez et al., 2008; Cosme et al., 2007; Karamanidou et al., 2011). Cosme et al. (2007) studied the interactions between protein fining agents and proanthocyanidins in white wine. They reported that the monomeric flavanols were significantly depleted by casein, and gelatin with low molecular weight (MW) distribution, and isinglass obtained from fish swim bladder. The degree of polymerization of polymeric proanthocyanidins that remained in the fined wine decreased significantly after addition of protein fining agents except when potassium caseinate was used. Casein and isinglass induced a significant decrease in wine color (A420nm), a decrease in browning potential and a decrease in turbidity. Cosme et al. (2012) focused their research on determining if non-allergenic pea protein or polyvinylpolypyrrolidone (PVPP) are possible alternatives for casein fining. The results indicate that flavonoid and non-flavonoid phenols decreased in the wines treated with potassium caseinate, pea protein, and PVPP. All fining agents decreased wine color. Potassium caseinate was the most effective fining agent for reducing browning potential. When applying the CIELaB chromatic characterization, they found that the value for b* (yellowness) decreased significantly with all fining agents assayed; however, the decrease was greater in all experiments fined with potassium caseinate, indicating a higher reduction in the yellow intensity of the fined wine. Chroma (C*) is a parameter that indicates the contribution of a*(redness) and b*(yellowness). The value of C* decreased significantly after addition of pea protein, potassium caseinate and formulations of pea protein with PVPP. They found that PVPP could be used alone or in combination with much smaller quantities of casein and still effectively reduce wine oxidation through removal of polyphenols in reduced and oxidized (quinones and quinone methides) forms, which includes simple phenolic acids and flavonoids. Recently in 2013, Oberholster et al. investigated the effect of gelatin and egg albumin on the phenolic composition of Pinotage wine. They found that both gelatin and egg albumin fining decreased the mean degree of polymerization (mDP) of tannin significantly by 26.4% and 25.2%, respectively, compared to the control. Egg albumin treatments significantly decreased the total pigment content compared to control.

CONCLUSION

This chapter reviewed the winemaking techniques which are admitted to affect the phenolic composition of wines. The maceration and the aging remain the most influencing steps while fermentation and filtration slightly impact the polyphenols content.

Despite all the progress made in this sector, some information remain contradictory. Moreover, the effects of some winemaking processes on wine polyphenols composition are still lacking. Therefore, more studies are required to elucidate the real impact of each step during the winemaking of a given wine.

REFERENCES

Alvarez, I., Aleixandre, J. L., García, M. J. & Lizama, V. (2006). Impact of prefermentative maceration on the phenolic and volatile compounds in Monastrell red wines. *Anal. Chim. Acta.*, *563*, 109-115.

Arapitsas, P., Matthias Scholz, M., Vrhovsek, U., Di Blasi, S., BiondiBartolini, A., Masuero, D., Perenzoni, D., Rigo, A. & Mattivi, F. (2012). A metabolomic approach to the study of wine micro-oxygenation. *Plos one*, *7*(5), e37783. doi:10.1371/journal.pone.0037783.

Atanasova, V., Fulcrand, H., Cheynier, V. & Moutounet, M. (2002). Effect of oxygenation on polyphenol changes occurring in the course of wine-making. *Anal. Chim. Acta*, *458*, 15–27.

Barra, C., Hackl, K., Durrschmid, K., Eder, R. & Dieter Kulbe, K. (2005). Effect of cold maceration, stem return, reheating and mannoprotein addition on color stabilization of Pinot noir wines. *Mitteilunngen Klosterneuburg*, *55*, 140-147.

Barrera-Garcia, V. D., Gougeon R. D., Di Majo, D., De Aguirre, C., Voilley, A. & Chassagne, D. (2007). Different sorption behaviors for wine polyphenols in contact with oak wood. *J. Agric. Food Chem.*, *55*, 7021–7027.

Baustita-Ortín, A. B., Fernández-Fernández, J. I., López-Roca, J. M. & Gómez-Plaza, E. (2007). The effects of enological practices in anthocyanins, phenolic compounds and wine colour and their dependence on grape characteristics. *J. Food Comp. Analy.*, *20*, 546-552.

Baustita-Ortín, A. B., Martínez-Cutillas, A., Ros-García, J. M., López-Roca, J. M. & Gómez-plaza, E. (2005). Improving colour extraction and stability in red wines: the use of maceration enzymes and enological tannins. *Int. J. Food Technol.*, *40*, 867-878.

Belleville, M. P., Brillouet, J. M., Tarado De La Fuente, B. & Moutonet, M. (1992). Fouling colloids during microporous alumina membrane filtration of wine. *J. Food Sci.*, *57*, 396-400.

Belleville, M. P., Brillouet, J. M., Tarado De La Fuente, B. & Moutonet, M. (1990). Polysaccharide effects on cross-flow microfiltration of two Red wines with a microporous alumina membrane. *J. Food Sci.*, *55*, 1598-1602.

Berger, J. L. & Cottereau, P. (2000). Winemaking of fruity red wines by pre-fermentation maceration under heat. *Bulletin de l'OIV*, 833-834, 473-480.

Bloem, A., Lonvaud-Funel, A. & De Revel, G. (2008). Hydrolysis of glycosidically bound flavour compounds from oak wood by Oenococcus Oeni. *Food Microbiol.*, *25*, 99-104.

Bloem, A., Bertrand, A., Lonvaud-Funel, A. & De Revel, G. (2007). Vanillin production from simple phenols by wine-associated lactic acid bacteria. *Lett. Appl. Microbiol.*, *44*, 62-67.

Bogianchini, M., Cerezo, A. B., Gomis, A., López, F. & García-Parrilla, M. C. (2011). Stability, antioxidant activity and phenolic composition of commercial and reverse osmosis obtained dealcoholised wines. *LWT- Food Sci. Technol.*, *44*, 1369-1375.

Borazan, A. & Bozan, B. (2013). The influence of pectolytic enzyme addition and prefermentive mash heating during the winemaking process on the phenolic composition of okuzgozu red wine. *Food Chem.*, *138*, 389-395.

Budić-Leto, I., Lovrić, T., Gajdoš Kljusurić, J., Pezo, I. & Vrhovšek, U. (2006). Anthocyanin composition of the red wine Babić affected by maceration treatment. *Eur Food Res. Technol.*, *222*, 397-402.

Busse-Valverde, N., Gómez-Plaza, E., López-Roca, J. M., Gil-MuÑoz, R. & Bautista-Ortín, A. B. (2011). The extraction of anthocyanins and proanthocyanidins from grapes to wine during fermentive maceration is affected by the enological technique. *J. Agric. Food Chem.*, *59*, 5450-5455.

Busse-Valverde N., Gómez-Plaza E., López-Roca J., Gil-Muηoz R., Fernandez-Fernandez J. & Baustita-Ortin A. (2010). Effect of different enological practices on skin and seed proanthocyanidins in three varietal wines. *J. Agric. Food Chem.*, *58*, 11333-11339.

Cabrita, M. J., Torres, M., Palma. V., Alves, E., Patão, R. & Costa Freitas, A. M. (2008). Impact of malolactic fermentation on low molecular weight phenolic compounds. *Talanta*, *74*, 1281-1286.

Cameira-dos-Santos, P. J., Brillouet, J. M., Cheynier, V. & Moutounet, M. (1996). Detection and partial characterization of new Anthocyanin-derived pigments in Wine. *J. Sci. Food Agric.*, *70*, 204-208.

Canal-Llaubères, R. M. & Pouns, J. P. (2002). Les enzymes de macération en vinification en rouge. Influence d'une nouvelle préparation sur la composition des vins. *Revue des Œnologues*, *104*, 29–31.

Cano-López, M., López-Roca, J. M., Pardo-Minguez, F. & Gómez Plaza, E. (2010). Oak barrel maturation vs. micro-oxygenation: Effect on the formation of anthocyanin-derived pigments and wine colour. *Food Chem.*, *119*, 191-195.

Caridi, A., Cufari, A., Lovino, R., Palumbo, R. & Tedesco, I. (2004). Influence of yeast on polyphenol composition of wine. *Food Technol. Biotechnol.*, *43*, 37-40.

Castellari, M., Arfelli, G., Riponi, C. & Amati, A. (1998). Evolution of phenolic compounds in red winemaking as affected by must oxygenation. *Am. J. Enol. Vitic.*, *49*, 91–94.

Castillo-Sánchez, J. X., García-Falcón, M. S., Garrido, J., Martínez-Carballo, E., Martins-Dias, L. R. & Mejuto, X. C. (2008). Phenolic compounds and colour stability of Vinhão wines: Influence of wine-making protocol and fining agents. *Food Chem.*, *106*, 18-26.

Castillo-Sánchez, J. J., Mejuto, J. C., Garrido, J. & García-Falcón, M. S. (2006). Influence of wine-making protocol and fining agents on the evolution of the anthocyanin content, colour and general organoleptic quality of Vinhao wines. *Food Chem.*, *97*, 130-136.

Castillo-Sánchez, J. X., García-Falcón, M. S., Garrido, J., Martínez-Carballo, E., Martins-Dias, L. R. & Mejuto, X. C. (2008). Phenolic compounds and colour stability of Vinhão wines: Influence of wine-making protocol and fining agents. *Food Chem.*, *106*, 18-26.

Cejudo-Bastante, M. J., Pérez-Coello, M. S. & Hermosím-Gutiérrez, I. (2011). Effect of wine micro-oxygenation treatment and storage period on colour-related phenolics, volatile composition and sensory characteristics. *LWT- Food Sci. Technol.*, *44*, 866-874.

Cheynier, V., Dueñas-Paton, M., Salas, E., Maury, C., Souquet, J. M., Sarni-Manchado, P. & Fulcrand, H. (2006). Structure and properties of wine pigments and tannins. *Am. J. Enol. Vitic.*, *57* (3), 298-305.

Cheynier, V., Flucrand, H., Brossaud, F., Asselin, C. & Moutounet, M. Phenolic composition as related to red wine flavor. In "Chemistry of wine Flavor", eds. Waterhouse, A., Ebeler, S,. (1999). American Chemical Society, Washington, DC. 124-141.

Cheynier, V., Prieur, C., Guyot, S., Rigaud, J. & Moutounet, M. (1997). The structures of tannins in grapes and wines and their interactions with proteins. American Chemical Society Symposium, *661*, 81-93.

Cosme, F., Capão, I., Filipe-Ribeiro, L., Bennett, R. N. & Mendes-Faia, A. (2012). Evaluating potential alternatives to potassium caseinate for white wine fining: Effects on physiochemical and sensory characteristics. *LWT- Food Sci. Technol.*, *46*, 382-387.

Cosme, F., Ricardo-Da-Silva, J. M. & Laureano, O. (2007). Protein fining agents: Characterization and red wine fining assays. *Ital. J. Food Sci.*, *19*, 39–56.

Cottereau, P., Aguera, E., Athès-Dutour, V., Bes, M., Caillé, S., Escudier, J. L., Mikolajczak, M., Roy, A., Sablayrolles, J. M., Samson, A., Souchon, I. & Vidal, J. P. (2010). Réduction de la teneur en alcool des vins: Étude comparative de différentes technologies. *Bulletin de l'OIV*, *83*, 31-42.

Couasnon, M. (1999). Une nouvelle technique: La macération préfermentaire à froid - Extraction à la neige carbonique 1re partie. *Revue des Œnologues*, *92*, 26-30.

Cunier, C. (1997). Ceppi di lievito e composizione fenolica dei vini rossi. *Vignevini*, 7-8, 39-42.

Czekaj, P., López, F. & Güell, C. (2000). Membrane fouling during microfiltration of fermented beverages. *J. Membr. Sci.*, *166*, 199-212.

Del Barrio-Galán, R., Pérez-Magariño, S., Ortega-Heras, M., Guadalupe, Z. & Ayestarán, B. (2012). Polysaccharide characterization of commercial dry yeast preparations and their effect on white and red wine composition. *LWT- Food Sci. Technol.*, *48*, 215-223.

Delcroix, A., Günata, Z., Sapis, J. C., Salmon, J. M. & Bayonove, C. (1994). Glycosidase activities of three enological yeast strains during winemaking: Effect on the terpenol content of Muscat wine. *Am. J. Enol. Vitic.*, *45*, 291-296.

De Revel, G., Bloem, A., Augustin, M., Lonvaud-Funel, A. & Bertrand, A. (2005). Interaction of Oenococcus oeni and oak wood compounds. *Food Microbiol.*, *22*, 569-575.

De Rosso, M., Panighel, A., Dalla Vedona, A., Stella, L. & Flamini, R. (2009). Changes in chemical composition of a red wine aged in Acacia, cherry, chestnut, mulberry, and oak wood barrels. *J. Agric. Food Chem.*, *57*, 1915-1920.

Drinkine, J., Lopes P., Kennedy J. A., Teissedre P. L. & Saucier C. (2007). Ethylidene bridged flavan-3-ols in red wine and correlation with wine age. *J. Agric. Food Chem.*, *55*, 6292–6299.

Ducasse, M. A., Canal-Llauberes, R. M., de Lumley, M., Williams, P., Souquet, J. M., Fulcrand, H., Doco, T. & Cheynier, V. (2010). Effect of macerating enzyme treatment on the polyphenol and polysaccharide composition of red wines. *Food Chem.*, *118*, 369–376.

Du Toit, W. j., Marais, J., Pretorius, I. S. & Du Toit, M. (2006). Oxygen in must and wine: A review. *S. Afr. J. Enol. Vitic.*, *27*, 76-94.

Dykes, S. (2007). The effect of oxygen dosage rate on the chemical and sensory changes occurring during micro-oxygenation of New Zealand red wine. Ph.D. Thesis, the University of Auckland, New Zealand.

El Rayess, Y., Albasi, C., Bacchin, P., Taillandier, P., Mietton-Peuchot, M. & Devatine, A. (2012). Analysis of membrane fouling during cross-flow microfiltration of wine. *Innov. Food Sci. Emerg.*, *16*, 398-408.

El Rayess, Y., Albasi, C., Bacchin, P., Taillandier, P., Mietton-Peuchot, M. & Devatine, A. (2011). Cross-flow microfiltration of wine: effect of colloids on critical fouling conditions. *J. Membr. Sci.*, 385-386, 177-186.

Escribano-Bailon T., Alvarez-Garcia M., Rivas-Gonzalo J. C., Heredia F. J. & Santos-Buelga, C. (2001). Color and stability of pigments derived from the acetaldehyde mediated condensation between malvidin 3–O-glucoside and (+)-catechin. *J. Agric. Food Chem.*, *49*, 1213–1217.

Es-Safi N. E., Le Guerneve C., Cheynier V. & Moutounet M. (2000). New phenolic compounds formed by evolution of (+)-catechin and glyoxylic acid in hydroalcoholic solution and their implication in color changes of grape-derived foods. *J Agric Food Chem.*, *48*, 4233–4240.

Feuillat, M., Escot, S., Dulau, L. & Charpentier, C. (2000). Release of polysaccharides by yeasts and the influence of released polysaccharides on colour stability and wine astringency. *Aust. J. Grape Wine R.*, *7*, 153-159.

Fornairon-Bonnefond, C., Camarasa, C., Moutonet, M. & Salmon, J. M. (2002). Newtrends on yeast autolysis and wine Aging on lees: A bibliographic review. *J. Int. Sci. Vigne Vin*, *36*, 49–69.

Francia-Aricha, E. M., Rivas-Gonzalo, J. C. & Santos-Buelga, C. (1998). Effect of malvidin-3-monoglucoside on the browning of monomeric and dimeric flavanols. *Zeitschrift für Lebensmitteluntersuchung und-Forschung A*, *207*, 223-228.

Fulcrand, H., Dueñas, M., Salas, E. & Cheynier, V. (2006). Phenolic reactions during winemaking and aging. *Am. J. Enol. Vitic.*, *57*, 289–297.

Fulcrand, H., Atanasova, V., Salas, E. & Cheynier, V. (2004). The fate of anthocyanins in wine: are there determining factors? In Red wine color: revealing the mysteries; Waterhouse, A. L, Kennedy, J, Eds.; ACS Symposium Series, 886; American Chemical Society: Washington, DC. 68-85.

Gambuti, A., Rinaldi, A., Lisanti, M. T., Pessina, R. & Moio, L. (2011). Partial dealcoholisation of red wines by membrane contactor technique: influence on colour, phenolic compounds and saliva precipitation index. *Eur. Food Res. Technol.*, *233*, 647-655.

Gao, L., Girard, B., Mazza, G. & Reynolds, A. G. (1997). Changes in Anthocyanins and Color Characteristics of Pinot Noir Wines during Different Vinification Processes. *J. Agric Food Chem.*, *45*, 2003-2008.

Gerbaux, V., Briffox, C. & Vincent, B. (2003). Optimisation de la macération finale à chaud, intérêt d'un enzymage et d'une macération sous chapeau immergé pour la vinification du Pinot noir. *Rev. Fr. Oenol.*, *201*, 16-21.

Gil-Muñoz, R., Moreno-Pérez, A., Vila-López, R., Fernádez-Fernández, J., Martínez-Cutillas, A. & Gómez-Plaza, E. (2009). Influence of low temperature prefermentative techniques on chromatic and phenolic characteristics of Syrah and Cabernet Sauvignon wines. *Eur. Food Res. Technol.*, *228*, 777–788.

Gómez-Benítez, J. Palacios Macías, V. M., Szekely Gorostiaga, P., Veas López, R. & Pérez Rodríguez. (2003). Comparison of electrodialysis and cold treatment on an industrial scale for tartrate stabilization of sherry wines. *J. Food Eng.*, *58*, 373-378.

Gómez-Cordoves, C., Gonzalez-San Jose, M., Junquera, B. & Estrella, I. (1995). Correlation between flavonoids and color in red wines aged in wood. *Am. J. Enol. Vitic.*, *46*, 295–298.

Gómez-Míguez, M., González-Miret, M. L. & Heredia, F. (2006). Evolution of colour and anthocyanin composition of Syrah wines elaborated with pre-fermentative cold maceration. *J. Food Eng.*, *79*, 271–278.

Gómez-Plaza, E. & Cano-López, M. (2011). A review on micro-oxygenation of red wines: Claims, benefits and the underlying chemistry. *Food Chem.*, *125*, 1131-1140.

Gómez-Plaza, E., Gil-Muñoz, R., López-Roca, J. M., De La Hera-Orts, M. L. & Martínez-Cultíllas, A. (2000). Effect of the addition of bentonite and polyvinylpolypyrrolidone on the colour and long-term stability of red wines. *J. Wine Res.*, *11*, 223-231.

González-Sanjosé, M. L., Ortega-Heras, M. & Pérez-Magariño, S. (2008). Microoxygenation treatment and sensory properties of young red wines. *Food Sci. Technol. Int.*, *14*, 123–130.

Gordillo B., Lopez-Infante M., Ramirez-Perez P., Gonzalez-Miret M. & Heredia F. (2010). Influence of prefermentive cold maceration on the color and anthocyanic copigmentation of organic Tempranillo wines elaborated in a warm climate. *J. Agric. Food Chem.*, *58*, 6797-6803.

Granès, D., Bouissou, D., Lutin, F., Moutounet, M. & Rousseau, J. (2009). L'élévation du pH des vins: causes, risques œnologiques, impacts de la mise en œuvre de moyens d'acidification. *Bulletin de l'OIV*, *82*, 57-70.

Guyot, S., Vercauteren, J. & Cheynier, V. (1996). Structural determination of colorless and yellow dimmers resulting from (+)-catechin coupling catalysed by grape polyphenol oxidase. *Phytochemistry*, *42*, 1279–1288.

Haslam E. (1980). In vino veritas: Oligomeric procyanidins and the Aging of red wines. *Phytochemistry*, *19*, 2577–2582.

Hayasaka, Y. & Kennedy, J. A. (2003). Mass spectrometric evidence for the formation of pigmented polymers in red wine. *Aust. J. Grape Wine Res.*, *9*, 210-220.

Hayasaka, Y., Birse, M., Eglinton, J. & Herderich, M. (2007). The effect of Saccharomyces cerevisiae bayanus yeast on colour properties and pigment profiles of a Cabernet Sauvignon red wine. *Aust. J. Grape Wine Res.*, *13*, 176-185.

Hermosín-Gutiérrez, I., Sánchez-Palomo Lorenzo, E. & Vicario Espinosa, A. (2005). Phenolic composition and magnitude of copigmentation in young and shortly aged red wines made from cultivars Cabernet Sauvignon, Cencibel and Syrah. *Food Chem.*, *92*, 269-283.

Hernández, T., Estrella, I., Carlavilla, D., Martín-Álvarez, P. J. & Victoria Moreno-Arribas, V. (2006). Phenolic compounds in red wine subjected to industrial malolactic fermentation and Aging on lees. *Anal. Chim. Acta*, *563*, 116-125.

Isabel Spranger, M., Cristina Clímaco, M., Sun, B., Eiriz, N., Fortunato, C., Nunes, A., Conceicao Leandro, M., Luisa Avelar, M. & Pedro Belchior, A. (2004). Differentiation of red winemaking technologies by phenolic and volatile composition. *Anal. Chim. Acta*, *513*, 151-161.

Kalkan Yildirim, H. (2011). Effects of fining agents on antioxidant capacity of red wines. *J. Inst. Brew.*, *117*, 55-60.

Kammerer, D., Claus, A., Schieber, A. & Carle, R. (2005). A novel process for the recovery of polyphenols from grape (vitis vinifera L.) pomace. *J. Food Sci.*, *70*, 157-163.

Karamanidou, A., Kallithraka, S. & Hatzidimitrio, E. (2011). Fining of red wines: Effects on their analytical and sensory parameters. *J. Int. Sci. Vigne Vin*, *45*, 47–60.

Kelebek, H., Canbas, A., Cabaroglu, T. & Selli, S. (2007). Improvement of anthocyanin content in the cv. Okuzgozu wines by using pectolytic enzymes. *Food Chem.*, *105*, 334–339.

Kontoudakis, N., Fernando, Z. M., Bosh, C. & Miquel, J. (2010). Grape phenolic maturity; determination methods and consequences on wine phenolic composition. Ph.D. Thesis, the University of Rovira i Virgili.

Koyama, K., Goto-Yamamoto, N. & Hashizume, K. (2007). Influence of maceration temperature in red wine vinification on extraction of phenolics from berry skins and seeds of grape (vitis vinifera). *Biosci. biotechnol. Biochem.*, *71*, 958-965.

Kovac, V., Alonso, E., Bourzeix, M. & Revilla, E. (1992). Effect of several enological practices on the content of catechins and proanthocyanidins of red wines. *J. Agric. Food Chem.*, *40*, 1953-1957.

Liao, H., Cai, Y. & Haslam, E. (1992). Polyphenols interactions. anthocyanins: copigmentation and colour changes in young red wines. *J. Sci. Food Agric.*, *59*, 299-305.

Liguori, L., Russo, P., Albanese, D. & Di Matteo, M. (2012). Effect of process parameters on partial dealcoholization of wine by osmotic distillation. *Food Bioprocess technol.*, *6*, 2514-2524.

Locascio, G. R., Millis, A. D. & Waterhouse, L. A. (2006). Reduction of catechin, rutin, and quercetin levels by interaction with food-related microorganisms in a resting state. *J. Sci. Food Agric.*, *86*, 2105-2112.

Lorincz, G., Kállay, M. & Pásti G. (1998). Effect of carbonic maceration on phenolic compounds of red wines. *Acta Aliment*, *27*, 341-355.

Mateus, N., Silva, A. M. S., Rivas-Gonzalo, J. C., Santos-Buelga, C. & Freitas, V. (2003). A new class of blue anthocyanin-derived pigments isolated from red wines. *J. Agric. Food Chem.*, *51*, 1919-1923.

Mauzaric, J. P. & Salmon, J. M. (2005). Interactions between yeast lees and wine polyphenols during simulation of wine aging: I. Analysis of remnant polyphenolic compounds in the resulting wines. *J. Agric. Food Chem.*, *53*, 5647-5653.

Mietton-Peuchot, M., Peuchot, C. & Milisic, V. (2008). Nanofiltration and reverse osmosis inwinemaking. *Desalination*, *231*, 283-289.

Mongas, M., Bartalome, B. & Gomez-Cordoves, C. (2005). Updated knowledge about the presence of phenolic compounds in wine. *Crit. Rev. Food Sci. Nutr.*, *45*, 85-118.

Morel-Salmi, C., Souquet, J. M., Bes, M. & Cheynier, V. (2006). Effect of Flash Release treatment on Phenolic extraction and wine composition. *J. Agric. Food Chem.*, *54*, 4270-4276.

Moreno-Arribas, M. V., Gómez-Cordovés, C. & Martín-Álvarez, P. J. (2008), Evolution of red wine anthocyanins during malolactic fermentation, postfermentive treatments and Aging with lees. *Food Chem.*, *109*, 149-158.

Moutounet, M. & Escudier, J. L. (2000). Prétraitement des raisins par Flash-détente sous vide. Incidence sur la qualité des vins. *Bull. O. I. V.*, *73*, 5-19.

Netzel, M., Strass, G., Bitsch, I., Konitz, R., Christmann, M. & Bitsch, R. (2003). Effect of grape processing on selected antioxidant phenolics in red wine. *J. Food Eng.*, *56*, 223-228.

Oberholster, A., Carstens, L. M. & Du Toit, W. J. (2013). Investigation of the effect of gelatine, egg albumin and cross-flow microfiltration on the phenolic composition of Pinotage wine. *Food Chem.*, *138*, 1275-1281.

Oszmianski, J., Romeyer, F. M., Sapis, J. C. & Macheix, J. J. (1986). Grape seed phenolics: Extraction as affected by some conditions occurring during wine processing. *Am. J. Enol. Vitic.*, *37*, 7-12.

Palomero, F., Morata, A., Benito, S., Calderón, F. & Suárez-Lepe, J. A. (2009a). Newgenera of yeasts for over-lees aging of red wine. *Food Chem.*, *112*, 432−441.

Pardo, F., Salinas, R., Alonso, G., Navarro, G. & Huerta, M. D. (1996). Effect of diverse enzyme preparations on the extraction and evolution of phenolic compounds in red wines. *Food Chem.*, *67*, 135–142.

Parley, A., Vanhanen, L. & Heatherbell, D. (2001). Effects of pre-fermentation enzyme maceration on extraction and colour stability in Pinot Noir wine. *Austr. J. Grape Wine R.*, *7*, 146–152.

Parish, M., Wollan, D. & Paul, R. (2000). Micro-oxygenation. A review. *The Australian Grape grower and Winemaker*, *438*, 47–50.

Parley, A. (1997). The effect of pre-fermentation enzyme maceration on extraction and colour stability in Pinot noir wine. Ph.D. Thesis, University of Lincoln, New Zealand.

Paul, R. (2002). Micro-oxygenation – Where now? In Proceedings of the ASVO seminar uses of gases in winemaking. Adelaide: Australian Society of Viticulture and Oenologie. 23-27.

Pérez-Serradilla, J. A. & Luque de Castro, M. D. (2008). Role of lees in wine production: A review. *Food Chem.*, *11*, 447-456.

Poncet-Legrand, C., Edelmann, A., Putaux, J. L., Cartalade, D., Sarni-Manchado, P. & Vernhet, A. (2006). Poly (L-proline) interactions with flavan-3-0ls units: influence of the molecular structure and the polyphenol/protein ratio. *Food Hydrocolloid*, *20*, 687-697.

Poirier, D. (1985). Contribution à l'étude de la microfiltration tangentielle sur membranes minérales et de son application à la clarification et la stabilisation des vins. Thèse de doctorat.

Poni, S., Bernizzoni, F., Gatti, M. & Civardi, S. (2009). Long-term performance of barbera grown under different training systems and within- row vine spacings. *Am. J. Enol. Vitic.*, *60*, 339-348.

Pozo-Bayón, M. A., Monagas, M., Polo, M. C. & Gómez-Cordovés, C. (2004). Occurrence of pyroanthocyanins in sparkling wines manufactured with red grape varieties. *J. Agric. Food Chem.*, *52*, 1300-1306.

Quideau, S., Jourdes, M., Lefeuvre, D., Montaudon, D., Saucier, C., Glories, Y., Pardon, P. & Pourquier, P. (2005). *Chem A Eur. J.*, *11*, 6503–6513.

Remy, S., Fulcrand, H., Labarbe, B., Cheynier, V. & Mounounet, M. (2000). First confirmation in red wine of products resulting from direct anthocyanin-tannin reactions. *J. Sci. Food Agric.*, *80*, 745-751.

Rentzsch, M., Schwarz, M. & Winterhalter, P. (2007). Pyroanthocyanins- an overview on structures, occurrence, and pathways of formation. *LWT- Food Sci. Technol.*, *18*, 526-534.

Revilla, I. & Gonzales-SanJose, M. (2003). Compositional changes during the storage of red wines treated with pectolytic enzymes: Low molecular-weight phenols and flavan-3-ol derivative levels. *Food Chem.*, *80*, 205–214.

Revilla, I. & González-Sanjosé, M. L. (2002). Multivariate evaluation of changes induced in red wine characteristics by the use of-extracting agents. *J. Agric. Food Chem.*, *50*, 4525–4530.

Revilla, I. & González-Sanjosé, M. L. (2001). Evolution during the storage of red wines treated with pectolytic enzymes: new anthocyanin pigments formation. *J. Wine Res.*, *12*, 183–197.

Ribéreau-Gayon, P., Dubourdieu, D., Doneche, B. & Lonvaud, A. (2006). *Handbook of Enology*, Volume *1*: The microbiology of winemaking and vinifications. Ed. Ribéreau-Gayon P. Wiley, Chichester, England.

Ribéreau-Gayon, P., Glories, Y., Maujean, A. & Dubourdieu, D. (2006). *Handbook of Enology*. The chemistry of wine stabilization and treatments. Ed. Ribéreau-Gayon P., Wiley, Chichester, England..

Ribéreau-Gayon, P., Pontallier, P. & Glories, Y. (1983). Some interpretations of colour changes in young red wines during their conservation. *J. Sci. Food Agric.*, *34*, 505–516.

Rieger, T. (2000). Micro-oxygenation presents promise with potential peril for quality winemaking. *Vineyard and Winery Management*, 83–88.

Romero-Cascales, I., Ros-Garcia, J. M., Lopez-Roca, J. M. & Gomez-Plaza, E. (2012). The effect of a commercial pectolytic enzyme on grape skin cell wall degradation and colour evolution during the maceration process. *Food Chem.*, *130*, 626–631.

Sarni-Manchado, P., Cheynier, V. & Moutounet, M. (1999). Interactions of grape seed tannins with salivary proteins. *J. Agri. Food Chem.*, *47*, 42-47.

Sánchez-Iglesias, M., González-Sanjosé, M. L., Pérez-Magariño, S., Ortega-heras, M. & González-Huerta, C. (2009). Effect of micro-oxygenation and wood type on the phenolic composition and color of an aged red wine. *J. Agric. Food Chem.*, *57*, 11498-11509.

Santos-Buelga, C., Francia-Aricha, E. M., Du Pascual-Teresa, S. & Rivas- Gonzalo, J. C. (1999). Contribution to the identification of the pigments responsible for the browning of anthocyanin-flavanol solutions. *Eur. Food Res. Technol.*, *209*, 411-415.

Schmidtke, M. L., Clark, A. C. & Scollary, G. R. (2011). Micro-oxygenation of red wine: Techniques, Applications, and outcomes, *Crit. Rev. Food Sci. Nutr.*, *51*, 115-131.

Serrano, M. & Paetzold, M. (1998). Incidences des filtrations sur la composition chimique et les qualités organoleptiques des *vins. J. Int. Sci. Vigne Vin*, hors-série, 53-57.

Sousa C., Mateus N., Silva A. M. S. & González-Paramás, A. M. (2007). Structural and chromatic characterization of a new Malvidin 3-glucoside–vanillyl–catechin pigment. *Food Chem.*, *102*, 1344–135.

Sousa C., Mateus N., Perez-Alonso J., Santos-Buelga C. & De Freitas V. (2005). Preliminary study of oaklins, a new class of brick-red catechin pyrylium pigments resulting from the reaction between catechin and wood aldehydes. *J. Agric Food Chem.*, *53*, 9249–9256.

Spayd, S. E., Tarara, J. M., Mee, D. L. & Ferguson, J. C. (2002). Separation of sunlight and temperature effects on the composition of vitis vinifera cv. Merlot berries. *Am. J. Enol. Vitic.*, *53*, 171-182.

Spranger, M. I., Clímaco, M. C., Sun, B., Eiriz, N., Fortunato, C., Nunes, A., Leandro, M. C., Avelar, M. L. & Belchior, P. (2004). Differenciation of red winemaking technologies by phenolic and volatile composition. *Anal. Chim. Acta*, *513*, 151-161.

Stankovic, S., Jovic, S., Zivkovic, J. & Pavlovic, R. (2012). Influence of age on red wine colour during fining with bentonite and gelatin. *Int. J. Food Prop.*, *15*, 326-335.

Sun, B. & Spranger, M. I. (2005). Changes in phenolic composition of Tinta Miúda red wines after 2 years of Aging in bottle: effect of winemaking technologies. *Eur. Food Res. Technol.*, *221*, 305-312.

Sun B., Spranger I., Roque-do-Vale F., Leandro C. & Belchior P. (2001). Effect of different winemaking technologies on phenolic composition in Tinta Miúda red wines. *J. Agric. Food Chem.*, *49*, 5809-5816.

Tanaka, T., Takahashi, R., Kouno, I. & Nonaka, K. (1994). Chemical evidence for the de-astringency (insolubilization of tannins) of persimmon fruit *J. Chem. Soc., Perkin Trans. 1*, *20*, 3013–3022.

Timberlake C. F. & Bridle P. (1976). Interactions between anthocyanins, phenolic compounds, and acetaldehyde and their significance in red wines. *Am. J. Enol. Vitic.*, *27*, 97–105.

Ulbricht, M., Ansorge, W., Danielzik, I., König, M. & Schuster, O. (2009). Fouling in microfiltration of wine: The influence of the membrane polymer on adsorption of polyphenols and polysaccharides. *Sep, Purif. Technol.*, *68*, 335-342.

Vidal, S. & Aagaard, O. (2008). Oxygen management during vinification and storage of Shiraz wine. *Aust. NZ Wine Ind. J.*, *23*, 56-63.

Vivas N. & Glories Y. (1996). Role of oak wood ellagitannins in the oxidation process of red wines during aging. *Am J. Enol. Vitic.*, *47*, 103–107.

Vrhovsek, U., Vanzo, A. & Nemanic, J. (2002). Effect of red wine maceration techniques on oligomeric and polymeric proanthocyanidins in wine, cv. *Blaufrankisch. Vitis*, *41*, 47-51.

Wildenradt H. L. & Singleton V. L. (1974). The production of aldehydes as a result of oxidation of polyphenolic compounds and its relation to wine aging. *Am J. Enol. Vitic.*, *25*, 119–126.

Wollan, D. (1997). Tannin extraction and manipulation – managing the options. In Allen, M., Wall, G., Bulleidm N. (eds.). ASVO Oenology seminar: phenolics and extraction. Adelaide, Australian Society of viticulture and oenology.

Yang Sun, S., Guang Jiang, W. & Ping Zhao, Y. (2011). Evaluation of different Saccharomyces cerevisiae strains on the profile of volatile compounds and polyphenols in cherry wines. *Food Chem.*, *127*, 547-555.

Zoecklein, B. W., Wolf, T. K., Pélanne, L., Kay Miller, M. & Birkenmaier, S. S. (2008). Effect of vertical shoot-positioned, smart-Dyson, and Geneva double-curtain training systems on viognier grape and wine composition. *Am. J. Enol. Vitic.*, *59*, 11-21.

In: Wine
Editor: Youssef El Rayess

ISBN: 978-1-63321-048-6

Chapter 5

RED WINE: A FUNCTIONAL BEVERAGE?

Giovanna Giovinazzo[1*], Ilaria Ingrosso[1], Mariana Tristezza[1], Francesco Grieco[1] and Youssef El Rayess[2,3,4]

[1]National Research Council. - Institute of Sciences of Food Production (ISPA), Unit of Lecce, via Prov. Lecce-Monteroni, Lecce, Italy

[2]Faculty of Agricultural and Food Sciences, Holy Spirit University of Kaslik, Jounieh, Lebanon

[3]Université de Toulouse, INPT, UPS, Laboratoire de Génie Chimique, 4 Allée Emile Monso, Toulouse, France.

[4]Centre de Viticulture et d'Œnologie de Midi-Pyrénées, Avenue de l'Agrobiopôle, Castanet-Tolosan, France

ABSTRACT

Phenolic compounds present in grape berries are extracted from the skin, seeds and pulp during the winemaking process. These substances have a potentially positive effect on human health, thus giving to red wine "bioactive properties" that can contribute to decrease the incidence of a number of human pathologies. However, basic and clinical sciences are showing that the reality is much more complex than this and that several issues, notably daily intake, bioavailability, or in vivo antioxidant activity are yet to be resolved. Viticulture and winemaking practices, which vary in different countries, determine the concentration of phenolic compounds in wine. In particular poor knowledge is available on the effects of different yeast strains on the final concentration of polyphenols in red wine. In this chapter, we summarize the recent findings concerning the effects of wine polyphenols on human health and propose future directions for research to increase the amount of these healthy compounds in wine.

Keywords: Wine, polyphenols, antioxidant, biological effects, cardiovascular disease, bioavailability

[*] Corresponding author: National Research Council. - Institute of Sciences of Food Production (ISPA), Unit of Lecce, via Prov. Lecce-Monteroni, 73100 Lecce, Italy. Email: giovanna.giovinazzo@ispa.cnr.it.

INTRODUCTION

Grapevine (*Vitis* spp) is the most cultivated fruit crop in the world, with an area dedicated to viticulture of 7.8 million hectares. 67.5 million tons of berries corresponding to 252 million hectoliters of wine were been produced during the 2012 vintage (http://www.oiv.int/). The berries are harvested primarily for winemaking (68%) but also to provide fresh table grapes (30%), raisins (2%) and other minor products.

Grape metabolism elaborates a vast array of secondary compounds, which confer selective advantages to the plant against environmental stresses. Among such products are the polyphenols, a large family of secondary metabolites, which are involved in plant responses to biotic and abiotic stresses. The most present polyphenols in plants are the flavonoids, the cinnamic and benzoic acids and the stilbenes. They derive from the phenylpropanoid metabolism, but flavonoids are ubiquitous in plants, whereas stilbenes are specific to certain plant families.

Other polyphenols, as the phytoalexins, are antimicrobial compounds synthesized in response to pathogen or herbivore attack. However, other roles have been described for stress-induced polyphenols, including the defense signaling of responses and protection against ultraviolet (UV) light damage (Chong et al., 2009; Jandet et al., 2002)

Phenolic compounds are secondary metabolites that are synthesized by grape (Chong et al., 2009) during normal development in response to stress conditions. Although structurally diverse, phenolics are classified into two groups: the flavonoids and the non-flavonoids. The flavonoid family includes the flavonols, such as myricetin, quercetin and kaempferol, which exist both as aglycons and sugar conjugates; the flavan-3-ols (such as (+)-catechin and (–)-epicatechin) and the anthocyanins, such as malvidin-3-glucoside. The non-flavonoids encompass hydroxybenzoic acids such as gallic acid; hydroxycinnamic acids, including *p*-coumaric, caffeic and caftaric acids and the stilbenes, such as *trans*-resveratrol and *cis*-resveratrol (Jandet et al., 2002). The synthesis of stilbenes in grape berry tissues is activated in response to fungal attack, to berry injury and also to ultraviolet irradiation (Jandet et al., 2002).

The anthocyanins and the proanthocyanidins are among the most important compounds for the quality of the red wine, since they are responsible for major characteristics of this beverage, i.e., color, bitterness, astringency and chemical stability towards oxidation. Red wine represents a concentrated source of anthocyanidins, catechins, and proanthocyanidins and other phenolics (Mattivi et al., 2002).

The anthocyanins, the resveratrol and the flavonols are especially contained in the berry skins. During winemaking, only a fraction of the grape flavonoids is selectively transferred to the wine and a final yield strongly depending on the management of the contact of the liquid must, containing berry skin and seeds, with the solid parts of the grape bunches and on the grape variety.

The data concerning the extractable phenolics of the grape cannot be simply generalized to predict the wine composition, since a high variability in the extraction yield from grape to wine is introduced by the technological factors governing the winemaking process (such as temperature, duration and intensity of the liquid-solid contact, final ethanol concentration). Moreover, many other factors (chemical microbiological and physical) have been demonstrated to further modify the phenolics structure and concentration during the fermentation, fining and storage of wine.

The various polyphenol families present in wine (Paixao et al., 2008; Arnous, et al., 2001) are important for a number of technological properties of wine such as clarity, hue, and palatal taste (Paixao et al., 2008). In particular, tannins confer astringency and structure to the beverage by the formation of complexes with the proteins of saliva. The knowledge about their qualitative and quantitative profile in grapes is very important to predict wine aging attitude and can help to solve problems related to color stability, especially in the case of red wines that are destined to long aging periods (Mattivi et al., 2002). The wine aging also changes the phenolic composition, as these compounds can suffer diverse transformations, like oxidation processes, condensation and polymerization reactions and extraction from wood, usually associated to the changes in color and colloidal stability (Saucier et al., 1997), flavor, bitterness and astringency (Sánchez-Iglesias et al., 2009). The polyphenolic fingerprint can be useful for the classification of wines, since it can give us information about the variety, the geographic and winery origin and even the applied winemaking technology (Spranger et al., 2004)

There is emerging evidence that a functional diet can help in modulating the immune system responses to the inflammation processes through a variety of mechanisms based on the absorption and utilization by the human metabolism of specific compounds.

Polyphenols are the principal compounds related to the wine consumption benefits due to their antioxidant and free radical scavenging properties. These physiological effects are especially associated to flavonoids and stilbenes (Paixao, et al., 2008), namely quercetin, (+)-catechin, gallic acid and *trans*-resveratrol (Soleas et al., 2001). Flavonoids have shown the inhibition of lower density lipoprotein (LDL) oxidation by macrophages and cupric ions, as increasing evidences shows that oxidized LDL and very LDL (VLDL) may be involved in the pathogenesis of atherosclerosis (Carluccio et al., 2007). Flavanols, such as (–)-epicatechin have many beneficial health effects such as anti-tumorgenic, anti-mutagenic, anti-pathogenic and anti-oxidative properties and has attracted interest because their concentrations in red wine are higher than other flavonoids (Tarola and Giannetti, 2007).

The stilbene *trans*-resveratrol has gained great attention and a number of scientific papers have appeared related to the moderate consumption of red wine the ability to inhibit platelet aggregation and LDL oxidation and its beneficial effects in health. Since trans-resveratrol is postulated to be involved in the health benefits associated with a moderate consumption of red wine, it is one of the most extensively studied natural products. Indeed, hundreds of studies have reported the beneficial effects of *trans*-resveratrol on neurological (Anekonda 2006) and cardiovascular systems (Lippi et al., 2010a-2010b). Soleas and coworkers had related that *trans*-resveratrol may be the most effective anticancer polyphenol present in red wine (Soleas et al. 2001).

One of the most striking biological activities of *trans*-resveratrol is its anticancer activity that prevents carcinogenesis in the three stages of tumor development (Cai et al., 2011). Further data provide interesting insights into the effect of this compound on the life span of different organisms, suggesting that *trans*-resveratrol might be regarded as a potential anti-aging agent in treating age-related human diseases (Pearson et al., 2008). In addition, the effects described in mice subjected to a high-calorie diet (Kundu et al., 2006) open up new approaches for treating not only age-related diseases but also obesity-related disorders (Subbaramaiah et al., 1998).

In this chapter, we attempt to summarize the recent findings concerning the effects of wine polyphenols on human health and propose future directions for research to increase the amount of these healthy compounds in wine.

TECHNOLOGICAL DETERMINANTS OF ANTIOXIDANT CONTENTS AND ACTIVITY IN GRAPES AND WINES

The grape cultivar plays an essential role in establishing the absolute amount of the flavonoids, thus determining the final content of polyphenols in red wines. Indeed, it has been proposed that the knowledge of the qualitative and quantitative analysis of polyphenolos in the berry allows the prediction of the final polyphenol content in the derived wine (Mattivi et al., 2002; Vinci et al., 2008). Mattivi et al. (2002) have studied the "phenolic potential" of twenty-five red grape cultivars (*Vitis vinifera* L.), including 21 Italian and 4 international red grape varieties, by analyzing the amount and the localization of the extractable flavonoids in the grape. The obtained results showed that the grape cultivar has an essential part in determining the absolute amount of the flavonoids, and the distribution between the berry skin and seeds of the flavan-3-ols. However, the terroir also detains a strong influence on the concentration of these compounds in the individual wines. Significant evidence about the influence of the terroir on the amount of polyphenols in red wines has been supplied by Kumšta et al. (2012), who analyzed grape growth under the same geographical region during four vintages. The Authors have studied the relationship between the properties of 43 wines of the cv. Riesling from six wine-growing sub-regions and their specific ecosystem, showing that the phenolic composition of grapes and wines is likely to be related to the terroir. A clear-cut association was detected in wines between wine-growing area and *trans*-resveratrol concentration, thus allowing the authors to demonstrate the capacity to distinguish different wine terroir by using the *trans*-resveratrol concentrations in the derived wine.

Because of the large influence of phenolic compounds on red wine quality, many winemaking techniques have been developed to influence the extraction of these compounds during vinification and most have been directed at enhancing extraction. Different factors limit the extraction of different classes of phenolic compounds. The impact of winemaking processes on phenolic extraction is well-known and several winemaking steps and practices are known to influence the phenolic composition of red wines.

Higher fermentation temperatures have been reported to increase phenolic extraction. A number of studies examined the effect of temperature on extraction of winemaking practices on phenolic compounds (Gao et al., 1997; Girard et al., 1997; 2001; Atanacković et al., 2012), showing that total phenolics content was greater at higher fermentation temperatures. In particular, Atanacković et al. (2012) have studied the influence of thermo-vinification on the resveratrol content, total phenolic content and antioxidant potential of red wines obtained from the cultivars Merlot, Cabernet Sauvignon, Pinot Noir and Prokupac. The obtained results indicated that thermo-vinificated wine samples resulted after thermo-vinification contained higher amount of phenolic compounds.

Several investigations have analyzed the effects of the maceration time, SO_2, pectolytic enzymes and fining agents addition on the final content of polyphenols in the produced wines.

The influence of different maceration stage on phenolic composition of Aglianico, Montepulciano, Nero di Troia and Sangiovese wines has been recently studied by Gambacorta (2011). The authors compared three different maceration technologies, i.e., the traditional (5 days of maceration at 25 °C with three daily punching-down), the prolonged maceration (10 days) and the cryomaceration (24 h at 5 °C), concluding that the phenols extraction from grapes was found to be mostly dependent on the grape variety rather than on the applied maceration technology.

Gambuti et al. (2007) have investigated the effect of the above parameters during vinification of grapes of the red the red cv. Aglianico. The authors concluded that the combination of the addition of SO_2 and pectolytic enzymes during the pre-fermentative phases enhanced the potential health protective effect of red wine, since these practices enhance the transfer of bioactive polyphenols from grape berries to the wine. A similar investigation has been recently carried out on the Vranec red grape cultivar (Ivanova et al., 2012; Kostadinović et al., 2012). The effects of maceration time and SO_2 and yeast on polyphenolic during vinification and wine aging were compared and the obtained results suggested that all parameters affected the content of total phenolics, anthocyanins, flavonoids and flavan-3-ols. Moreover, during aging, wines produced after 6 and 10 days of maceration showed no changes in their phenolic content, whereas a slight reduction of total phenolics was detected in wines obtained after 3 days of grape maceration.

The clarification stage is performed during the vinification process in order to obtain limpid and bright wines by the eliminating substances in suspense as well as the wine-protein instability. However the introduction into the wine of different fining agents affects the quality of the final product. In fact, fining agents such as PVPP, gelatin, egg albumin and casein have demonstrated that they can decrease the concentration of a large number of phenolic compounds and can alter the color in treated (Castellari et al., 1998; Maury et al., 2001; Castillo-Sanchez et al., 2008).

An interesting comparison on the effects of nine different winemaking procedures (traditional, delestage, saignée, delayed punching-down, addition of grape-seed tannins, addition of ellagic-skin-seed tannins addition, must-wine heating, cryomaceration, prolonged maceration) on the phenolic content and antioxidant activity of musts and wines derived from the economically relevant red grape cultivar "Primitivo" has been documented by Baiano et al. (2009). A number of evidence was revealed by the authors: i) tannins addition raised the phenolic content of musts and wines in a greater amount when compared to the other technologies; ii) wine aging, independently on the winemaking procedure, determined an enhancement of the antioxidant activity: iii) wine obtained by traditional technology, saignée and tannins addition were denoted by the highest antioxidant activities; iv) wines produced through traditional winemaking had the highest content in anthocyanins.

There has also been an increasing interest in understanding the potential of wine yeast and their derivate to influence the phenolic profile of red wines. Yeast lees has been demonstrated to adsorb anthocyanins (Morata et al. 2003) and to reduce the polyphenolic content (Mazauric and Salmon, 2005; Lesko et al., 2011) of wines. Polyphenols are adsorbed by the macromolecules released during yeast autolysis in a similar manner to protein-fining agents.

A number of studies on the effect of different yeast strains used for vinification of Pinot noir (Girard et al., 2001; Lorenzini 2001), Cabernet franc and Merlot (Mazza et al., 1999),

Vranec wines (Ivanova et al., 2012) and Sangiovese (Zini et al., 2002) showed that the influence of the yeast starter on the phenolic content of produced wines was not significant.

However, several other investigations produced positive evidences about the potential correlation between yeast starter cultures and total antioxidant capacity of wines. A study on the phenolic profiles of wines produced from Pinot noir cultivar indicated that the dissimilarity of their polyphenolic profile was due to the different yeast strain utilized (Dumont et al., 1993). Cuinier (1997) compared results obtained from Gamay grape must fermentations carried out using eighteen different yeast strains. The obtained wines showed differences up to 100 mg/L in anthocyanins and 10 absorbance units in total phenolics content. Caridi et al. (2004) described a correlation between the yeast strain employed in the fermentative process and the polyphenolic composition of wine, highlighting that strain properties are able to alter chromatic properties, phenolic profile and antioxidant activity of the final product. Similar results were obtained in a recent study using Vranec and Merlot wines form Macedonia, where the increase of *trans*-resveratrol concentration and antioxidant activity in the produced wines was influenced by the yeast strain used during fermentation (Kostadinović et al., 2012).

Brandolini et al. (2007) carried out a large scale analysis testing the effect of 19 strains of *Saccharomyces cerevisiae,* isolated from different grapes and wines, on the total polyphenol content and the antioxidant capacity of obtained wines. The authors demonstrated that the tested parameters were significantly influenced by the yeast strain, and contributing positively to the final quality of the wine and to its functional properties.

Polyphenols Biological Properties

The epidemiological data from France (French Paradox) strongly suggests a protective effect of red wine despite a high-fat diet. The conclusive studies showed that the de-alcoholized red wine had cardiovascular protective effects in short-term studies on humans (Karatzi et al., 2005; Papamichael et al., 2006). Both flavonoids and non-flavonoid phenolic compounds have been implicated in the protective effects of wine on the cardiovascular system. These compounds were also described as potent antioxidants as they reduce low-density lipoprotein (LDL) cholesterol oxidation, modulate cell signaling pathways, reduce platelet aggregation, inhibit the growth of some tumor types, and exhibit anti-inflammatory, antibacterial, antifungal, antiviral, neuroprotective, anti-proliferative and anti-angiogenic activities. However, the beneficial effects of moderate wine consumption may be attributed to the overall mix of all of its components and not to a specific action of one.

Oxidative stress results in direct or indirect reactive oxygen species ROS-mediated damage of nucleic acids, proteins, and lipids, and has been implicated in carcinogenesis, neurodegenerative disease, atherosclerosis, diabetes, and aging. As antioxidants, polyphenols may protect cell constituents against oxidative damage and, therefore, limit the risk of various degenerative diseases associated to oxidative stress. The antioxidant effect of polyphenols limits the damage by acting directly on ROS.

Numerous studies in vitro as well as in animals and humans demonstrate beneficial effects of grape polyphenols on traditional cardiovascular risk factors as: hypertension, hypercholesterolemia, diabetes and smoking leading to atherosclerosis (Auger et al., 2005;

Bernatova et al., 2002; Peng et al., 2005; Pinent el al., 2004). Wine polyphenols can reduce the risk of heart disease by their antioxidant and anti-inflammatory activities, by decreasing the platelet aggregation and by improving the endothelial function and increasing fibrinolysis.

It is well-known that the LDL oxidation is the key mechanism in atherosclerosis. Red wine intake exhibit positive effect on biomarkers of atherosclerosis including a decrease of LDL/HDL ratio, oxidized LDL levels, lipoprotein, clotting factors and fibrinogen levels (Avellone et al., 2006; Guilford and Pezzuto, 2011). Polyphenols may exert also antithrombotic effect by inhibition of patelet aggregation. A dysfunction of endothelial cells may alter the balance between the synthesis and interaction of proteins that promote clot formation and fibrinolytic proteins that facilitate fibrinolysis. Short-term ingestion of red wine improved enthothelial function in patients with coronary artery disease (whelan et al., 2004). This is could be explained by the promotion of endothelium nitric oxide production which have a vasorelaxation effect. This last effect can reduce also the development of hypertension.

In recent years, a large number of studies have been devoted to illustrate the potential cancer chemopreventive activities of red wine polyphenols (Soleas et al., 1997; Yang et al., 2001; Femia et al., 2005). These latter have been shown to block carcinogenesis and to inhibit the growth of tumor in whole animals or in cell culture. Different mechanisms have been suggested to explain the anti-carcinogenic effects and are reviewed by Scalbert et al. (2005) and Rodriguo et al. (2011), Guilford and Pezzuto (2011). Nowadays, literature shows that moderate wine consumption and due to the presence of polyphenols may decrease the risk of several cancers: colon, ovarian, prostate, lung and basal cell carcinoma. The moderate wine consumption was associated with a decreased risk of developing esophageal adenocarcinoma (Kubo et al., 2009) and also lung cancer (Chao, 2007). Hudson et al. (2007) reported that grape polyphenols induced prostate tumor cell lines apoptosis. Oak et al. (2005) showed that red wine polyphenols inhibit angiogenesis by reducing the proliferation and migration of endothelial and vascular smooth muscle cells. Eng et al. (2002) found that red wine polyphenols have an inhibitory activity against aromatase (a cytochrome P450) involved in breast tumor growth.

Wine polyphenols were shown to exhibit anti-inflammatory effect. This effect appears to be associated with the inhibition of nitric oxide production. A critical step in both inflammation and atherosclerosis is the adhesion of circulating monocytes to vascular endothelial cells, which involves vascular cell adhesion molecule-1 (VCAM-1) and intercellular adhesion molecule-1 (ICAM-1) (Hidalgo et al., 2012). Studies showed that polyphenols contribute to the attenuation of inflammatory diseases by decreasing ICAM-1 and VCAM-1 levels (Kwon et al., 2005; Lotito and Frei, 2006). The ani-inflammation effect of polyphenols has been mainly ascribed to immunomodulation and antioxidative action (Chacona et al., 2009). In fact, polyphenols modulate cytokine expression as basic pathway to anti-inflammation. Also, wine polyphenols can exhibit anti-inflammatory activities by their capacities to scavenge nitric oxide and to decrease tne nitric oxide synthase activity.

Epidemiological and animal studies have shown that wine polyphenols can reduce the risk of developing of neurological disorders such as dementia, Parkinson's and Alzheimer's diseases (Basli et al. 2012). Oxidative stress resulting in ROS generation is responsible of many forms of cellular deterioration leading to various chronic pathologies like neurodegenerative disorder. The antioxidant effect of polyphenols protects cell constituents against oxidative alteration and thus limits the risk of developing degenerative disorders. Also, wine polyphenols may regulate the nitric oxide activity at the level of endothelial nitric oxide synthase (eNOS) protein expression in endothelial cells (Wallerath et al., 2003). A preventive approach to reduce tissue injury associated with the risk of cerebral ischemia is constituted by eNOS up-regulation by wine polyphenols.

Type 2 diabetes is characterized by insulin resistance, hyperglycemia, overproduction of glucose by the liver and defection of β-cells in the pancreas. Long term effects of diabetic patients include an increased risk for cardiovascular disease, blindness, nerve and kidney damage, and limb amputations. Several studies have reported the anti-diabetic effect of grape and wine polyphenols (Marfella et al., 2006; Kar el al., 2009; Zumino, 2009). Wine polyphenols may affect glycemia through different mechanisms including: the inhibition of glucose adsorption in the gut or its uptake by peripheral tissues; the inhibition of β-glucosidase, α-amylase and sucrose in rats; the inhibition of gluconeogenesis, adregenic stimulation of glucose intake or the stimulation of insulin by pancreatic β-cells; the modulation of SIRT1 gene improving whole-body glucose homeostasis and insulin sensitivity in rats. Also, the antioxidant and the anti-inflammatory properties of wine polyphenols may be responsible for the positive response in type 2 diabetes.

Phenolic acids: This group of phenols is defined non-coloured phenols abundant in white grape berries and responsible for the bitter and astringent properties of wine (Tian et al., 2009). In wine, there are two groups of phenolic acids; hydroxybenzoic acids and hydroxycinnamic acids (Cabrita et al., 2008). Hydroxybenzoic acids, including gallic acid, protocatechuic acid, gentisic acid, *p*-hydroxybenzoic acid, vanillic acid and syringic acid, derive from benzoic acid. Chlorogenic acid, as the main constituent of the hydroxycinnamic acid derivative group, increased with harvest time delay, and the same occurred with sinapic acid, while the converse was the true of caffeic acid and ferulic acid, which were also esterified with tartaric acid as the known compounds caftaric acid and fertaric acid, respectively (Gil-Muñoz et al., 1999; Cabrita et al., 2008). Hydroxycinnamic acids have gained an increasing interest in health because they are known to be potent antioxidants. These compounds have been described as chain-breaking antioxidants acting through radical scavenging activity that is related to their hydrogen or electron donating capacity and to the ability to delocalize/stabilize the resulting phenoxyl radical within their structure. The free radical scavenger ability of antioxidants can be predicted from standard one-electron potentials.

Phenolic acids represent important fraction of wine phenolics, but their biological effects have been scarcely investigated. The interrelationship between antioxidative capacity and vasodilatory activity, two potentially beneficial biological effects, of phenolic acids from wine were examined. Antioxidative and vasodilatory effects of phenolic acids relate to the number of hydroxyl groups in the phenyl ring, degree of compactness and branching of molecules, and three-dimensional distributions of atomic polarisability of the tested molecules (Teixeira et al., 2013). Caffeic acid has been shown to have neuroprotective effects against injury induced by 5-S-cysteinyl-dopamine, against Aβ-induced neurotoxicity and by inhibiting peroxynitrite-induced neuronal injury (Sul et al., 2009; Vauzour et al., 2010). Ferulic acid has been cited as an anti-diabetic effect by lowering blood glucose and by increasing plasma insulin (Jung et al., 2007)

Anthocyanins: Anthocyanins are plant pigments belonging to a subset of flavonoids with a particularly high antioxidant capacity and concomitantly strong health-promoting effects

(Bártíková et al., 2013; Yoo et al., 2010). Anthocyanins have been credited with capacity to modulate cognitive and motor function, to enhance memory, and to have a role in preventing age-related declines in neural function (Youdim et al., 2002).

As part of the human diet, they offer protection against cancer, inhibiting the initiation and progression stages of tumor development (Martin et al., 2013). They also reduce inflammatory inducers of tumor initiation, suppress angiogenesis, and minimize cancer-induced DNA damage in animal disease models. Anthocyanins also protect against cardiovascular diseases and age-related degenerative diseases associated with metabolic syndrome (Aggarwal and Shishodia, 2004; Renaud and de Lorgeril, 1992). Dell'Agli et al. (2005) have shown that wine anthocyanins inhibit the activity of phosphodiesterase-5 reducing the risk of cardiovascular disease by vasorelaxation. Grape juice and wine anthocyanins in synergy with other flavonoids have been cited as responsible for antiplatelet activity in human and dog systems (Shanmuganayam et al., 2002). Anthocyanins have also neuroprotective benefits by reducing age-associated oxidative stress and by improving cognitive brain function (Tarozzi et al., 2007).

Flavonols: Dietary flavonols inhibit LDL oxidation and so reduce the primary risk factor for atherosclerosis and related diseases. Long-term dietary administration of flavonols (such as quercetin) offers cardio-protection and improves the levels of CVD risk factors in animals (Vina et al., 2007), and the animal studies are supported by human epidemiological studies, which show inverse correlations between the occurrence of CVD, certain cancers, and age-related degenerative diseases and the consumption of flavonol rich diets (Mink et al., 2007; Seeram et al., 2004; Soobrattee et al., 2006).

Flavonols have been linked to protective effects against several specific cancers, including leukemia and pancreatic, breast, cervical, prostate, uterine, and urinary tract cancers. Subjects with regular flavonol intake have a 10–60% lower incidence of these types of cancer compared with subjects with low flavonol intake.

This protective activity results from both the action of flavonols as stimulators of antioxidant defenses and their direct inhibitory effects on cellular proliferation. Quercetin consumption has been reported to be inversely associated with breast cancer incidence (Fink et al., 2007).

Resveratrol: Resveratrol is a stilbene phytoalexin produced by specific plant species in response to biotic and abiotic challenges. It is thought to be one of the principal agents in the health-promoting effects of red wine (Baur and Sinclair, 2006). Resveratrol is a stilbene phenolic compound, it being in the same family of molecules that includes the *cis* or *trans* isomers of glucosides (piceid) and polymers (viniferins). Resveratrol and piceid are present mainly in grape and grape products, but also in lower amounts in peanuts, pistachios, some berries, and chocolate. Results of clinical studies show that the most important source of resveratrol and piceid are wines (98.4%), grape and grape juices (1.6%), whereas peanuts, pistachios, and berries contribute to less than 0.01% (Zamora-Ros et al., 2008). Wine is the major source of resveratrol and piceid to the diet, ranging from 95 and 98.7% for *trans*-resveratrol and *trans*-piceid to 99.9% and 99.7% for *cis*-resveratrol, *cis*-piceid, respectively. Other foods items such as grapes contribute by amount of 3.8% of total *trans*-resveratrol, whereas other contributors such as pistachios or berries provide less than 1% of the dietary total amount of *trans*-resveratrol and *trans*-piceid.

Resveratrol, possess numerous important bioactivities, including anti-inflammatory, antioxidant, anti-aggregatory functions, and modulation of lipoprotein metabolism (Gerard and Mullin, 2011; D'Introno et al., 2009; Frei, 2004). It has also been shown to detain chemopreventive properties against certain forms of cancer, and cardiovascular disorders and to have positive effects on longevity (Kundu and, Surh, 2008; Vergara et al., 2012).

Aside from cardiovascular disease, resveratrol has been reported to potentially benefit a number of conditions, including cancer (Kaminski et al., 2011). Anticancer activity of this compound is mainly due to induction of apoptosis via several pathways, as well as alteration of gene expressions, all leading to a decrease in tumor initiation, promotion, and progression. (Udenigwe et al., 2008) Resveratrol blocks the growth of lymphoma cells and also increases their rate of cell death (Bruno et al., 2003).

Resveratrol sensitizes chemotherapy-resistant lymphoma cells to treatment with paclitaxel-based chemotherapy. (Jazirehi et al., 2004) Resveratrol also reduces the production of growth factors, such as vascular endothelial growth factor and interleukin- (Baur et al., 2006), and may reduce the ability of lymphoma cells to spread to other organs (Dulak et al., 2005) Resveratrol has been studied for its effect on cell cycle signaling for colon cancer. (Fonar et al., 2011; Nguyen et al., 2009) Finally, was demonstrated that in vitro administration of resveratrol favorably altered gene expression in the androgen axis and in cell cycle regulators, providing potential anticancer benefit for prostate cancer (Jones et al., 2005). Moreover, *trans*-resveratrol appears to protect against diabetes (Sharma et al., 2007) and neurodegenerative disorders (Anekonda, 2006), due to induction of Sirtuin-1 genes (Wood et al., 2004). *Trans*-resveratrol might also contribute to increasing the lifespan of metazoans and mice by miming the effect of caloric restriction and thus decreasing age-related signs (Park et al., 2012; Pearson et al., 2008). Experimental studies have shown that resveratrol exhibits both an anti-inflammatory and cardio-protective potential by inhibiting the expression of inflammatory mediators and the monocyte adhesion to vascular endothelial cells (Carluccio et al., 2007; Csiszar et al., 2006). Although resveratrol exhibits potent anticancer activities against transformed cells, its effectiveness is limited by its poor bioavailability and as a dietary phytonutrients it is most effective against tumors with which it comes in direct contact, such as skin cancers and tumors of the gastrointestinal tract. Furthermore inhibition of sirtuin 1 by both pharmacological and genetic means abolished protein de-acetylation and autophagy as stimulated by resveratrol, but not by piceatannol, indicating that these compounds act through distinct molecular pathways. In support of this notion, resveratrol and piceatannol synergized in inducing autophagy as well as in promoting cytoplasmic protein d-eacetylation. (Pietrocola et al., 2012)

Flavanols: Flavanols are present either as monomers, (-)-epicatechin, (+) catechin and gallocatechin gallate and as oligomers and polymers also called condensed tannins or proanthocyanidins. Catechin and proanthocyanidins have proved to be potent antioxidants in different *in vitro* models, and in some *in vivo* intervention studies. However their potential beneficial effect on cardiovascular health is not merely due to this property but includes the different mechanisms implicated on cardiovascular conditions or problems, *i.e.*, hypertension, inflammation, cellular proliferation, trombogenesis, hyperglycemia and hypercholesterolemia.

Catechin and proanthocyanidins have been shown to have free radical scavenging and antioxidant activities. It seems that the proanthocyanidin dimer have the most antioxidant effect. These effects confer to the flavanols a cardioprotection action by limiting the oxidative stress factors.

In breast cancer cell lines, epicatechin inhibits metastatic cell, hepatocyte growth factor signaling and cell motility; causes cell arrest in S phase; modulates NO signaling; induces killer caspases; and inhibits NF-κB signaling (Butt and Sultan 2009). Grape seed proanthocyanidins exert their anti-cancer effects through the inhibition of the constitutive expression of various NF-κB responsive genes/ proteins such as cyclooxygenase-2, inducible nitric oxide synthase, proliferating cell nuclear antigen and MMP-9 in human epidermoid carcinoma A431 cells. Furthermore, treatment of human epidermal keratinocytes with grape seed proanthocyanidins inhibits UV-induced oxidative stress-mediated activation of ERK1/2, JNK and p38 of MAPK family (Nandakumar et al., 2008). Also, grape seeds proanthocyanidins have anit-inflammatory effects by modulating the adipokine and cytokine gene expression. Wine catechins have strong antimicrobial activity against *Porphyromonas gingivalis* and *Prevotella intermedia.*

Polyphenol Mechanisms of Action against Chronic Diseases

Reactive oxygen and nitrogen species are considered major determinants of many degenerative pathologies, including CVD, certain cancers, neurodegenerative diseases, chronic inflammation and tissue aging (Closa and Folch-Puy, 2004; Valko et al., 2006). Essentially, these molecules can act by damaging biological structures and molecules, but can also influence cell signaling either indirectly (by changing the redox cellular status) or directly (by participating in intracellular signaling) (Virgili and Marino, 2008). Although the antioxidant properties of phytonutrients have been a major focus of research, there is an emerging view that they may act not only by scavenging reactive oxygen and nitrogen species or suppressing their production but also by enhancing the endogenous antioxidant capacity of cells/tissues (e.g., glutathione synthesis) or by influencing signaling pathways through interaction with proteins, enzymes, and nuclear receptors (Virgili and Marino, 2008).

Red wine contains varying levels of resveratrol, flavonols, anthocyanins, and catechins, which are all effective antioxidants. However, the cardio-protection afforded by these polyphenols is likely the result of multiple biological activities independent of their antioxidant activities, including improvement of endothelial function, reduction in LDL uptake, oxidation and aggregation, reduction of blood pressure and inhibition of platelet aggregation (Bose et al., 2008), Interestingly, dietary polyphenols have been suggested to prevent LDL oxidation by direct binding to specific sites in LDLs (Wang and Goodman, 1999).

The chemopreventive activity of phytonutrients has been suggested to be associated with the ability to block the progression of latent tumors through different mechanisms (Beliveau and Gingras 2007). Polyphenols can modulate the action of these enzymes that detoxify carcinogens or proteins that export them from cells (Conney, 2003), such as the platelet-derived growth factor receptor (for example, several anthocyanin) (Martin et al., 2011). In particular, the resveratrol promotes apoptosis and cell death of tumor cells (Jung et al., 2006) vascularizes the microtumors by blocking the metalloproteases and reduces inflammation by inhibiting cyclooxygenase 2 (D'Introno et al., 2009; Aggarwal and Shishodia, 2004).

Polyphenols have been shown to modulate energy metabolism and consequently reduce insulin resistance, type 2 diabetes and metabolic syndrome (Hanhineva et al., 2011). AMP-activated protein kinase (AMPK) is a conserved sensor of cellular energy (Khan et al., 2005). Low energy stores determined by caloric restriction increase the AMP/ATP ratio and activate AMPK activity, which in turn activate sirtuin 1 (SIRT1), a class III histone deacetylase (HDAC), and influence glucose/lipid metabolism and age-related diseases (Wood et al., 2004). Therefore, activation of AMPK has been proposed as a strategy for treating metabolic syndrome and delaying aging (Curtis et al., 2005).

Recent studies have demonstrated that many dietary polyphenols can act as mimics of caloric restriction and activate AMPK, thus suppressing hepatic gluconeogenesis and inducing hepatic fatty-acid β-oxidation. Wine polyphenols can stimulate the glucose transporters in muscle and adipose tissues, with an overall reduction of the glucose level in the blood and the lipid content of the liver as well as an improvement in insulin sensitivity (Hanhineva et al., 2011).

Resveratrol, quercetin and anthocyanins also extend life span in different species (Baur and Sinclair, 2006) and, in particular, resveratrol directly inhibits cAMP-degrading phosphodiesterases, leading to an increase in cAMP and subsequent activation of AMPK (Park et al., 2012).

Polyphenol Bioavailability and Delivery

The biological effects of grape polyphenols in human and experimental models demonstrate antioxidant properties closely associated with the maintenance of endothelial function, increase in antioxidant capacity, protection against LDL oxidation and neuroprotective effects. Recent patents regarding grape polyphenols show a tendency to return to natural products with a minimum use of severe extraction processes and organic solvents. Moreover, the recent patents regarding human health show more pharmaceutical use of grape juice and other polyphenol-rich products. The application of such products in clinical trials as a substitute or co-adjuvant with drugs may be useful in future research. (Gollücke and Ribeiro, 2012)

Several investigations have proposed that untargeted metabolomics serve as a powerful approach for providing new clues about metabolism, bioavailability, and excretion of wine polyphenols-intake biomarkers. To our knowledge, there are a limited number of studies applying metabolomics to wine intake. In this context, Grun et al. (2008) have identified ten metabolites that were significantly increased after consumption for four weeks, of 800 mg of red per week, wine and juice of grape polyphenols. Indeed, metabolomic approaches represent an interesting tool for unraveling the complex relations between the consumption of polyphenols and human health.

Bioavailability is a challenging subject to study, because the various biotic factors that may affect it. In particular, little information are available on the effect of the food matrix on the bioavailability of polyphenols, which may affect their absorption (Ortega et al., 2009; Perez-Jimenez et al., 2009). One of the main factors influencing polyphenols bioavailability can be found in their chemical structure. In foods, most polyphenols exist as polymers or in glycosylated forms and aglycon. In these native forms, polyphenols cannot be absorbed and have to be hydrolyzed by the intestinal enzymes or by the colonic micro flora before absorption. Anthocyanins represent an exception, because the intact glycosides can be absorbed and detected in the circulation (Nurmi et al., 2009). The specific chemical structures of polyphenols, as well as the type of the sugar in the glycoside, determine their rate and extent of intestinal absorption.

Following the ingestion of polyphenols, the absorption of part of them occurs in the small intestine, where the glycosides are hydrolyzed. The fraction of polyphenols that is not absorbed in the small intestine reaches the colon, where it undergoes to substantial structural modifications. In fact, the colonic micro flora hydrolyzes glycosides into aglycones and degrades them to simple phenolic acids. This microbial activity is of great importance for the production of specific active metabolites.

It is important to underline that there is great inter-individual variability in producing these active metabolites (Lu and Anderson, 1998; Morton et al., 2002). Once absorbed, and prior to the passage into the blood stream, polyphenols undergo other structural modifications due to conjugation processes, that mainly include methylation, sulfation and glucuronidation. These modifications could affect the bioavailability of polyphenols and, consequently, their biological activity. The conjugation mechanisms are highly efficient and free aglycones are generally either absent or present in low concentrations in plasma after consumption of nutritional doses.

The therapeutic potential of resveratrol can only be adopted *in vivo* only if the limitations related to its bioavailability can be overcome. Research programs are currently exploring other means of enhancing resveratrol bioavailability, including: i) co-administration with metabolism inhibitors in order to prolong its presence *in vivo*, ii) use of resveratrol analogous endowed with a better bioavailability, iii) drug delivery system employing nanotechnology (Johnson et al., 2011; Goldberg et al., 2003). As far as concerns the first approach, some authors have evaluated the possibility of enhancing the pharmacokinetic parameters of resveratrol by partially inhibiting its glucuronidation by means of co-administration with specific inhibitors (Hoshino et al., 2010).

The second strategy consists in the evaluation of new naturally-occurring and/or synthetic analogous of resveratrol endowed with the same structural backbone and some chemical modifications resulting in better efficacy (Cai et al., 2011). In this context, the biological activity exerted by resveratrol metabolites is still debated. Therefore, further studies are needed, including the possibility to induce deglycosylation to release resveratrol into the target tissues (Hassan-Khabbar et al., 2008). However, conventional formulations alone are probably inadequate to overcome the physicochemical and pharmacokinetic limitations of resveratrol bioavailability. To solve this problem, a number of novel drug delivery systems have been proposed to protect and stabilize resveratrol and to enhance its bioavailability: nano- and micro-formulations for resveratrol encapsulation that include liposomes, polymeric nanoparticles, solid lipid nanoparticles, lipospheres, cyclodextrins, polymeric microspheres, yeast cell carriers and calcium or zinc pectinate beads (Amri et al., 2012). At the present time, the research in this field is mainly focused on exploring multiparticulate forms from the millimeter to micrometer range and colloidal carriers in the nanometer range in order to improve the oral intake as these offer the possibility of easy chronic administration. For all current formulation attempts, the main issue is to determine whether the drug delivery system is efficient in improving the pharmacokinetic profile of resveratrol and promoting the *in vivo* therapeutic effects. Perspectives now lie in the development of innovative formulations able to overcome all the limitations governing resveratrol bioavailability as a pre-requisite to securing efficient and sustained delivery *in vivo* (Amri et al., 2012).

CONCLUSION

The precise nature of the role played by polyphenols in human health has been largely highlighted in these last years. The grape is one of the principal sources for the human intake of healthy polyphenols and their concentration in the berry is mainly cultivar-dependent.

A better knowledge concerning the composition and dynamics of polyphenols profile in red grape will help vine-dresser and wine makers in producing grape-derived products and wines with high content of phenolic antioxidants and considerable antioxidant activity, maintaining optimal organoleptic properties and a significant link with the original terroir.

Even though, a better understanding is still request about the several different cellular mechanisms and complex pathways involved in wine polyphenols metabolism, the present findings suggest that the contribute of antioxidant phenols through a reasonable daily drinking of red wines may offer additional protection against *in vivo* oxidation and other damages of human cell components.

ACKNOWLEDGMENT

This chapter was supported by a grant from the Regione Puglia Project PS_008 - INNOWINE- "Biotecnologie innovative per il miglioramento della qualità e sicurezza dei vini tipici pugliesi".

REFERENCES

Aggarwal, B. B. & Shishodia, S. (2004). Suppression of the nuclear factor-κB activation pathway by spice derived phytochemicals, reasoning for seasoning. *Ann. NY. Acad. Sci.*, *1030*, 434–441.

Amri, A., Chaumeil, J. C., Sfar, S. & Charrueau, C. (2012). Administration of resveratrol, What formulation solutions to bioavailability limitations? *J. Control Rel.*, *158*, 182–193.

Anekonda, T. S. (2006). Resveratrol a boon for treating Alzheimer's disease? *Brain Res. Rev.*, *52*, 316-326.

Arnous, A., Makris, D. P. & Kefalas, P. (2001). Effect of principal polyphenolic components in relation to antioxidant characteristics of aged red wines. *J. Agr. Food Chem.*, *49*, 5736 – 5742.

Atanacković, M., Petrović, A., Jović, S., Gojković- Bukarica, L., Bursać, M. & Cvejić, J. (2012). Influence of winemaking techniques on the resveratrol content, total phenolic content and antioxidant potential of red wines. *Food Chem.*, *131*, 513–518.

Auger, C., Teissedre, P. L., Gérain, P., Lequeux, N., Bornet, A., Serisier, S., Besancon, P., Caporiccio, B., Cristol, J.-P. & Rouanet, J. M. (2005). Dietary wine phenolics catechin, quercetin, and resveratrol efficiently protect hypercholesterolemic hamsters against aortic fatty streak accumulation. *J. Agric. Food Chem.*, *53*, 2015-2021.

Avellone, G., Di Garbo, V., Campisi, D., De Simone, R., Raneli, G., Scaglione, R. & Licata, G. (2006). Effects of moderate Sicilian red wine consumption on inflammatory biomarkers of atherosclerosis. *Eur. J. Clin. Nutr.*, *60*, 41-47.

Baiano, A., Terracone, C., Gambacorta, G. & La Notte, E. (2009). Phenolic content and antioxidant activity of Primitivo wine: comparison among winemaking technologies. *J. Food Sci.*, *74*, C258-267.

Bártíková, H., Skálová, L., Dršata, J. & Boušová, I. (2013). Interaction of anthocyanins with drug-metabolizing and antioxidant enzymes. *Curr Med Chem.*, *20*, 4665-79.

Basli, A., Soulet, S., Chaher, N., Mérillon, J. M., Chibane, M., Monti, J. P. & Richard, T. (2012). Wine polyphenols: potential agents in neuroprotection. *Oxidative medicine and cellular longevity*, http://dx.doi.org/10.1155/2012/805762

Baur, J. A. & Sinclair, D. A. (2006). Therapeutic potential of resveratrol, the in vivo evidence. *Nat. Rev.*, *5*, 493–506.

Beliveau, R. & Gingras, D. (2007). Role of nutrition in preventing cancer. *Can. Fam. Physician*, *53*, 1905–1911.

Bernátová, I., Pechánová, O., Babál, P., Kyselá, S., Stvrtina, S. & Andriantsitohaina, R. (2002). Wine polyphenols improve cardiovascular remodeling and vascular function in NO-deficient hypertension. *Am. J. Physiol.-Heart C.*, *282*, 942-948.

Bose, M., Lambert, J. D., Ju, J., Reuhl, K. R., Shapses, S. A. & Yang, C. S. (2008). The major green tea polyphenol, (-)-epigallocatechin-3-gallate, inhibits obesity, metabolic syndrome, and fatty liver disease in high-fat-fed mice. *J. Nutr.*, *138*, 1677–1683.

Brandolini, V., Fiore, C., Maietti, A., Tedeschi, P. & Romano, P. (2007). Influence of *Saccharomyces cerevisiae* strains on wine total antioxidant capacity evaluated by photochemiluminescence. *World J. Microbiol. Biotechnol.*, *23*, 581–586.

Bruno, R., Ghisolfi, L., Priulla, M., Nicolin, A. & Bertelli, A. (2003). Wine and tumors: study of resveratrol. *Drugs Exp Clin Res.*, *29*, 257-261.

Butt, M. S. & Sultan, M. T. (2009). Green tea: nature's defense against malignancies. *Crit. Rev. Food Sci. Nutr.*, *49*, 463–473.

Cai, H., Sale, S., Britton, R. G., Brown K., Steward, W. P. & Gescher, A. J. (2011). Pharmacokinetics in mice and metabolism in murine and human liver fractions of the putative cancer chemopreventive agents 3,4,5,5,7-pentamethoxyflavone and tricin (4,5,7-trihydroxy-3,5-dimethoxyflavone). *Cancer Chemother. Pharmacol.*, *67*, 255-263.

Cabrita, M. J, Torres, M., Palma, V., Alves, E., Patão, R. & Costa Feritas, A. M. (2008) Impact of malolactic fermentation on low molecular weight phenolic compounds. *Talanta*, *74*, 1281-1286.

Caridi, A., Cufari, A., Lovino, R., Palumbo, R. & Tedesco, I. (2004). Influence of yeast on polyphenols in wine. *Food Technol. Biotechnol.*, *42*, 37–40.

Carluccio, M. A., Ancora, M. A., Massaro, M., Carluccio, M., Scoditti, E., Distante, A., Storelli, C. & De Caterina, R. (2007). Homocysteine induces VCAM-1 gene expression through NF-kappaB and NAD(P)H oxidase activation, protective role of Mediterranean diet polyphenolic antioxidants. *Am. J. Physiol. Heart Circ. Physiol.*, *293*, 2344-2354.

Castellari, M., Spinabelli, U., Riponi, C. & Amati, A. (1998). Influence of some technological practices on the quantity of resveratrol in wine. *Zeitschrift Fur Lebensmittel-Untersuchung Und-Forschung A*, *206*, 151–155.

Castillo-Sanchez, J. X., Garcia-Falcon, M. S., Garrido, J., Martinez-Carballo, E., Martins-Dias, L. R. & Mejuto, X. C. (2008). Phenolic compounds and colour stability of Vinhao wines: influence of wine-making protocol and fining agents. *Food Chem.*, *106*, 18–26.

Chacón, M. R., Ceperuelo-Mallafré, V., Maymó-Masip, E., Mateo-Sanz, J. M., Arola, L., Guitiérrez, C., Fernandez-Real, J. M., Ardevol, A., Simon, I. & Vendrell, J. (2009). Grape-seed procyanidins modulate inflammation on human differentiated adipocytes< i> in vitro</i>. *Cytokine*, *47*, 137-142.

Chao, C. (2007). Associations between beer, wine, and liquor consumption and lung cancer risk: a meta-analysis. *Cancer Epidem. Biomar.*, *16*, 2436-2447.

Chong, J. L., Poutaraud, A. & Hugueney, P. (2009). Metabolism and roles of stilbenes in plants. *Plant Sci.*, *177*,143–155.

Closa, D. & Folch-Puy, E. (2004). Oxygen free radicals and the systemic inflammatory response. *IUBMB Life*, *56*, 185–191.

Conney, A. H. (2003). Enzyme induction and dietary chemicals as approaches to cancer chemoprevention: the seventh DeWitt S. Goodman lecture. *Cancer Res.*, *63*, 7005-7031.

Csiszar, A., Smith, K., Labinsky, N., Orosz, Z., Rivera, A. & Ungvari, Z. (2006). Resveratrol attenuates TNF-alpha-induced activation of coronary arterial endothelial cells, role of NF-kappaB inhibition. *Am. J. Physiol. Heart Circ. Physiol.*, *291*, 1694-1699.

Cuinier, C. (1997). Ceppi di lievito e composizione fenolica dei vini rossi. *Vignevini*, 7/8, 39-42.

Curtis, R., Geesaman, B. J. & Di Stefano, P. S. (2005). Ageing and metabolism, drug discovery opportunities. *Nat. Rev. Drug Discov.*, *4*, 569–580.

D'Introno, A., Paradiso, A., Scoditti, E., D'Amico, L., De Paolis, A., Carluccio, M. A., Nicoletti, I., DeGara, L., Santino, A. & Giovinazzo, G. (2009). Antioxidant and anti-inflammatory properties of tomato fruit synthesizing different amount of stilbenes. *Plant Biotech. J.*, *7*, 422- 429.

Dell'Agli, M., Galli, G. V., Vrhovsek, U., Mattivi, F. & Bosisio, E. (2005). In vitro inhibition of human cGMP-specific phosphodiesterase-5 by polyphenols from red grapes. *J. Agric. Food Chem.*, *53*, 1960-1965.

Dulak, J. (2005) Nutraceuticals as anti-angiogenic agents: hopes and reality. *J Physiol Pharmacol.*, *56*, 51-67.

Dumont, A., Watson, B. T., McDaniel M. R. & Price, S. F. (1993). Evaluation of the effects of commercial yeast strains on the composition, aroma, and flavor of Pinot noir. *Abstr. Am. J. Enol. Vitic.*, *44*, 355.

Eng, E. T., Williams,D., Mandava, U., Kirma, N., Tekmal, R., R. & Chen, S. (2002). Anti-Aromatase Chemicals in Red Wine. *Ann. NY.Acad. Sci.*, 963, 239-246.

Femia, A. P., Caderni, G., Vignali, F., Salvadori, M., Giannini, A., Biggeri, A., Gee, J., Przybylska, K., Cheynier, V. & Dolara, P. (2005). Effect of polyphenolic extracts from red wine and 4–OH–coumaric acid on 1, 2–dimethylhydrazine–induced colon carcinogenesis in rats. *Eur. J. Nutr.*, *44*, 79-84.

Fink, B. N., Steck, S. E., Wolff, M. S., Britton, J. A., Kabat, G. C., Schroeder, J. C., Teitelbaum, S. L., Neugut, A. I. & Gammon, M. D. (2007). Dietary flavonoid intake and breast cancer risk among women on Long Island. *Am. J. Epidemiol.*, *165*, 514–523.

Fonar, Y. & Frank, D. (2011). FAK and WNT signaling: the meeting of two pathways in cancer and development. *Anticancer Agents Med. Chem.*, *11*, 600-606.

Frei, B. (2004), Efficacy of dietary antioxidants to prevent oxidative damage and inhibit chronic disease. *J. Nutr.*, *134*, 3196-3198.

Gambacorta, G., Antonacci, D., Pati, S., la Gatta, M., Faccia, M., Coletta, A. & La Notte, E. (2011). Influence of winemaking technologies on phenolic composition of Italian red wines. *Eur. Food Res. Technol.*, *233*, 1057–1066.

Gambuti, A., Strollo, D., Erbaggio, A., Lecce, L. & Moio, L. (2007). Effect of winemaking practices on color indexes and selected bioactive phenolics of Aglianico wine. *J. Food Sci.*, *72*, S623-S628.

Gao, L., Girard, B., Mazza G. & Reynolds, A. G. (1997). Changes in anthocyanins and color characteristics of Pinot noir wines during different vinification processes. *J. Agric. Food Chem.*, *45*, 2003-2008.

Gerard, E. & Mullin, M. D. (2011). Red Wine, Grapes, and Better Health-Resveratrol. *Nutr. Clin. Pract.*, *26*, 722-723.

Gil-Muñoz, R., Gómez-Plaza, E., Martínez, A. & López-Roca, J. M. (1999) Evolution of phenolic compounds during wine fermentation and post-fermentation: Influence of grape temperature. *J. Food Comp. Anal.*, *12*, 259-272.

Girard, B., Kopp, T. G., Reynolds, A. G. & Cliff, M. (1997). Influnence of vinification treatments on aroma constituents and sensory descriptors of Pinot noir wines. *Am. J. Enol. Vitic.*, *48*, 198-206.

Girard, B., Yuksel, D., Cliff, M. A, Delaquis, P. & Reynolds A. G. (2001). Vinification effects on the sensory, colour, and GC profiles of Pinot noir wines from British Colombia. *Food Res. Int.*, *34*, 483-499.

Goldberg, D. M., Yan, J. & Soleas, G. J. (2003). Absorption of three wine-related polyphenols in three different matrices by healthy subjects. *Clin. Biochem.*, *36*, 79-87.

Gollücke, A. P. & Ribeiro, D. A. (2012). Use of grape polyphenols for promoting human health: a review of patents. *Recent Pat. Food Nutr. Agric.*, *4*, 26-30.

Grun, C. H., Van Dorsten, F. A., Jacobs, D. M., Le Belleguic, M., van Velzen, E. J., Bingham, M. O., Janssen, H. G. & Van Duynhoven, J. P. (2008). GCMS methods for metabolic profiling of microbial fermentation products of dietary polyphenols in human and in vitro intervention studies. *J. Chromatogr. B.*, *871*, 212–219.

Guilford, J. M. & Pezzuto, J. M. (2011). Wine and health: a review. *Am. J. Enol. Vitic.*, *62*, 471-486.

Hanhineva, K., Torronen, R., Bondia-Pons, I., Pekkinen, J., Kolehmainen, M., Mykkanen, H. & Poutanen, K. (2011). Impact of dietary polyphenols on carbohydrate metabolism. *Int. J. Mol. Sci.*, *11*, 1365–1402.

Hassan-Khabbar, S., Cottart, C. H., Wendum, D., Vibert, F., Clot, J. P., Savouret, J. F., Conti, M. & Nivet-Antoine, V. (2008). Post ischemic treatment by trans-resveratrol in rat liver ischemia–reperfusion, a possible strategy in liver surgery. *Liver Transpl.*, *14*, 451-459.

Hidalgo, M., Martin-Santamaria, S., Recio, I., Sanchez-Moreno, C., de Pascual-Teresa, B., Rimbach, G. & de Pascual-Teresa, S. (2012). Potential anti-inflammatory, anti-adhesive, anti/estrogenic, and angiotensin-converting enzyme inhibitory activities of anthocyanins and their gut metabolites. *Genes Nutr.*, *7*, 295-306.

Hoshino, J., Park, E. J., Kondratyuk, T. P., Marler, L., Pezzuto, J. M., van Breemen, R. B., Mo, S., Li, Y. & Cushman, M. (2010). Selective synthesis and biological evaluation of sulfate-conjugated resveratrol metabolites. *J. Med. Chem.*, *53*, 5033–5043.

Hudson, T. S., Hartle, D. K., Hursting, S. D., Nunez, N. P., Wang, T. T., Young, H. A., Arany, P. & Green, J. E. (2007). Inhibition of prostate cancer growth by muscadine grape skin extract and resveratrol through distinct mechanisms. *Cancer Res.*, *67*, 8396-8405.

Ivanova, V., Vojnoski, B. & Stefova, M. (2012). Effect of winemaking treatment and wine aging on phenolic content in Vranec wines. *J. Food Sci. Technol.*, *49*(2), 161-172.

Jandet, P., Douillet-Breuil, A. C., Bessis, R., Debord, S., Sbaghi, M. & Adrian, M. (2002). Phytoalexins from the vitaceae, Biosynthesis, phytoalexin gene expression in transgenic plants, antifungal activity, and metabolism. *J. Agric. Food Chem.*, *50*, 2731-2741.

Jazirehi, A. R. & Bonavida, B. (2004) Resveratrol modifies the expression of apoptotic regulatory proteins and sensitizes non-Hodgkin's lymphoma and multiple myeloma cell lines to paclitaxel-induced apoptosis. *Mol Cancer Ther.*, *3*, 71-84.

Johnson, J. J., Nihal, M., Siddiqui, I. A., Scarlett, C. O., Bailey, H. H., Mukhtar, H. & Ahmad, N. (2011). Enhancing the bioavailability of resveratrol by combining it with piperine *Mol. Nutr. Food Res.*, *55*, 1169-1176.

Jones, S. B., DePrimo, S. E., Whitfield, M. L. & Brooks, J. D. (2005) Resveratrol induced gene expression profiles in human prostate cancer cells. *Cancer Epidemiol Biomarkers Prev*, *14*, 596-604.

Jung, E. M., Park, J. W., Choi, K. S., Park, J. W., Lee, H. I., Lee, K. S. & Kwon, T. K. (2006). Curcumin sensitizes tumor necrosis factor related apoptosis-inducing ligand (TRAIL)-mediated apoptosis through CHOP-independent DR5 upregulation. *Carcinogenesis*, *27*, 2008–2017.

Jung, E. H., Ran Kim, S., Hwang, I. K. & Youl Ha, T. (2007). Hypoglycemic effects of a phenolic acid fraction of rice bran and ferulic acid in C57BL/KsJ-db/db mice. *J. Agric. Food Chem.*, *55*, 9800-9804.

Kaminski, B. M., Steinhilber, D., Stein, J. M. & Ulrich, S. (2011) Phytochemicals resveratrol and sulforaphane as potential agents for enhancing the anti-tumor activities of conventional cancer therapies. *Curr Pharm Biotechnol.*, *67*, 1167-1178.

Kar, P., Laight, D., Rooprai, H. K., Shaw, K. M. & Cummings, M. (2009). Effects of grape seed extract in Type 2 diabetic subjects at high cardiovascular risk: a double blind randomized placebo controlled trial examining metabolic markers, vascular tone, inflammation, oxidative stress and insulin sensitivity. *Diabetic Med.*, *26*, 526-531.

Karatzi, K. N., Papamichael, C. M., Karatzis, E. N., Papaioannou, T. G., Aznaouridis, K. A., Katsichti, P. P., Stamatelopoulos, K., Zampelas, A., Lekakis, J. & Mavrikakis, M. E. (2005). Red wine acutely induces favorable effects on wave reflections and central pressures in coronary artery disease patients. *Am. J. hypertens.*, *18*, 1161-1167.

Khan, B. B., Alquier, T., Carling, D. & Hardie, D. G. (2005). AMP-activated protein kinase: ancient energy gauge provides clues to modern understanding of metabolism. *Cell Metab.*, *1*, 15–25.

Kostadinović, S., Wilkens, A., Stefova, M., Ivanova, V., Vojnoski, B., Mirhosseini, H. & Winterhalter, P. (2012). Stilbene levels and antioxidant activity of Vranec and Merlot wines from Macedonia: effect of variety and enological practices. *Food Chem.*, *135*, 3003-3009.

Kubo, A., Levin, T. R., Block, G., Rumore, G. J., Quesenberry Jr, C. P., Buffler, P. & Corley, D. A. (2009). Alcohol types and sociodemographic characteristics as risk factors for Barrett's esophagus. *Gastroenterology*, *136*(3), 806-815

Kumšta, M., Pavloušek, P. & Kupsa, J. (2012). Influence of terroir on the concentration of selected stilbenes in wines of the cv. Riesling in the Czech Republic, *Hort. Sci.*, *39*, 38–46.

Kundu, J. K. & Surh, Y. J. (2008). Cancer Chemopreventive and therapeutic potential of resveratrol, mechanistic perspectives. *Cancer Lett.*, *269*, 243-261.

Kundu, J. K., Shin, Y. K., Kim, S. H. & Surh, Y. J. (2006). Resveratrol inhibits phorbol ester-induced expression of COX-2 and activation of NF-kappaB in mouse skin by blocking IkappaB kinase activity. *Carcinogenesis*, *27*, 1465-1474.

Kuwajerwala, N., Cifuentes, E., Gautam, S., Menon, M., Barrack, E. R. & Reddy, G. P. V. (2002). Resveratrol induces prostate cancer cell entry into S phase and inhibits DNA synthesis. *Cancer Res.*, *62*, 2488–2492.

Kwon, H. M., Choi, Y. J., Jeong, Y. J., Kang, S. W., Kang, I. J., Lim, S. S. & Kang, Y. H. (2005). Anti-inflammatory inhibition of endothelial cell adhesion molecule expression by flavone derivatives. *J. Agric. Food Chem.*, *53*, 5150-5157.

Leskó, A., Kállay, M., Nyúl-Pühra, B. & Nyitrai-Sárdy, D. (2011). The change of polyphenolic composition and tyrosol content of the wine as an effect of sur lie method. *Acta Alim.*, *40*, 79–90.

Lippi, G., Franchini, M., Favaloro, E. J. & Targher, G. (2010). Moderate red wine consumption and cardiovascular disease risk, beyond the "French paradox". *Semin. Thromb. Hemost.*, *36*, 59-70.

Lippi, G., Franchini, M. & Guidi, G. C. (2010). Red wine and cardiovascular health, The "French Paradox" revisited. *Int. J. Wine Res.*, *2*, 1-7.

Lorenzini, F. (2001). Comparaison de six preparations de levures oenologiques sur du Pinot noir. *Rev. Suisse Vitic. Arboric. Hortic.*, *33*, 73-76.

Lotito, S. B. & Frei, B. (2006). Dietary Flavonoids Attenuate Tumor Necrosis Factor α-induced Adhesion Molecule Expression in Human Aortic Endothelial Cells structure-function relationships and activity after first pass metabolism. *J. Biol. Chem.*, *281*, 37102-37110.

Lu, L. J. & Anderson, K. E. (1998). Sex and long-term soy diets affect the metabolism and excretion of soy isoflavones in humans. *Am. J. Clin. Nutr.*, *68*, 1500S-1504S.

Marfella, R., Cacciapuoti, F., Siniscalchi, M., Sasso, F. C., Marchese, F., Cinone, F., Musacchio, E., Marfella, M. A., Ruggiero, L., Chiorazzo, G., Liberti, D., Chiorazzo, G., Nicoletti, G. F., Saron, C., D'Andrea, F., Ammendola, C., Verza, M. & Coppola, L. (2006). Effect of moderate red wine intake on cardiac prognosis after recent acute myocardial infarction of subjects with Type 2 diabetes mellitus. *Diabetic Med.*, *23*, 974-981.

Martin, C., Butelli, E., Petroni, K. & Tonelli, C. (2011). How Can Research on Plants Contribute to Promoting Human Health? *Plant Cell*, *23*, 1685–1699.

Martin, C., Zhang, Y., Tonelli, C. & Petroni K. (2013). Plants, Diet, and Health. *Annu. Rev. Plant Biol.*, *64*, 19–46.

Mattivi, F., Zulian, C., Nicolini, G. & Valenti, L. (2002). Wine, Biodiversity, Technology, and Antioxidants. *Ann. N.Y. Acad. Sci*, *957*, 37–56.

Maury, C., Sarni-Manchado, P., Lefebvre, S., Cheynier, V. & Moutounet, M. (2001). Influence of fining with different molecular weight gelatins on proanthocyanidin composition and perception of wines. *Am. J. Enol. and Vitic.*, *52*, 140–144.

Mazauric, J. P. & Salmon, J. M. (2005). Interactions between yeast lees and wine polyphenols during simulation of wine aging: I. Analysis of remnant polyphenolic compounds in the resulting wines. *J. Agric. Food Chem.*, *53*, 5647-5653.

Mazza, G., Fukumoto, L., Delaquis, P., Girard, B. & Ewert B. (1999). Anthocyanins, phenolics, and color of Cabernet franc, Merlot, and Pinot noir wines from British Colombia. *J. Agric. Food Chem.*, *47*, 4009-4017.

Mink, P. J., Scrafford, C. G., Barraj, L. M., Harnack, L., Hong, C. P., Nettleton, J. A. & Jacobs Jr, D. R. (2007). Flavonoid intake and cardiovascular disease mortality, a prospective study in postmenopausal women. *Am. J. Clin. Nutr.*, *85*, 895–909.

Morata, A., Gomez-Cordoves, C., Subervolia, J., Bartolome, B., Colomo, B. & Suarez J. A. (2003). Adsorption of anthocyanins by yeast cell walls during fermentation of red wines. *J. Agric. Food Chem.*, *51*, 4084-4088.

Morton, M. S., Arisaka, O., Miyake, N., Morgan, L. D. & Evans, B. A. (2002). Phytoestrogen concentrations in serum from Japanese men and women over forty years of age. *J. Nutr.*, *132*, 3168-3171.

Nandakumar, V., Singh, T. & Katiyar, S. K. (2008). Multi-targeted prevention and therapy of cancer by proanthocyanidins. *Cancer lett.*, *269*, 378-387

Nguyen, A.V., Martinez, M., Stamos, M. J., Moyer, M. P., Planutis, K., Hope, C. & Holcombe, R. F. (2009) Results of a phase I pilot clinical trial examining the effect of plant-derived resveratrol and grape powder on Wnt pathway target gene expression in colonic mucosa and colon cancer. *Cancer Manag Res.*, *1*, 25-37.

Nurmi, T., Mursu, J., Heinonen, M., Nurmi, A., Hiltunen, R. & Voutilainen, S. (2009). Metabolism of berry anthocyanins to phenolic acids in humans. *J. Agric. Food Chem.*, *57*, 2274–2281.

Oak, M. H., El Bedoui, J. & Schini-Kerth, V. B. (2005). Antiangiogenic properties of natural polyphenols from red wine and green tea. *J. Nutr. Biochem.*, *16*, 1-8.

Ortega, N., Reguant, J., Romero, M. P., Macia, A. & Motilva, M. J. (2009). Effect of fat content on the digestibility and bioaccessibility of cocoa polyphenol by an in vitro digestion model. *J. Agric. Food Chem.*, *57*, 5743–5749.

Paixao, N., Pereira, V., Marques, J. C. & Camara, J. S. (2008). Quantification of polyphenols with potential antioxidant properties in wines using reverse phase HPLC. *J. Sep. Sci.*, *31*, 2189-2198.

Papamichael, C., Karatzi, K., Karatzis, E., Papaioannou, T. G., Katsichti, P., Zampelas, A. & Lekakis, J. (2006). Combined acute effects of red wine consumption and cigarette smoking on haemodynamics of young smokers. *J. hypertens.*, *24*, 1287-1292.

Park, S. J., Ahmad, F., Philp, A., Baar, K., Williams, T., Luo, H., Ke H., Rehmann, H., Taussig, R., Brown, A. L., Kim, M. K., Beaven, M. A., Burgin, A. B., Manganiello, V. & Chung, J. H. (2012). Resveratrol ameliorates aging-related metabolic phenotypes by inhibiting cAMP phosphodiesterases. *Cell*, *14*, 421-433.

Pearson K. J. Baur, J. A., Lewis, K. N., Peshkin, L., Price, N. L., Labinskyy, N., Swindell, W. R., Kamara, D., Minor, R., Perez, E., Jamieson, H., Zhang, Y., Dunn, S., Sharma, K., Pleshko, N., Woollett, L., Csiszar, A., Ikeno, Y., Le Couteur, D., Elliott, P., Becker, K., Navas, P., Ingram, D., Wolf, N., Ungvari, Z., Sinclair, D. & de Cabo, R. (2008).

Resveratrol delays age-related deterioration and mimics transcriptional aspects of dietary restriction without extending life span. *Cell Metab.*, *8*, 157-168.

Peng, N., Clark, J. T., Prasain, J., Kim, H., White, C. R. & Wyss, J. M. (2005). Antihypertensive and cognitive effects of grape polyphenols in estrogen-depleted, female, spontaneously hypertensive rats. *Am. J. Physiol.-Reg. I*, *289*, 771-775.

Perez-Jimenez, J., Serrano, J., Tabernero, M., Arranz, S., Diaz-Rubio, M. E., Garcia-Diz, L., Goni, I. & Saura-Calixto, F. (2009). Bioavailability of phenolic antioxidants associated with dietary fiber, plasma antioxidant capacity after acute and long-term intake in humans. *Plant. Foods Hum. Nutr.*, *64*, 102–107.

Pietrocola, F., Mariño, G., Lissa, D., Vacchell, i. E., Malik, S. A., Niso-Santano, M., Zamzami, N., Galluzzi, L., Maiuri, M. C. & Kroemer, G. (2012) Pro-autophagic polyphenols reduce the acetylation of cytoplasmic proteins. *Cell Cycle*, *11*, 3851-60.

Pinent, M., Blay, M., Blade, M. C., Salvado, M. J., Arola, L. & Ardevol, A. (2004). Grape seed-derived procyanidins have an antihyperglycemic effect in streptozotocin-induced diabetic rats and insulinomimetic activity in insulin-sensitive cell lines. *Endocrinology*, *145*, 4985-4990.

Renaud, S. & de Lorgeril, M. (1992). Wine, alcohol, platelets and the French paradox for coronary heart disease. *Lancet*, *339*, 1523–1526.

Rodrigo, R., Miranda, A. & Vergara, L. (2011). Modulation of endogenous antioxidant system by wine polyphenols in human disease. *Clin. Chim. Acta*, *412*, 410-424.

Sánchez-Iglesias, M., González-Sanjosé, M. L., Pérez-Magariño, S., Ortega-Heras, M. & González-Huerta, C. (2009). Effect of micro-oxygenation and wood type on the phenolic composition and color of an aged red wine. *J Agric Food Chem.*, *57*, 11498–11509.

Saucier, C., Little, D. & Glories, Y. (1997). First evidence of acetaldehyde-flavanol condensation products in red wine. *Am. J. Enol. Vitic.*, *48*, 370–373.

Scalbert, A., Manach, C., Morand, C., Rémésy, C. & Jiménez, L. (2005). Dietary polyphenols and the prevention of diseases. *Crit. Rev. Food Sci. Nutr.*, *45*, 287-306.

Seeram, N. P., Adams, L. S., Hardy, M. L., & Heber, D. (2004). Total cranberry extract versus its phytochemical constituents, antiproliferative and synergistic effects against human tumor cell lines. *J. Agric. Food Chem.*, *52*, 2512–2517.

Shanmuganayagam, D., Beahm, M. R., Osman, H. E., Krueger, C. G., Reed, J. D. & Folts, J. D. (2002). Grape seed and grape skin extracts elicit a greater antiplatelet effect when used in combination than when used individually in dogs and humans. *J. Nutr.*,*132*, 3592-3598.

Sharma, S., Chopra, K. & Kulkarni, S. K. (2007). Effect of insulin and its combination with resveratrol or curcumin in attenuation of diabetic neuropathic pain, participation of nitric oxide F TNF-alpha. *Phytother. Res.*, *21*, 278-283.

Soleas, G. J., Diamandis, E. P. & Goldberg, D. M. (2001). The world of resveratrol. *Adv. Exp. Med. Biol.*, *492*, 159-182.

Soleas, G. J., Diamandis, E. P. & Goldberg, D. M. (1997). Wine as a biological fluid: history, production, and role in disease prevention. *J. Clin. Lab. Anal.*, *11*, 287-313.

Soobrattee, M. A., Bahorun, T. & Aruoma, O. I. (2006). Chemopreventive actions of polyphenolic compounds in cancer. *Biofactors*, *27*, 19–35.

Spranger, M. I., Clìmaco, M. C., Sun, B., Eiriz, N. Fortunato, C., Nunes, A., Leandro, M. C., Avelar, M. L. & Belchior, P. A. (2004). Differentiation of red winemaking technologies by phenolic and volatile composition. *Anal. Chim. Acta*, *513*, 151–161.

Subbaramaiah, K., Chung, W. J., Michaluart, P., Telang, N., Tanabe, T., Inoue, H., Jang, M., Pezzuto, J. M. & Dannenberg, A. J. (1998). Resveratrol inhibits cyclooxygenase-2 transcription and activity in phorbol ester-treated human mammary epithelial cells. *J Biol Chem.*, *273*, 21875-21882.

Sul, D., Kim, H. S., Lee, D., Joo, S. S., Hwang, K. W. & Park, S. Y. (2009). Protective effect of caffeic acid against beta-amyloid-induced neurotoxicity by the inhibition of calcium influx and tau phosphorylation. *Life Sci.*, *84*, 257-262.

Tarola, A. M. & Giannetti, V. (2007). Determination by LC of Polyphenols in Italian Red Wine. *Chromatographia*, *65*, 367–371.

Tarozzi, A., Morroni, F., Hrelia, S., Angeloni, C., Marchesi, A., Cantelli-Forti, G. & Hrelia, P. (2007). Neuroprotective effects of anthocyanins and their in vivo metabolites in SH-SY5Y cells. *Neurosci. Lett.*, *424*, 36-40.

Teixeira, J., Gaspar, A., Garrido, E. M., Garrido, J. & Borges, F. (2013) Hydroxycinnamic Acid Antioxidants: An Electrochemical Overview. *Bio. Med. Res. Int.* Article ID 251754, 11 pages.

Tian, R. R., Pan, Q. H., Zhan, J. C., Li, J. M., Wan, S. B., Zhang, Q. H. & Huang, W. D. (2009) Comparison of phenolic acids and flavan-3-ols during wine fermentation of grapes with different harvest times. *Molecules*, *14*, 827-838.

Udenigwe, C. C., Ramprasath, V. R., Aluko, R. E. & Jones, P. J. (2008) Potential of resveratrol in anticancer and anti-inflammatory therapy. *Nutr Rev.*, *66*, 445-454.

Valko, M., Rhodes, C. J., Moncol, J., Izakovic, M. & Mazur, M. (2006). Free radicals, metals and antioxidants in oxidative stress-induced cancer. *Chem. Biol. Interact.*, *160*, 1–40.

Vauzour, D., Houseman, E. J., George, T. W., Corona, G., Garnotel, R., Jackson, K. G., Sellier. C., Gillery, P., Kennedy, O. B., Lovegrove, J. A. & Spencer, J. P. (2010). Moderate Champagne consumption promotes an acute improvement in acute endothelial-independent vascular function in healthy human volunteers. *Brit. J. Nutr.*, *103*, 1168-1178.

Vergara, D., Simeone, P., Toraldo, D., Del Boccio, P., Vergaro, V., Leporatti, S., Pieragostino, D., Tinelli, A., De Domenico, S., Alberti, S., Urbani, A., Salzet, M., Santino, A. & Maffia, M. (2012). Resveratrol down regulates Akt/GSK and ERK signalling pathways in OVCAR-3 ovarian cancer cells. *Mol Biosyst.*, *8*, 1078-1087.

Viña, J., Gomez-Cabrera, M. C. & Borras, C. (2007). Fostering antioxidant defences, up-regulation of antioxidant genes or antioxidant supplementation? *Br. J. Nutr.*, *98*, S36–S40.

Vinci, G., Eramo, S. L., M., Nicoletti, I. & Restuccia, D. (2008). influence of environmental and technological parameters on phenolic composition in red wine. *J. Commodity Sci. Technol. Qual.*, *47*, 245-266

Virgili, F. & Marino, M. (2008). Regulation of cellular signals from nutritional molecules, a specific role for phytochemicals, beyond antioxidant activity. *Free Radic. Biol. Med.*, *45*, 1205–1216.

Wallerath, T., Poleo, D., Li, H. & Förstermann, U. (2003). Red wine increases the expression of human endothelial nitric oxide synthasea mechanism that may contribute to its beneficial cardiovascular effects. *J. Am. Coll. Cardiol.*, *41*, 471-478.

Wang, W. & Goodman, M. T. (1999). Antioxidant property of dietary phenolic agents in a human LDL oxidation ex vivo model, interaction of protein binding activity. *Nutr. Res.*, *19*, 191–202.

Whelan, A. P., Sutherland, W. H. F., McCormick, M. P., Yeoman, D. J., De Jong, S. A. & Williams, M. J. A. (2004). Effects of white and red wine on endothelial function in subjects with coronary artery disease. *Internal Med. J.*, *34*, 224-228.

Wood, J. G., Rogina, B., Lavu, S., Howitz, K., Helfand, S. L., Tatar, M. & Sinclair, D. (2004). Sirtuin activators mimic caloric restriction and delay ageing in metazoans. *Nature*, *430*, 686-689.

Yang, C. S., Landau, J. M., Huang, M. T. & Newmark, H. L. (2001). Inhibition of carcinogenesis by dietary polyphenolic compounds. *Annu. Rev. Nutr.*, *21*, 381-406.

Yoo, Y. J., Saliba, A. J. & Prenzler, P. D. (2010) Should Red Wine Be Considered a Functional Food? *Compr. Rev. Food Sci.*, *9*, 530-551.

Youdim, K. A., McDonald, J., Kalt, W. & Joseph, J. A. (2002). Potential role of dietary flavonoids in reducing microvascular endothelium vulnerability to oxidative and inflammatory insults. *J. Nutr. Biochem.*,*13*, 282-288.

Zamora-Ros, R., Andres-Lacueva, C., Lamuela-Raventos, R. M., Berenguer, T., Jakszyn, P., Martinez, C., Sanchez, M., Navarro, C., Chirlaque, M., Tormo, M.-J., Quiros, J., Amiano, P., Dorronsoro, M., Larranaga, N., Barricarte, A., Ardanaz, E. & Gonzalez, C. (2008). Concentrations of resveratrol and derivatives in foods and estimation of dietary intake in a Spanish population, European Prospective Investigation into Cancer and Nutrition (EPIC)-Spain cohort. *Br. J. Nutr.*, *100*, 188–196.

Zini, S., Siliani, A., Vagnoli, P. & Rosi, I. (2002). Un nuovo ceppo selezionato di *Saccharomyces cerevisiae* nella fermentazione di uve Sangiovese. *Enologo*, *38*, 95-100.

Zunino, S. (2009). Type 2 diabetes and glycemic response to grapes or grape products. *J. Nutr.*, *139*, 1794-1800.

In: Wine
Editor: Youssef El Rayess

ISBN: 978-1-63321-048-6

Chapter 6

WINE PHENOLS AND LACCASE: AN AMBIVALENT RELATIONSHIP

Harald Claus[1,*], Andrea Sabel[1] and Helmut König[1]
[1]Institute of Microbiology and Wine Research,
Johannes Gutenberg-University,
Mainz, Germany

ABSTRACT

Wine grapes possess a very high phenolic content compared to many other fruits and crops. The phenolic compounds significantly affect the color, odor and taste of the wine. In addition to the visual and gustatory properties, they serve as free radical scavengers and have antioxidant effects. Some polyphenols have been associated with anticarcinogenic, anti-inflammatory, antibacterial and antihepatotoxic properties. Therefore, they have a high impact on the quality of the wine and are of great public health importance. Wine phenols are extremely sensitive to oxygen and thus lead to the development of brown oxidation products accompanied by loss of nutritive values of the wine. Enzymatic oxidation takes place under the influence of polyphenoloxidases including tyrosinase and laccase, copper-containing enzymes that catalyse the oxidative conversion of phenols to highly reactive quinones.

In contrast to tyrosinase which originates from the grapes themselves, laccase is produced by the phytopathogenic fungus *Botrytis cinerea* and enters the must with contaminated berries. Effective inactivation of laccase can be achieved by short time heating or by high doses of sulphite, whereas bentonite additions are not very useful. Although uncontrolled action of *Botrytis*-laccase has a dramatic impact on wine quality, the oxidative power of the enzyme has been tested for beneficial biotechnological applications in vinification: Treatment of must with laccase promotes wine stabilization by removal of phenols and avoidance of subsequent browning reactions; hazardous waste waters from wine-related industries can be detoxified by the use of fungal laccases; and enzymatic oxidation of polyphenols *in vitro* yields oligomeric products with enhanced pharmacological properties.

---

* E-mail: hclaus@uni-mainz.de.

Keywords: Laccase, wine browning, phenolic compounds, resveratrol, antioxidants, antimicrobial activity, *Botrytis cinerea*

INTRODUCTION

A vast of polyphenols (i.e., phenols with at least two hydroxyl groups linked to the aromatic ring) contribute to the red pigmentation, the bitter and astringent components, and to the taste of grapes and wine (Eder and Wendelin, 2002; Adams, 2006; Landete, 2012). The primary constituents of the phenolic wine compounds are flavonoids which share a basic C_6-C_3-C_6 structure. They make up approximately 85 % of the total phenols and enclose the anthocyanins, flavonols, flavanols and condensed tannins. The non-flavonoid phenolic acids include the hydroxycinnamates (C_6-C_3), hydroxybenzoates (C_6-C_1) and stilbenes (C_6-C_2-C_6). A major part of the polyphenols in nature is conjugated with sugars or organic acids (Landete, 2012), rendering them more hydrophilic. Derivates of cinnamic acid (including caffeic, p-coumaric, and ferulic acids) can form ester bonds with tartaric acid. The hydroxybenzoates, comprising p-hydroxybenzoic, protocatecheic, vanillic, gallic, and syringic acids, are more prevalent in maturing wines. Hydrolysable tannins, polyphenols descending from gallic acid, are leached out from oak barrels (barrique). Especially in red wines they are responsible for some bitterness and astringency (Dietrich and Pour-Nikfardjam, 2009; Moreno-Arribas and Polo, 2010).

The phenolic composition of wines is dependent on the grape variety and time of harvesting (Tian et al., 2009). The total polyphenol contents of red wines is in the range of 300 – 5000 mg/l, and thus about ten times higher than that of white wines, which ranges between 60 -200 mg/l (Eder and Wendelin, 2002). The analytical methods for determination of polyphenols from grapes, wines and extracts have been recently reviewed by Lorrain et al. (2013). Actual surveys of the polyphenol composition of wines are presented by Dietrich and Pour-Nikfardjam (2009), and can be found in the book of Morena-Arriba and Polo (2010) or online (http://www.phenol).

Due to their presumptive beneficial impact on human health, polyphenols in red wine, especially resveratrol, have gained increasing scientific interest. They are believed to reduce the effects of cell-damaging free radicals in the body. In particular, these antioxidants slow down the removal of nitric oxide from the blood, lowering blood pressure and therefore reducing the risk of heart problems and strokes. There is increasing evidence for similar health promoting properties of white wine and champagne in spite of the fact that the total phenolic content is significantly lower than in red wines. The non-flavoid phenolics tyrosol and hydroxytyrosol, present in white wine, seem to exert analogous effects as resveratrol (Zinnai et al., 2013).

Biotic and abiotic reactions alter the phenol composition of must and wine. *Saccharomyces cerevisiae* cells adsorb phenolic compounds or cause color loss by enzymatic hydrolysis involving a periplasmatic anthocyanin-β-D-glucosidase (Caridi et al., 2004). Lactic acid bacteria such as *Lactobacillus plantarum* are able to metabolize cinnamic acids to 4-vinylphenol and 4-vinylguaiacol (Landete et al., 2007). The off-flavors ethylphenol and ethylguaiacol result from the enzymatic conversion of p-coumaric acid and ferulic acid by *Brettanomyces* strains (Dittrich and Großmann, 2010). Glycosidases produced mainly by non-

Saccharomyces yeasts split off sugar moieties from glycosylated polyphenols and terpenes to yield the corresponding aglycons (Sabel et al., 2013).

Most of all, oxidative reactions, both spontaneously and enzymatically catalyzed (Du Toit et al., 2006; Li et al., 2008; Oliveira et al., 2011), have dramatic effects on the final phenol composition from the grape berry up to the bottled wine. Once the berry integrity has been destroyed, oxidative enzymes (phenoloxidases) and their phenolic substrates are exposed to air, resulting in enzymatic browning. Phenoloxidase activities are particularly high in fruits and vegetables, which contain significant levels of phenolic compounds including different grape varieties (Zocca et al., 2007). The responsible enzymes are the copper containing tyrosinase (from grapes) and laccase (from *Botrytis cinerea*) which use molecular oxygen for substrate oxidation (Claus, 2010, Mayer and Staples, 2002). Tyrosinase (E.C. 1.14.18.1) hydroxylates monophenols to ortho-diphenols, and oxidizes the latter to ortho-quinone intermediates which polymerize to form brown pigments (Claus and Decker 2006). Tyrosinases are supposed to be sensitive to must and wine conditions (low pH, presence of alcohol and sulphite) and thus exert only a minor influence on the nutritional and organoleptic properties of wine.

Laccase (E.C. 1.10.3.2) has no monohydroxylase activity, but oxidizes a wide spectrum of different polyphenols and other compounds by a radical mechanism. This enzyme is very stable under wine conditions and has serious impacts on the phenolic composition and quality of must and wine. Some typical phenolic laccase substrates are shown in Figure 1.

This chapter summarizes facts on possible detrimental effects of (*Botrytis*) laccase on wine polyphenols but also rules out the enzyme´s positive potential for wine-related applications. The various roles of the stilbene resveratrol as health-promoting antioxidant and as a phytoalexin are highlighted.

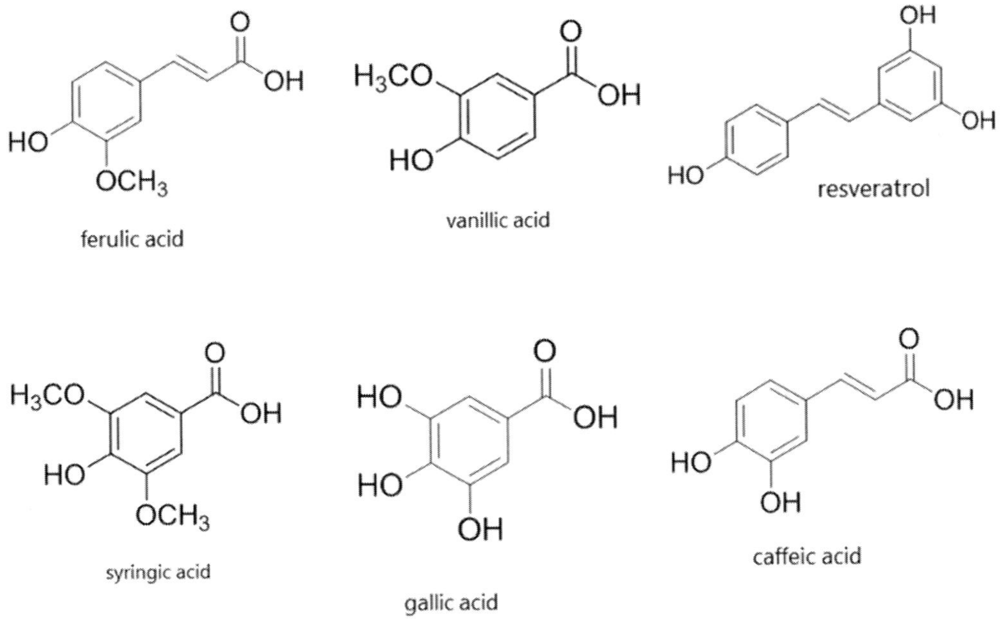

Figure 1. Some typical phenolic laccase substrates.

POLYPHENOLS

Wine Organoleptic Properties

Among polyphenolic compounds, two types of flavonoids, the anthocyanins and flavanols (i.e., catechins, proanthocyanidins, condensed tannins), are particularly relevant to the quality of red wines, as they are key compounds for color definition and astringency. Other flavonoids such as flavonols may have some influence on color and bitterness, although they are present in red wines in much lower amounts. Phenolic acids and hydrolysable tannins, released from barrel wood, may also have an influence on wine taste and color, and hydroxycinnamoyl derivatives from grape must are involved in the oxidative browning of wines together with flavanols. Besides, some of these perceptions may be modified by other sensory characteristics (e.g., sourness, sweetness) related to other wine components. Wine phenolic composition depends on the original grape and on oenological practices and storage conditions. Anthocyanins and tannins are located in the solid parts of the berry from where they are released during winemaking. Furthermore, they undergo various enzymatic and chemical reactions as the wine is made and aged. The new compounds formed often exhibit sensory properties different from those of their precursors, hence modifying the quality of the wines (Moreno-Arribas and Polo, 2010).

Health Promoting Properties

Due to their positive impact on human health, dietary polyphenols have got increasing scientific interest. Cardiovascular effects, tumor cell apoptosis, cytoprotective enzyme induction, antitumoral action, or activation of oxidative-stress-responsive transcription factors have been reported (Han et al., 2007; Stevenson and Hurst, 2007; He et al., 2008; Hanhineva et al., 2010; Arranz et al., 2012; Landete, 2012; Li and Förstermann, 2012; Li et al., 2012; Stuart and Robb, 2013). The positive effects of moderate red wine consumption have been mainly attributed to the antioxidant capacity of polyphenols (de Beer et al., 2002) which is generally measured in *vitro* by radical scavenging assays (de Beer et al., 2002; Campanella et al., 2004; Huang et al., 2005). These methods give hints, but no proofs for antioxidant effects *in vivo*. Polyphenols are apparently absorbed from the human upper gastrointestinal tract by mechanisms which are not fully characterized. Some polyphenolic glycosides may interact with active sugar transporters, lactase pholorizin hydrolase or cytosolic β-glucosidase (Thilakarathna and Rupasinghe, 2013). The more hydrophobic aglycones appear to undergo passive diffusion. As a result only 5-10 % of the polyphenols enters the bloodstream and is further transformed by enzymes in the kidneys and the liver to methyl-, sulphate- or glucuronide-conjugates (de Beer et al., 2002). The maximum steady-state concentration of conjugated and unconjugated forms in plasma is in the range of 1 -5 μM, even with a diet high in polyphenolics (Stevenson and Hurst, 2007). It should be mentioned that correct determination of polyphenols in human blood should not be restricted to the traditional measurements in plasma only, because they may be bound by human erythrocytes and other blood cells (Koren et al., 2010). Nevertheless, the physiological concentration of polyphenols appears below the level to exert direct antioxidant effects (Soleas et al., 1997). However, recent studies indicate that they stimulate in *vivo* the expression of antioxidant enzymes (e.g., superoxide dismutase, catalase, glutathione peroxidase, NAD(P)H-quinone oxidoreductase). Furthermore, polyphenols were found to influence signaling pathways involved in inflammatory reactions and cell proliferation (Stevenson and Hurst, 2007; Stuart and Robb, 2013).

The main portion of the polyphenols (especially oligomeric and polymeric polyphenols such as tannins, reaching molecular masses close to 40 kDa) passes almost unchanged in the colon, where bacteria hydrolyze β-glycosidic bonds to release aglycones, which are more active antioxidants than the glycosides (de Beer et al., 2002; Stevenson and Hurst, 2007; Landete, 2012; Cardono et al., 2013). The colonic microorganisms also cleave the modified phenolic conjugates and break down flavonoids and procyanidines into simpler phenolic compounds. The latter possess antioxidant and antimicrobial activities which contribute to the biological activity after absorption (de Beer et al., 2002; Stevenson and Hurst, 2007). On the other hand, recent studies suggested that the phenolic substrates and their metabolites produced may in turn modulate the composition of the microbial populations through selective prebiotic effects and antimicrobial activities against pathogenic bacteria (Cardono et al., 2013).

As the low bioavailability of flavonoids limits their healthy effects, various attempts have been made to enhance their bioactivity in *vivo* such as: improving the intestinal absorption by use of absorption enhancers, novel delivery systems, improving metabolic stability and changing the site of absorption from large to small intestine (Thilakarathna and Rupasinghe, 2013).

Antimicrobial Activities

Polyphenols kill microorganisms or inhibit the growth of bacteria, fungi or protozoa (Landete, 2012). Several mechanisms of action in the growth inhibition of bacteria are involved, such as permeabilization and destabilization of the plasma membrane, inhibition of extracellular microbial enzymes, direct actions on microbial metabolism, and deprivation of substrates required for microbial growth (Campos et al., 2009; Dietrich and Pour-Nikfardjam, 2009). Studies have shown that certain phenolic compounds present in wine affect bacterial growth and metabolism (Vivas et al., 1997; Reguant et al., 2000; Campos et al., 2003; 2009; García-Ruiz et al., 2008; 2009; 2011; 2012). Theobald et al. (2008) reported that epigallocatechine gallate (EGCG) exerts a concentration-dependent impact on the growth of the wine bacterium *Oenococcus oeni* strain B2. It was shown that EGCG had a stimulating effect at 400–500 mg l^{-1} but an inhibitory effect when present above 540 mg l^{-1} in the culture media. García-Ruiz et al. (2011) investigated the inhibitory potential of 18 phenolic compounds (including hydroxybenzoic acids, hydroxycinnamic acids, phenolic alcohols, stilbenes, flavan-3-ols and flavonols) on wine-related lactic acid bacteria of the species *Ooenococcus oeni*, *Lactobacillus hilgardii* and *Pediococcus pentosaceus*. In general, flavonols and stilbenes showed the greatest inhibitory effects on the growth of the tested strains. Hydroxycinnamic acids and hydroxybenzoic acids and their esters exhibited medium inhibitory effect, and phenolic alcohols as well as flavanol-3-ols showed the lowest effect on the growth of the strains studied. In comparison to the antimicrobial additives used in winemaking the inhibitory impact of most phenolic compounds were stronger than those of the common used potassium metabisulphite and lysozyme.

In a subsequent study García-Ruiz et al. (2012) determined whether phenolic extracts with antimicrobial activity may be considered as an alternative to the use of sulfur dioxide (SO_2) for controlling malolactic fermentation (MLF) in winemaking. Growth inhibition of six oenological strains of *Lactobacillus hilgardii, Lactobacillus casei, Lactobacillus plantarum, Pediococcus pentosaceus* and *Oenococcus oeni* by phenolic extracts from different origins (spices, flowers, leaves, fruits, legumes, seeds, skins, agricultural by-products and others) was evaluated. A total of 24 extracts were found to significantly inhibit the growth of at least two of the strains studied. Some of these extracts were also active against the acetic acid bacteria *Acetobacter aceti* and *Gluconobacter oxydans*. Transmission electron microscopy of the bacterial cells after incubation with the phenolic extract confirmed damage of the integrity of the cell membrane. These results demonstrated the potential applicability of phenolic extracts as antimicrobial agents in winemaking.

In a similar approach Pastorkova et al. (2013) investigated the antimicrobial potential of 15 grape phenolic compounds of various chemical classes (phenolic acids, stilbenes and flavonoids) against yeasts and acetic acid bacteria. Pterostilbene, resveratrol and luteolin were among six active compounds that possessed the strongest inhibitory effects against all microorganisms tested. In the case of phenolic acids, myricetin, p-coumaric and ferulic acids exhibited selective antimicrobial activity depending upon the species of yeasts and bacteria tested. In comparison with potassium metabisulphite, all microorganisms tested were more susceptible to the phenols.

In addition to oenological-relevant microorganisms, wine extracts have been found to be effective against strains of pathogenic bacteria and yeasts like *Staphylococcus aureus, Escherichia coli* and *Candida albicans* (Papadopoulou et al., 2005). Cueva et al. (2012) investigated the effect of seven wine phenolic compounds and six oenological phenolic extracts on the growth of pathogenic bacteria associated with respiratory diseases (*Pseudomonas aeruginosa*, *Staphylococcus aureus*, *Moraxella catarrhalis*, *Enterococcus faecalis*, *Streptococcus sp.* Group F, *Streptococcus agalactiae* and *Streptococcus pneumoniae*). Gallic acid and ethyl gallate were the compounds that showed the highest antimicrobial activity. Gram-negative bacteria were more susceptible than Gram-positive bacteria to the action of phenolic compounds and extracts.

The results revealed antibacterial and antifungal effects against wine spoilage and pathogenic microorganisms of several polyphenols naturally occurring in grapes. Recent studies suggested the possible application of plant phenolic extracts to reduce the use of sulphites in winemaking (Gonzales-Rompinelli et al., 2013). Nevertheless, this optimistic assessment must be taken with care: strains of the common wine bacterium *Lactobacillus plantarum* were able to grow in the presence of high concentrations of some phenolic compounds (Landete et al., 2007) and Guzzo et al. (2011) reported that some red wine phenolics compromise the activity of lysozyme, which is used in winemaking to control microbial contamination.

BOTRYTIS CINEREA

Plant Infection

Botrytis cinerea (teleomorphic form: *Botryotinia fuckeliana*) has been identified as a pathogen of more than 235 plant species including grapes, lettuce, tomatoes, tobacco and strawberries. Related species such as *Botrytis allii*, *Botrytis byssoides*, *Botrytis squamosa*, *Botrytis fabae* and *Botrytis gladioli* are pathogens in onions, beans and flowers. In viticulture, *Botrytis cinerea* may cause both serious loss and enhancement of quality, depending not only on the stage of ripening in which berries are infected but also on weather conditions. Under dry and warm conditions, infections of ripe berries may raise the quality especially of white cultivars. In this case, berry ingredients are concentrated due to the perforation of the berry skin by the fungus. In addition, *Botrytis cinerea* produces gluconic acid which confers a pronounced tastiness to the wine. Consequently, late infections of mature berries facilitate the production of dessert wines like “Trockenbeerenauslesen”, “Sauternes” and “Tokay”. On the other hand, berry infection at an early stage of ripening and during long lasting wetness, reduces the quality, because of berry decay (Kassemeyer and Berkelmann-Löhnertz, 2009).

Botrytis cinerea and other members of the phytopathogenic *Sklerotiniaceae* (such as *Sklerotinia sclerotium*, *Sklerotinia minor* and *Monilinia fructigena*) are able to produce an extracellular β-glucan polymer which is attached onto growing hyphae. The presence of pathogenic-related exoenzymes (such as laccase, polygalacturonase, pectin methylesterases) in the glycan sheath surrounding the hyphae of *Botrytis* is well documented (Gil-ad et al., 2001). The penetration of the host cuticle by *Botrytis cinerea* mediated by cuteolytic enzymes triggers a programmed cell death in the epidermis and the underlying cells before they are invaded by hyphae. Several secondary metabolites showing phytototoxic and apoptotic activities have been identified in culture filtrates of *Botrytis cinerea*. The fungus produces two series of phytotoxic metabolites: a family of polyketide lactones known as the botcinolides and botrylactones, and secondly the sesquiterpenoid botrydial (Choquer et al., 2007; Collado et al., 2007).

Natural Defense Mechanisms and Control Measures

Grape berries have developed several strategies in response to fungal attack. Large amounts of pathogenic-related (PR) defense proteins (chitinase, β-1,3-glucanase, osmotin, and thaumatin-like proteins) are produced (Choquer at al., 2007; Wigand et al., 2009; Claus et al., 2014), which might in turn be degraded by several *Botrytis cinerea* proteases (Cichewicz et al., 2000). Resveratrol, a stilbene produced by grapes, inhibits the spread of *Botrytis cinerea* infection. This compound produces cytological changes in the fungus such as development of additional germ tubes in conidia, cytoplasmic granulations and retraction of cytoplasm in the hyphae tip (Cotoras et al., 2004). Laccase produced by the fungus is assumed to oxidize and to inactivate resveratrol. In contrast, it was stated that a specific laccase (stilbene oxidase) converts resveratrol into compounds which are more toxic to the fungus. Thus, resveratrol has been considered to act as a natural profungicide (Schouten et al., 2002).

Botrytis cinerea has traditionally been controlled by commercial fungicides (dicarboximides and benzimidazoles). Unfortunately, the use of these fungicides has caused serious problems such as the appearance of highly resistant strains and the contamination of soil and water. Natural products isolated from plants could be alternatives to chemical fungicides (Cotoras et al., 2004). Cucurbitacins (triterpenoids) protect cucumber tissue against infection by *Botrytis cinerea* and inhibit laccase production by the fungus (Viterbo et al., 1993a). The compounds 5,7-dihydroy-3,8-dimethoxyflavone and 3-β-hydroxy-kaurenoic acid, isolated from plant resinous exudates, showed fungitoxic activity (Cotoras et al., 2004).

Biological control is becoming more popular as alternative to synthetic fungicides because of environmental concerns and the possible selection of resistant pathogen populations. Several biological control agents have been developed and investigated for use against fungal pathogens of grapes, such as *Botrytis cinerea, Aspergillus sp.*, and *Rhizopus sp.* Among 80 commercial biological control agents tested, only a few proved to be effective against *Botrytis cinerea*. A strain of *Bacillus (*UYBC38) and the yeast *Hanseniaspora uvarum* (UYNS13) showed high antagonistic capability against the pathogen and were effective in controlling rot of grape (Rabosto et al., 2006). Furthermore, a strain FY-102 of *Pichia anomala* (renamed *Wickerhamomyces anomalus*) was found to inhibit grey mold symptoms on the grape-vine, suggesting that the yeast could control expression of the disease (Masih et al., 2000).

Commercial applications of UYBC38 in vineyards showed that this bacterium could survive on leaves and fruits for 45 to 70 days after spraying. No antagonistic effect was found between this strain and wine yeasts. Utkhede and Mathur (2002) reported that commercial products of *Gliocladium catenulatum* J1446 (Prestop®), *Bacillus subtilis* (QRD137), and *Rhodosporidium diobovatum* S33 could be useful for biological control of stem canker of greenhouse tomatoes caused by *Botrytis cinerea*.

Lessons from the Genome

Botrytis cinerea and *Sclerotinia sclerotiorum* are closely related plant pathogenic fungi notable for their wide host ranges and environmental persistence. The genomes of one strain of *Sclerotinia sclerotiorum* and two strains of *Botrytis cinerea* have recently been sequenced by Amselem et al. (2011). Their 38–39 Mb genomes include 11,860–14,270 predicted genes, which share 83 % amino acid identity on average between the two species. The *Sclerotinia sclerotiorum* genome comprises 16 chromosomes and reveals large-scale co-linearity with the *Botyrtis cinerea* genomes. Seven percent of the *Sclerotinia sclerotiorum* genome comprises transposable elements compared to 1 % of *Botrytis cinerea*. The arsenal of genes associated with the infection processes is similar between the species, including genes involved in plant cell wall degradation and oxalic acid production.

Analysis of secondary metabolism gene clusters revealed an expansion in number and diversity of *Botrytis cinerea*–specific compounds relative to *Sclerotinia sclerotiorum*. The potential diversity in secondary metabolism might be involved in adaptation to specific ecological niches. This genomic resource should contribute to a better understanding to what makes these fungi such successful and persistent pathogens of agronomic crops (Amselem et al., 2011).

Impact on Must and Wine

Botrytis cinerea infections drastically affect the chemical composition of must and subsequently the alcoholic fermentation by altering species heterogeneity and succession during the time course (Mills et al., 2002; Cilindre et al.,2007; Nisiotou et al., 2007; Dittrich and Großmann, 2010; Bokulich et al., 2012; Ky et al., 2012). As a consequence, the phenolic and organoleptic properties of the derived wines are very different from those of healthy grapes (Table 1). However, it should be kept in mind that the altered aroma profile by "noble-rot" is desired for production of famous sweet white wines (Tosi et al., 2012).

Table 1. Impact of *Botrytis cinerea* on must and wine*

Increasing	Decreasing
glucose, fructose, galactose, arabinose	glucose/fructose
polyols (glycerol, mannitol, sorbitol, inositol, erythrol)	
gluconic acid, malic acid, acetic acid, citric acid, galactaric acid, galacturonic acid, 2-ketoglutaric acid, pyruvic acid	tartaric acid
polysaccharides (β-glucan), viscosity	filtrability
laccase, glucosidases, esterases, pectinases	polyphenol content color (anthocyanins)
proteases	nitrogen content (amino acids, proteins); foaming properties of sparkling wine
ethanol, methanol	vitamins (thiamin)
botrycins osmolarity	
microbial diversity osmotolerant and fructophilic yeasts acetic and lactic acid bacteria	*Oenococcus oeni*
SO_2 demand	

* Modified from Dittrich and Großmann, 2010.

An Italian sweet *Passito* wine presented higher contents of organic acids (particularly malic acid), main metabolites derived from *Botrytis cinerea* (laccase, glycerol, gluconic acid and benzaldehyde) and low contents of total polyphenols. Among the volatile components, normal fatty acid ethyl esters, branched-chain esters and benzaldehyde characterized this wine (Giordano et al., 2009).

Pronounced effects of *Botrytis cinerea* infections on wine quality are caused by the production of exoenzymes. For example, glycosidases cleave glycosylated flavono- and anthocyanin-3-glucosides. Hydrolysis of flavonol-glycosides has been reported to result in haze development as the insoluble aglycones precipitate (Moreno-Arribas and Polo, 2010). However, the most dramatic impact has to be blamed on the action of the oxidative enzyme laccase.

FUNGAL LACCASES

Laccase [EC 1.10.3.2] or laccase-like enzymes have been detected in all domains of life (Claus, 2004; Claus and Strong, 2010). The best studied laccases are those of the lignin-degrading white-rot fungi (Baldrian, 2006). The enzyme is a prominent member of the blue multi-copper oxidases which carry four copper ions in the polypeptide chain. The T1 copper has a trigonal coordination, with two histidines and a cysteine as conserved ligands. It is the site of substrate oxidation and it has been widely argued that hydrophobic residues (Phe, Ile) in the near surroundings of the T1 copper strongly influence the oxidation potential of the enzyme which varies in the range between E_0 +400 mV to +800 mV, depending on the individual laccase (Giardina et al., 2010). The T2 and T3 copper atoms form a trinuclear cluster, where reduction of molecular oxygen to water takes place. The T2 copper is coordinated by two histidines and one water molecule and each of the two T3 copper atoms by three histidines. A number of 3D structures of laccases from basidiomycetes have been reported so far. On the other hand, crystal structures of ascomycetes have been resolved only for *Melanocarpus albomyces* and recently for *Botrytis aclada* (Kittl et al., 2012). The structures show that the protein monomer is normally organized in three sequentially arranged cupredoxin domains. Each of the three domains displays similar ß-barrel type architecture, related to the small blue copper proteins such as azurin or plastocyanin. Disulphide bonds link domain one with domains two and three, while the trinuclear cluster bridges the first and third domains. The T1 copper located in domain 3 is the primary substrate electron acceptor site and is connected to the oxygen-reducing T2/T3 trinuclear cluster by a His-Cys-His tripeptide.

The following simplified catalytic cycle has been proposed: the reducing substrate is bound in a cleft at the enzyme surface and is oxidized by the T1 copper site in domain three. Electrons donated by four equivalents of the reducing substrate are transferred via a strongly conserved His-Cys-His tripeptide and progressively leads to the reduction of all four Cu(II) ions in the polypeptide to the Cu(I) state. Reoxidation of the cuprous ions occurs at trinuclear T2/T3 cluster with the concomitant reduction of molecular oxygen, resulting in the formation of two water molecules. Reduction of oxygen by laccase occurs in 2-electron steps involving a peroxide intermediate, bridging the trinuclear copper site. The free radicals generated from laccase oxidation are very reactive and undergo further non-enzymatic cross-linking reactions (Figure 2). Laccases show relaxed specificity for their reducing substrates but a strong affinity for oxygen. Basically, any compound with characteristics similar to a diphenol will be oxidized by laccase as long as its redox potential is not too high (E_o < +1000 mV). Classical substrates of fungal laccases include various lignin-derived phenols and aromatic amines. Currently, about 100 natural and artificial compounds are known to be oxidized by laccases (Reiss et al. 2013). Chromogenic compounds commonly used for detection and photometric measurement of laccase activities are 4-hydroxy-3,5-dimethoxy-benzaldehyde azine [syringaldazine], 2,2´-azino-di-(-3-ethylbenzothiazoline-6-sulfonic acid) [ABTS] or 2,6-dimethoxyphenol. Another method to quantify laccase activities is the polarographic determination of oxygen consumption in the course of substrate oxidation.

Often, several high glycosylated isoforms are produced by the fungi, as well constitutively and/or after addition of inducers (copper, phenolic acids). Although usually active as monomeric proteins, some laccases consist of several subunits, forming hetero- or homo-dimers (Giardina et al., 2010).

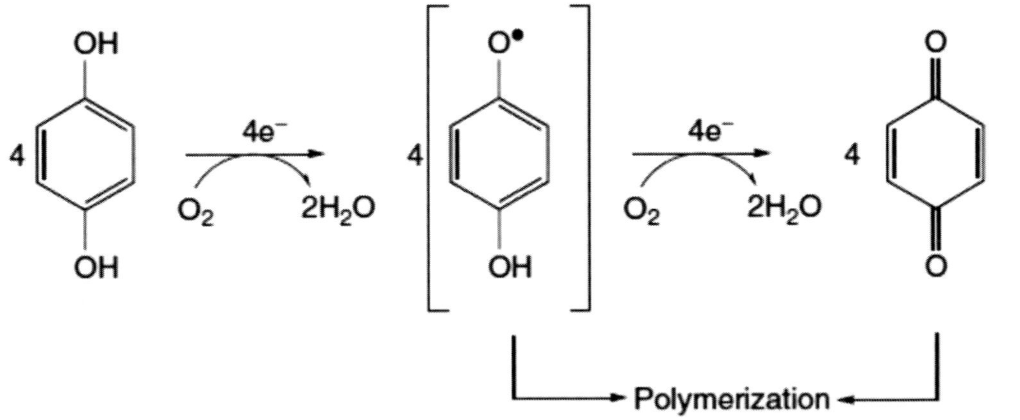

Figure 2. Typical reaction mechanism of laccase showing oxidation of a para-diphenol.

Fungal laccases are rather resistant against detergents like SDS, but due to their reactive copper centers, are inhibited by halides, carbonates, and heavy metals. In addition to the general inhibitors of metal-containing oxidases like cyanide, sodium azide or fluoride, hitherto no specific inhibitor has been described (Johannes and Majcherczyk, 2000).

Due to the abundance of laccase and laccase-like enzymes, there are numerous and diverse natural functions for these oxidoreductases (Reiss et al., 2013). Fungal laccases probably play roles in spore pigmentation and morphogenesis, fungal plant-pathogen/host interactions and stress defense. Furthermore they are involved in degradation of lignin and turnover of humic matter (Claus and Filip, 1998; Baldrian, 2006). Bollag et al. (1988) showed that laccase of *Rhizoctonia praticola* reversed the inhibitory effects of a number of phenolic compounds on fungal growth. This allows phytopathogenic fungi, such as *Botrytis cinerea* to detoxify phytoalexins and tannins, thereby increasing fungal virulence (Mayer and Staples, 2002).

Botrytis Laccase

Botrytis laccase has been intensively studied for decades by the group of Alfred Mayer (Mayer and Staple, 2002; Mayer, 2006). Several glycosylated isoenzymes with rather different molecular masses have been purified and biochemically characterized from different *Botrytis cinerea* strains (Table 2). Obviously when talking about *Botrytis* laccase, not necessarily the same enzyme is meant. It might be argued, whether *Botrytis* species determinations in earlier studies could be verified by modern molecular approaches. Nevertheless, strain specific variations, different culture conditions and purification procedures deliver different laccase isoenzymes which may assemble to oligomers. The situation is even more complicated as the extracellular *Botrytis* laccase is partly entrapped in a glycan sheath (Gil-ad et al., 2001). It might be worth to mention that a laccase-like stilbene oxidase of only 32 kDa has been purified by Pezet (1998). This does not mean that "normal-sized" ≈ 60 kDa laccases, which were also present in the culture filtrate of this *Botrytis cinerea* strain did not oxidize stilbenes, but *vice versa*, the small enzyme did apparently not oxidize the typical laccase substrate syringaldazine.

Table 2. Biochemical properties of purified laccases from *Botrytis sp.*

Species/Strain	Molecular Features	Reference
B. cinerea/unnamed	MW: 56 kDa pI: 2.2 pH opt: 4.0-4.7 (depending on substrate) T opt: 30 – 55°C (depending on substrate) Km (mM) for quinol: 0.19; 4-methylcatechol: 0.045; p-cresol: 0.67; ascorbic acid: 0.77 Inhibitors: KCN, EDTA	Dubernet et al., 1977
B. cinerea/unnamed	Glycoprotein MW: 35.5 kDa pI 2.5 – 3.5 pH opt: 4.5 Km (mM) for O_2: 0.25	Gigi et al., 1980; 1981
B. cinerea/strain 841 isolated from grapes	Two glycosylated isoforms MW: 72.4 kDa (SDS-PAGE) 102 kDa (SEC sephacryl S-200) pI: 2.8 (Lac 1), 2.6 (Lac2) pH opt: 2.5 – 3.0 (Lac1) 4.0 (Lac2) Topt: 55°C (Lac1) 60°C (Lac2) Km (mM) for quinol: (0.133 Lac1; 0.533 Lac2) Inhibition: 30 s at 60°C or 20 s at 80 °C Inhibitors: KCN, not EDTA up to 20 mM	Zouari et al., 1987
B. cinerea/strain 61-34 isolated from onion	Glycoprotein MW: 74 kDa Oligomerisation occurs pH opt: 3.5 T opt: 60°C Km (mM) for 2,6-dimethoxyphenol: 0.1 Inhibitors: KCN, EDTA, Na azide Active in 4 % SDS and 2 M urea	Slomczynski et al., 1995
B. cinerea/strain P-69 isolated from grapes	Six glycosylated isoforms MW: 64.5 – 97.6 kDa pI: 3.5 – 4.7 pH opt: 5.0 – 6.0 All isoforms oxidize hydroxystilbenes, but some do not react with syringaldazine	Pezet et al., 1991
B. cinerea/strain P-69 isolated from grapes	MW: 32 kDa Km (mM) for resveratrol: 0.02 Highest activity with pterostilbine Lowest activity with syringaldazine Main product: ε-viniferin	Pezet, 1998
B. cinerea/strain F226 isolated from chestnut leaf	Crude enzyme preparation pH opt: 5.5 T opt: 40°C	Fortina et al., 1996
Botrytis aclada/strain DSM 62081; laccase gene was recombinant expressed in *Pichia pastoris*	Glycoprotein MW: 100 kDa (glycosylated) - 73 kDa (deglycosylated) pI: 2.4 pH opt: 3 – 5 (depending from substrate) Km (mM) for ABTS: 0.02; 2,6-dimethoxyphenol: 0.019; guaiacol: 0.073; catechol: 0.139; hydroquinone: 0.091; p-anisidine: 0.592	Kittl et al., 2012

A more homogenous picture appears when looking in the genome of *Botrytis cinerea* strains and related species (Table 3). Usually, two laccase genes are found with typical molecular sizes for fungal laccases around 60 kDa. Eight to eleven sequon structures (*Asn-Xaa-Ser/Thr*) indicate possible N-glycosylations sites.

Table 3. Molecular features of laccase genes from *Botrytis cinerea* and related phytopathogenic ascomycetes

Species/Strain	AS	MW (Da)[1]	pI[1]	N-glycosylation sites[1]	Size of signal peptide (AS)[1]	Genbank Acc. No.	Reference
B. aclada/DSM 62081*	580	63629	4.46	9	19	AFC76164	Kittl et al., 2012
B. cinerea/ SA556							
Laccase 1	561	60337	5.89	8	20	AAK77952	Schouten et al., 2002
Laccase 2	581	63434	4.47	8	19	AAK77953	Schouten et al., 2002
B. cinerea/T4							
Laccase 1	565	60788	5.89	8	20	CCD45861	Amselem et al., 2011
Laccase 2	581	63434	4.47	9	19	CCD44233	Amselem et al., 2011
B. cinerea/BcDW1							
Laccase 1	663	71134	5.72	9	19	EMR81386	-
Laccase 2	565	60788	5.89	8	20	EMR84219	
Monilinia fructigena/DSM 2677 Laccase 2	584	63748	4.72	11	19	ABM21605	Hirschhäuser and Fröhlich, 2007
Sclerotinia minor/DSM 65016 Laccase 2	580	63463	4.94	9	19	ABM21603	Hirschhäuser and Fröhlich, 2007
Sclerotinia sclerotiorum/DSM 1946 Laccase 2	580	63421	4.78	10	19	ABM2160	Hirschhäuser and Fröhlich, 2007
Sclerotinia sclerotiorum/1980 UF-70 Laccase precursor	600	65661	5.64	9	19	EDO02635	-

*Crystal structure available at a solution of 1.67 Å (PDB: 3SQR_A).

[1] Calculated using the Expasy Bioinformatics platform (http://www.expasy.org/).

A signal peptide of 19–20 amino acids underlines that the gene products are probably secreted in the environment. The phylogenetic relatedness of the *Sclerotiniacae* (Amselem et al., 2011) is also manifested in the similarity of their laccase genes (Kittl et al., 2012).

For *Botrytis cinerea* it has been reported, that phenolic compounds, especially gallic acid or a product of gallic acid metabolism can induce laccase production (Gigi et al., 1980; Viterbo et al., 1993 b). Pectin acts as a second inducer of extracellular laccase production, in the presence of a phenolic substance as a first inducer. Pectin alone fails to induce enzyme formation (Marbach et al., 1985). Furthermore, terpenoids can stimulate laccase production (Cotoras et al., 2004). On the other hand, cucurbitacins have been found to inhibit laccase formation (Viterbo et al., 1993 a;b).

Botrytis laccases oxidize in *vitro* a wide spectrum of polyphenolics. Table 4 gives an impression about the substrate diversity which surely could be extended when further substrates are tested. Many of these phenolic compounds are converted to water-insoluble colored polymers.

Table 4. Substrate specificity of a crude laccase from *Botrytis cinerea* (Dubernet et al., 1977)

Substrate	Relative activity*
phloroglucinol	143
caffeic acid	132
protocatechuic acid	119
3-methylcatechol	110
gallic acid	109
p-cresol	109
ferulic acid	109
catechol	104
(+)-catechin	100
chlorogenic acid	100
3,3-dimethylphenol	100
4-methylcatechol	100
quinol	100
3,4-dihydroxyphenylalanine	97
anthocyanins	97
ascorbic acid	95
N,N-dimethyl-p-phenylenediamine	95
p-phenylenediamine	95
p-coumaric acid	90
leucoanthocyanins	84
vanillic acid	33
tyrosine	7
p-hydroxybenzoic acid	0

* Enzyme activity was determined by measuring oxygen consumption using a polarographic O_2 electrode. The standard substrate quinol was taken as 100 % (10 mM, pH 4.75, 25 °C).

Botrytis Laccase and Wine Browning

The browning of wine, primarily due to the enzymatic and chemical oxidation of the phenolic compounds, represents one of the most feared processes which may arise during winemaking. The release of laccase into must from grapes infected with *Botrytis cinerea* may cause a significant reduction of the content of phenolic compounds (Mayer, 2006). Polyphenolics, including the important classes of wine phenolics (phenolic acids, catechins, anthocyanins, tannins and stilbenes) are converted into the corresponding quinones, which often react further to dark colored polymers (Walker, 1975; Fowler et al., 2011). The latter are generally insoluble in water and precipitate from must and wine.

Furthermore, the oxidation of phenolic compounds may adversely affect the sensory and nutritional properties of wine and other foods (Oliveira et al., 2011; Zinnai et al., 2013). However in this respect, with the exception of resveratrol, currently only few studies focus on the fate of individual wine phenolics.

Due to its intrinsic stability and activity at low pH values, considerable laccase activities (3.0 U/ml) can persist in must (Zivkovic et al., 2011). The prevention of browning reactions by phenoloxidases has always been a challenge to food scientists (De Leonardis et al., 2010). In winemaking, sulfur dioxide is the generally used antioxidant to control wine browning. However, concerns about its ability to induce severe pseudoallergic reactions have created a great need for its reduction or replacement in recent years (Postolache et al., 2012). Bentonite is a common oenological additive to remove undesirable grape wine proteins. In our recent study (Zivkovic et al., 2011), the binding and inhibition of a laccase from the white-rot fungus *Polyporus pinisitus* in wine, must, and grape juice by two types of bentonites was investigated in a model system. Depending on the matrix tested the bentonites had very different effects on the laccase activity: (1) No inactivation in grape juice and must. (2) Low inhibition in white wine. (3) Stimulation of the activity in red wine. The different effects are probably due to different levels of protein and phenolic concentrations in the samples.

Conclusively, bentonite treatment was not effective for the removal of a model laccase from must. The probable reason is that laccases are acidic glycoproteins which are not efficiently bound to the negatively charged bentonite under wine conditions. Figure 3 shows the effect of a laccase from *Polyporus pinisitus* on the Vis-spectrum of a *Dornfelder* red wine. Due to the oxidation by laccase, the color-typical peaks of the red wine around 500 nm have vanished. Addition of even a high-dose of bentonite did not inhibit the reaction but induced color loss itself.

Unlike grape oxidases, which have been reported to be sensitive to sulphites even at low concentrations, fungal laccases tend to be more resistant. In our study, we found an irreversible inhibition of a "natural" laccase activity in a white must by addition of as much as 200 mg/l sodium bisulphite (Figure 4).

The most effective weapon to eliminate the laccase activity in the must is heat treatment. After 2 min at 75 °C the enzyme was completely inhibited, whereas lower temperatures afford longer incubation times (Figure 5).

Due to the dramatically impact on wine quality there is a strong demand to monitor *Botrytis* and its laccase activities in must and wine. Methods have been developed to quantify laccase activities by using chromogenic substrates such as syringaldazine (Grassin and Dubourdieu, 1989; Cuadrado et al., 2005), use of *Botrytis*-specific antibodies (Sivertsen at al.,

2005; Dewey et al., 2008), and detection of laccase genes by PCR methods (Hirschhäuser and Fröhlich, 2007).

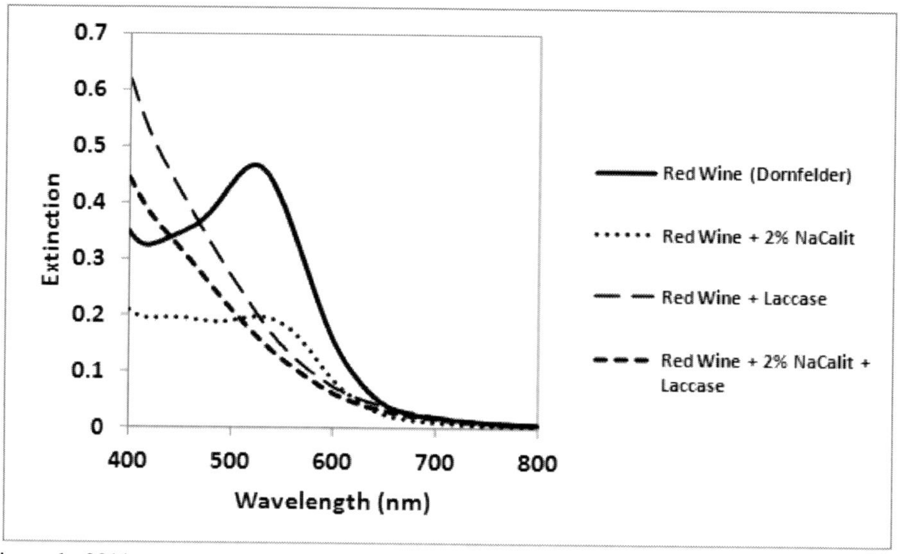

Zivkovic et al., 2011.

Figure 3. Vis-spectra of a *Dornfelder* red wine after incubation with a model laccase from *Polyporus pinisitus* and in the presence of a Na/Ca Bentonite.

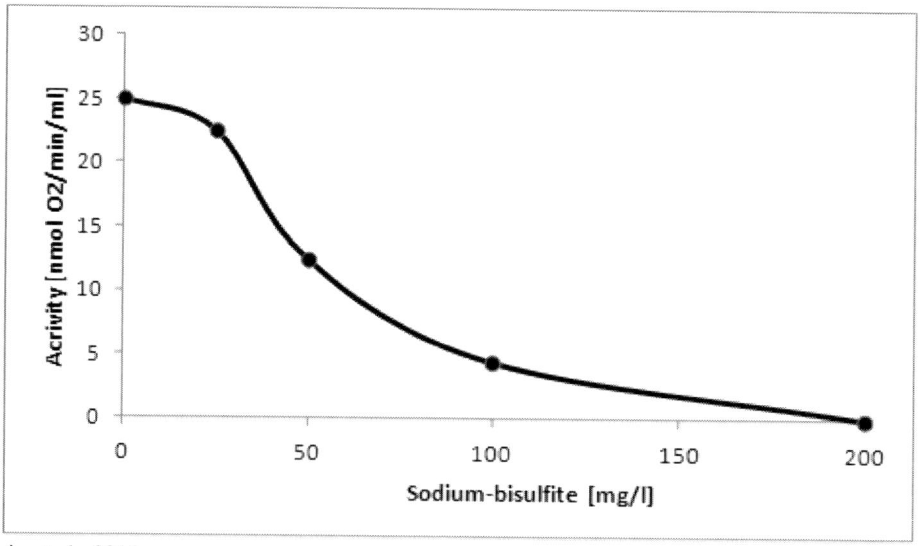

Zivkovic et al., 2011.

Figure 4. Inhibition of laccase activity in must by sulphite.

Botrytis Laccase and Resveratrol

Stilbenes such as resveratrol are biosynthesized by green plants of various taxonomic groups. The soil nematode-associated bacterium *Photorhabdus* is the only known stilbene producer outside the plant kingdom (Bode, 2009).

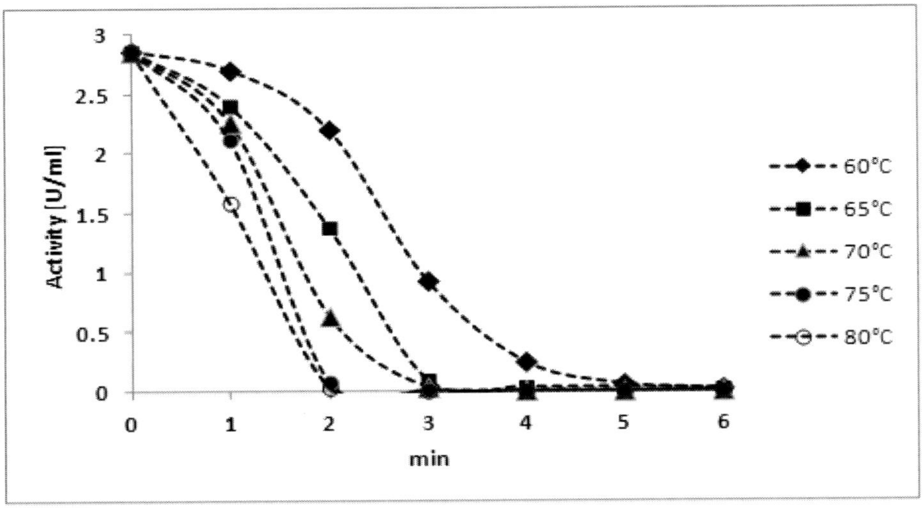

Zivkovic et al., 2011.

Figure 5. Inhibition of laccase activity in must by heat treatment.

The hydroxystilbene resveratrol is one of the best analyzed compounds in wine, especially because of its beneficial influence on human health. The positive antioxidant, cardioprotective, neuroprotective or anticancer effects have been discussed in numerous publications (for example by Soleas et al., 1997; Stevenson and Hurst, 2007; Li et al., 2012; Stuart and Robb, 2013).

The original well investigated function of the molecule is that of a phytoalexin. The latter are secondary metabolites produced by plants in response to biotic and abiotic stress factors and have antifungal and antimicrobial activities. Resistance of *Vitis* species to fungal infection is generally correlated with their ability to produce stilbene phytoalexines. Exogenic application of trans-resveratrol to grape berries is an effective means of enhancing their resistance to fungal infection. In laboratory experiments, trans-resveratrol inhibits the downy mildew fungus *Plasmopara viticola* and a number of other fungal pathogens, including *Cladosporium cucumerinum*, *Oidium tuckeri*, *Pyricularia oryzae*, *Sphaeropsis sapinea* and *Botrytis cinerea*.

Trans-resveratrol is composed of p-coumaryl-CoA and three molecules of malonyl-CoA in a reaction catalyzed by stilbene synthase. Trans-resveratrol is further converted into different derivates:

- Piceid or polydatin (glycosylation)
- Pterostilbene (methoxylation)
- ε- and δ- viniferin (oxidative oligomerization)

It should be noted that some resveratrol derivates are actually present in must and wine at higher levels than resveratrol itself. Piceid levels of red wines were found to be as much as tenfold higher. Pterostilbene and ε-viniferin are more potent inhibitors of fungal growth than trans-resveratrol, though they may not accumulate to levels required for physiological activity (Stuart and Robb, 2013).

The mechanism of action of stilbenes against *Botrytis cinerea* is not well understood. It has been suggested that resveratrol inhibits the respiration of fungal cells, probably by acting as an uncoupling agent or by membrane lipid peroxidation. Pterostilbene is more active than the less hydrophobic resveratrol, due to its increased diffusion through the cytoplasmic membrane (Caruso et al., 2011). *In vitro* experiments suggest that phytopathogenicity of *Botrytis cinerea* strains is associated with their ability to degrade stilbenes. In *Botrytis cinerea* cultures, trans-resveratrol induces a specific laccase gene (*Bclcc2*), and trans ε-viniferin has been identified as oxidation product of laccase-activity (Pezet, 1998; Schouten et al., 2002). Thus, *Botrytis cinerea* would commit suicide, because the dimer has been found to be more toxic than the monomer (Schouten et al., 2002). The situation is puzzling, because the small stilbene oxidase has been postulated as the resveratrol converting enzyme, whereas the gene product of *Bclcc2* is a normal-sized laccase (Table 3). Other researchers have suggested that *Botrytis cinerea* detoxifies trans-resveratrol by laccase activity (Pezet et al., 1991). The role of *Botrytis cinerea* laccase in grape infection thus remains unresolved (Favaron et al., 2009). The situation is even more complicated as *Botrytis cinerea* cultures convert trans-resveratrol not solely into ε-viniferin, but also into a number of other dimers such as resveratrol trans-dehydrodimer, leachinol F, and pallidol (Cichewicz et al., 2000). Like resveratrol, trans-pterostilbene undergoes oxidative dimerization by the laccase-activity of *B. cinerea* (Breuil et al., 1999).

BIOTECHNOLOGICAL ASPECTS

Resveratrol Enhanced Wine

Resveratrol levels in wines depend on a number of factors such as grape variety, geographical location, climate conditions, fungal infections of grapes, UV light exposure and oenological techniques (González-Candelas et al., 2000). Various methods have been tried to increase the resveratrol level in wine, which is in the range of 1 - 2 mg/l and 3 - 6 mg/l for normal white and red wines, respectively.

In order to achieve as high phenolic accumulation as possible, the interactive effects of storage conditions (temperature and time) and UV-C radiation on the polyphenols content in a postharvest *Redglobe* table grape variety were investigated by Crupi et al. (2013). During a storage time longer than 48 h, both cold storage and UV-C exposure positively enhanced the content of cis- and trans-piceid, together with quercetin-3-O-galactoside and quercetin-3-O-glucoside up to three fold compared to control grape samples. The authors concluded that combined postharvest treatments could lead to possible "functional" grapes, responding to the rising consumers demand to have foods that support and promote health.

Philip Norrie (2009) presented a patented technique which involves extraction of resveratrol from the grape skins and adding it back into the bulk wine at a desired concentration. Todaro et al. (2008) tested β-glucosidases (EC 3.2.1.21) from different sources to increase the trans-resveratrol content in some Sicilian wines by hydrolyzing glycosylated precursor isomers. An enzyme preparation of *Aspergillus niger* increased trans-resveratrol up to 75 %, with no change of the physical-chemical properties and bouquet of the wine. Transgenic yeasts overexpressing glycosyl-hydrolases have been shown to increase resveratrol content in wine (González-Candelas et al., 2000).

Novel Bioactive Polyphenols

The natural role of stilbenes in defending the producers against microbial attack is widely accepted. Thus, they can be important precursors for new antimicrobial therapeutics (Kumar et al., 2012). Laccase has been used to synthesize new molecules from resveratrol and other polyphenols with enhanced antioxidant and other beneficial features (Ponzoni et al., 2007). As mentioned above, laccase-catalyzed oxidations of polyphenols in more or less aqueous systems, like must and wine, yields mainly insoluble polymers which are no longer bioavailable.

However, when the enzymatic reactions proceed in organic solvents, substrates and products keep in solution. In recent years, many attempts have been made to synthesize *in vitro* oligomeric polyphenols with new beneficial properties (Table 5).

The question arises whether in wine (> 10 % v/v ethanol) similar beneficial phenolic oligomers are produced by the action of *Botrytis* laccase.

Under certain conditions, polyphenolics may display prooxidant rather than antioxidant properties and this may lead to undesirable consequences such as inhibition of mitochondrial respiration, direct DNA damage, lipid peroxidation and mutagenicity. A prooxidant effect was demonstrated for flavonoid phenoxy radicals generated by laccase from *Sclerotinia sclerotiorum* (Moţ et al., 2014).

Table 5. Laccase-catalyzed synthesis of oligomeric polyphenols with enhanced beneficial properties*

Polyphenol	Laccase source	Main products (as far identified)	Biological features investigated
Caffeic acid	*Trametes versicolor*	Oligomers/Polymers; Conjugate with Chitosan	Antioxidant activity Antimicrobial activity
Catechin	Not described	Conjugate with gelatine	Antioxidant activity
Catechin	Not described	Conjugate with poly(ε-lysine)	Inhibition of xanthin oxidase, hyaluronidase, collagenase
Catechin		Polycatechin	Antioxidant activity; Inhibition of xanthin oxidase
2,6-Dimethoxyphenol	*Trametes pubescens*	Dimer	Antioxidant
Epigallocatechin gallate	*Myceliophtora*	Oligomer	Antioxidant activity; Inhibition of xanthin oxidase

Table 5. (Continued)

Polyphenol	Laccase source	Main products (as far identified)	Biological features investigated
Ferulic acid	*Trametes hirsuta*		Antibacterial activity
Ferulic acid	*Tramtes pubescens*	Dimers	Antioxidant activity
Gallic acid	*Trametes versicolor*	Oligomers/polymers; Conjugate with chitosan	Antioxidant activity Antimicrobial activity
Kaempferol, quercitin, rutin, gallic acid, caffeic acid, vanillin	*Ustilago maydis* (also source of tyrosinase)	Polymer	Antioxidant activity Inhibition of lipidperoxidation/ Effect on ROS with hepatitic cell line WRL-68
Resveratrol	*Pyricularia oryzae*		Antioxidant activity
Resveratrol-related stilbenoids	*Trametes versicolor*	Dehydrodimers	Antiproliferative/Antitumor activities (carcinoma cell line SW580)
Rutin	*Myceliophtora*	Polymer	Antioxidant activity
Rutin, esculin	*Trametes versicolor*	Oligomers	Anti-mutagenic activity
Rutin	*Pycnococcus sp. Myceliophtora thermophyla (Suberase©)*	Oligomers (mainly dimers)	Antioxidant, Anti-inflammatory, Anti-aging activities (Inhibition of matrix metalloproteinase)
trans-Resveratrol	*Myceliophtora thermophyla; Trametes pubescens*	Dehydrodimer different from ε-viniferin	Antioxidant activity
Tyrosol	*Trametes trogii*	Dimer	Antibacterial activity Antifungal activity
Vanillin, tannin acid	*Trametes hirsuta*	Non-identified polymer	Antifungal activity

* As reviewed by Uyama and Kobayahi, 2006; Witayakran and Ragauskas 2009; Jeon et al., 2012.

Wine Stabilization

Although the uncontrolled oxidation of must and wine by *Botrytis* laccase is a nightmare for wine-makers, attempts have been made for a directed laccase treatment of beverages.

Due to a complex sequence of events, where the polyphenols play an important role, oxidative reactions can occur in musts and wines causing flavor and color alterations. This phenomenon of oxidation is known as maderization. Different methods have been used in order to prevent these reactions in wines, such as the removal of phenolics with polyvinylpolypyrrolidone (PVPP), and the use of sulfur dioxide to block oxidizers, among others.

An alternative is the use of phenoloxidases that selectively target specific polyphenols causing oxidative browning (Servilli et al., 2000; Minussi et al., 2002; 2007). These polyphenolic substances are oxidized by the laccases, polymerized and then removed by clarification.

Enzymatic treatment is interesting for its specific action and as a ‘‘mild technology’’, with less drastic effects on the characteristics of the wine. Several publications in the literature report that laccase treatment promotes wine stabilization. The enzyme is consistent with some important requirements when used for the elimination of polyphenols from wines, such as activity and stability in acid media (pH 2.5 – 4.0).

A laccase of *Trametes versicolor* removed more than 90 % of ferulic acid from a model solution and 34 % of phenolic compounds from wines. A mutant laccase from *Trametes versicolor* (optimum pH 2.7) eliminated up to 70 % catechin and 90 % of anthocyanidins in a model solution in 3 h of treatment and removed 50 % of total polyphenols from a black grape juice.

Maier et al. (1990) evaluated the polyphenol content, color, haze stability and sensorial quality of Riesling wines prepared with and without enzymatic treatment of the must. The results showed that wines made by laccase treatment were the best, suggesting that a stable and high quality wine can be made with little or no added SO_2. Since the use of laccase as a food additive is still not allowed, this enzyme has been tested for wine production in an immobilized form, to ensure its elimination from the must and its reutilization (Brenna and Bianchi 1994). The results obtained suggest the potential of this immobilized enzyme for continuous enzymatic catalysis in food processing.

In recent years, the moderate consumption of red wines has been shown to benefit human health. As already mentioned, the effect has mainly been attributed to the antiradical properties of polyphenols that are present in red wines. In view of this, it is necessary to analyze the possible interactions of laccase with these beneficial wine compounds which in principle are the substrates for the enzyme. According to Minussi et al. (2007), the treatment of a red must with laccase from *Trametes versicolor* mainly affects the phenolic compounds responsible for the must antioxidant properties. However, the treatment of white musts with the same laccase showed higher reduction in total phenol than in total antioxidant potential, so the treatment of white musts seems feasible. In experiments of Espín and Wichers (2000), laccase of *Pyricularia oryzae* oxidized resveratrol but did not change its antiradical capacity.

The oxidative power of laccase has been used in biosensors for a rapid and reliable amperometric estimation of the total content of polyphenolic compounds in wines (Gomez et al., 2004; Gamella et al., 2006). Another application of a fungal laccase (“suberase”) is the treatment of cork stoppers. The enzymatic oxidation of the phenols avoids evolvement of the off-flavor tetra-chloroanisol (Sponholz, 2000).

Transformation of Biogenic Amines

Increasing health concerns are related to the presence of biogenic amines in must and wine (König et al., 2013). Kaschak et al. (2009) detected tyramine, phenylethylamine, ethanolamine and isopentylamine most frequently in German wines from discount supermarkets. Aromatic amines like tyramine are readily oxidized by phenoloxidases (Claus and Strong, 2010) and it is challenging to investigate the influence of laccase on biogenic amines in wine. Indeed, laccase-like multicopper oxidases from *Lactobacillus plantarum* and

Pediococcus acidilactici have been recently described which are able to degrade biogenic amines in wine (Callejon et al., 2014).

Detoxification of Wine-Related Waste Waters

Many industrial wastewaters contain harmful phenolic compounds that are toxic and have high color content. The potential use of phenoloxidases (peroxidases, laccase, tyrosinases) to detoxify such effluents has already been recognized in the 1980s (Claus et al., 2002; Strong and Claus, 2011; Mukherjee et al., 2013).

The alcohol fermentation industry includes brewing, distilling, wine manufacture and production of bioethanol as fuel. Each of these categories produces wastewaters with common characteristics, such as acidic pH values and high biochemical oxygen demand (BOD) and chemical oxygen demand (COD). The brown color of wine-distillery wastewater is generally attributed to phenolic compounds, although in some distillery wastewaters, such as those derived from molasses. The brown color is attributed to nitrogenous polymers known as melanoidins. These complexes are formed as a result of condensation reactions between reducing saccharides and amines or amino acids, which is known as the Maillard reaction. Concentrations of phenolic compounds in such effluents are ranging from 29 to 766 mg/l. Certain phenolic compounds have been implicated in the inhibition of biological treatment by anaerobic digesters.

The treatment of such wastewaters with cultures of the laccase producing fungus *Trametes pubescens* MB 89 decreased the chemical oxygen demand of up to 83 ± 2.1 %, phenolic compounds of 87 ± 1.6 % and color of 88 ± 4.7 %. *Trametes pubescens* MB 89 greatly improved the quality of all six wastewaters tested, although two wastewaters had to be diluted to below 50 % to allow bioremediation by the submerged fungal culture. Laccase synthesis greater than 1500 U/l was obtained in all wastewaters, with a maximum of 8997 U/l. The complete fungal system was found to be superior to enzymatic treatment alone: a cell-free laccase preparation of this fungus lowered total phenolics by up to 61 ± 0.5 %, however the color was increased by up to 160 ± 5 %.

Exploitation of Wine-Related Wastes

Vine (*Vitis vinifera*) is the most cultivated fruit crop in the world, with a global production of around 69 million tons, of which 80 % are used in wine-making (Martínez-Ávila et al., 2012). These processes generate a massive quantity of grape waste products, considerably increasing environmental pollution due to high content of organic substances. In total, more than 20 % of wine production is waste (Landete et al., 2007). On the other hand, grape waste is a potential source of phenolic compounds that have been shown to be beneficial to human health. Solid by-products from wine industry were recently subjected to evaluation as potential antioxidant phytochemicals. Scola et al. (2010) have shown that it is possible to obtain flavan-3-ol compounds from wine wastes. The main phenolic compounds found in the extracts were catechin and epicatechin, followed by procyanidin B3, procyanidin B1, procyanidin B2, gallic acid, epigallocatechin, and procyanidin B4. All flavan-3-ol extracts showed significant *in vitro* and *in vivo* activities. It was found that the extracts were able to prevent lipid and protein oxidative damage in the cerebral cortex, cerebellum and hippocampus tissues of rats. Grape marc, a byproduct of winemaking, contains significant amounts of substances that are not efficiently used (Negro et al., 2003; De Campos et al., 2008). Augustin-Salazar et al. (2014) investigated the effect of different solvent systems on the extraction of phenolic substances and antioxidant capacity from the Carignan grape (*Vitis vinifera L.*) marc, as well as from the peduncle, skin and seed of Carignan. The main phenolic substances that remained in wine marc after fermentation were myricetin, catechin and epicatechin. The results of that study suggested that grape marc, considered as industrial waste, is a natural resource of antioxidants. Vine leaves with high polyphenolic potential and significant antioxidant and antimicrobial activities, which remain in the must after the grape harvest, could be also interesting for further exploitation (Fernandes et al., 2013; Katalinic et al., 2013).

More preliminary studies are promising and exploitation of wine industry waste as a source of valuable products may be cost-effective and merits profounder investigation (Makris et al., 2007; Chamorro et al., 2012; Martínez-Ávila et al., 2012).

CONCLUSION

The presence of phenolic compounds in wine is positively related with its sensory characteristics, such as flavor, astringency and color. Furthermore, polyphenols are beneficial to human health due to their antioxidant, anti-inflammatory, vasodilating and antimicrobial properties. Uncontrolled oxidative browning reactions catalyzed by *Botrytis* laccase have a dramatic impact on wine quality as they reduce the polyphenol content. However, the same enzymatic mechanisms can be biotechnologically used for wine stabilization, detoxification of wine-related waste or synthesis of new pharmacological compounds.

ACKNOWLEDGMENT

The authors want to thank the *Stiftung Rheinland-Pfalz für Innovation* (Rhineland-Palatinate, Germany) for financial support (Project-No. 961-386261/1051).

REFERENCES

Adams, D. O., (2006). Phenolics and ripening in grape berries. *Am. J. Enol. Vitic.*, 57, 249-256.

Amselem, J., Cuomo, C. A., van Kan, J. A. L, et al. (2011). Genomic analysis of the necrotrophic fungal pathogens *Sclerotinia sclerotiorum* and *Botrytis cinerea*. *PLoS Genetics,* 8, e1002230.

Arranz, S., Chiva-Blanch, G. Valderas-Martínez, P., Medina-Remón, A., Lamuela-Raventós, R.M., Estruch, R., (2012). Wine, beer, alcohol and polyphenols on cardiovascular disease and cancer. *Nutrient,* 4, 759-781.

Augustin-Salazar, S., Medina-Juárez, L. A., Soto-Valdez, H., Manzaneres-López, F., Gámez-Meza, N. (2014) Influence of the solvent system on the composition of phenolic substances and antioxidant capacity of extracts of grape (*Vitis vinifera L.*) marc *Austr. J. Grape Wine. Res.*, Doi: 10.1111/ajgw.12063.

Baldrian, P., (2006). Fungal laccases – occurrence and properties. *FEMS Microbiol. Rev.,* 30, 215-242.

Bode, H. B., (2009). Entomopathogenic bacteria as a source of secondary metabolites. *Curr. Opin. Chem. Biol.*, 13, 224-230.

Bokulich, N. A., Joseph, C. M. L., Allen, G., Benson, A. K., Mills, D. A., (2012). Next-generation sequencing reveals significant bacterial diversity of botrytized wine. *PLoS One,* 7, e36357, 1-9.

Bollag, J. M., Shuttleworth, K. L., Anderson, D. H., (1988). Laccase-mediated detoxification of phenolic compounds. *Appl. Environ. Microbiol.,* 54, 3086-3091.

Brenna, O., Bianchi, E., (1994). Immobilised laccase for phenolic removal in must and wine. *Biotechnol. Lett.,* 16, 35-40.

Breuil, A. C., Jeandet, P., Adria, M., Chopin, F., Pirio, N., Meunier, P., Bessis, R., (1999). Characterization of a pterostilbene dehydrodimer produced by laccase of *Botrytis cinerea. Phytopathol.,* 89, 298-302.

Callejon, S., Sendra, R., Ferrer, S., Pardo, I., (2014). Identification of a novel enzymatic activity from lactic bacteria able to degrade biogenic amines in wine. *Appl. Microbiol. Biotechnol.,* 98, 185-198.

Camanella, L., Bonanni, A., Finotti, E., Tomassetti, M., (2004). Biosensors for determination of total and natural antioxidant capacity of red and white wines: comparison with other spectrophotometric and fluorimetric methods. *Biosens. Bioelectron.,* 19, 641-651.

Campos, F. M., Couto, J. A., Hogg, T. A., (2003). Influence of phenolic acids on growth and inactivation of *Oenococcus oeni* and *Lactobacillus hilgardii. J. Appl. Microbiol.*, 94, 167-174.

Campos, F. M., Couto, J. A., Figueiredo, A. R., Tóth, I. V., Rangel, A. O. S. S., Hogg, T. A., (2009) Cell membrane damage induced by phenolic acids on wine lactic acid bacteria. *Int. J. Food Microbiol.*, 135, 144-151.

Cardono, F., Andrés-Lacueva, C., Tulipani, S., Tinahones, F. J., Queipo-Ortuño, M. I., (2013). Benefits of polyphenols on gut microbiota and implications in human health. *J. Nutr. Biochem.*, 24, 1415-1422.

Caridi, A., Cufari, A., Lovino, R., Palumbo, R., Tedesco, I., (2004). Influence of yeast on polyphenol composition of wine. *Food Technol. Biotechnol.*, 42, 37-40.

Caruso, F., Mendoza, L., Castro, P., Cotoras, M., Aguirre, M., Matsuhiro, B., Isaacs, M., Rossi, M., Viglianti, A., Antonioletti, R., (2011). Antifugal activity of resveratrol against *Botrytis cinerea* is improved using 2-furyl derivates. *PLoS ONE,* 6, e25421.

Chamorro, S., Viveros, A., Alvarez, I., Vega, E., Brenes, A., (2012). Changes in polyphenol and polysaccharide content of grape seed extract and grape pomace after enzymatic treatment. *Food Chem.*, 133, 308-314.

Choquer, M., Fournier, E., Kunz, C., Levis, C., Pradier, J. M., Simon, A., Viaud, M., (2007). *Botrytis cinerea* virulence factors: new insights into a necrotrophic and polyphageous pathogen. *FEMS Microbiol. Lett.*, 277, 1-10.

Cichewicz, R. H., Kouzi, S. A., Hamann, M. T., (2000). Dimerization of resveratrol by the grapevine pathogen *Botrytis cinerea. J. Nat. Prod.*, 63, 29-33.

Cilindre, C., Castro, A. J., Clément, C., Jenadet, P., Marchal, R., (2007). Influence of *Botrytis cinerea* infection on Champagne wine proteins (characterized by two-dimensional electrophoresis/immunodetection) and wine foaming properties. *Food Chem.,* 103, 139-149.

Claus, H., (2004). Laccases: structure, reactions, distribution. *Micron,* 35, 93-96.

Claus, H., (2010). Copper-containing oxidases: occurrence in soil microorganisms, properties and applications. In: Shermeti, I., Varma, A., Editors. Soil heavy metals, Soil Biology Vol. 19, Springer-Verlag Berlin Heidelberg, 281-313.

Claus, H., Filip, Z., (1998). Degradation and transformation of aquatic humic substances by laccase-producing fungi *Cladosporium cladosporioides* and *Polyporus versicolor. Acta Hydrochim. Hydrobiol.,* 26, 180-185.

Claus, H., Decker, H., (2006). Bacterial Tyrosinases. *Sys. Appl. Microbiol.*, 29, 3-14.

Claus, H., Strong, P. J., (2010). Laccase. In: Flickinger, M. C., Editor. Encyclopedia of Industrial Biotechnology: Bioprocess, Bioseparation, and Cell Technology. John Wiley and Sons, DOI 10.1002/9780470054581.eib379.

Claus, H., Faber, G., König, H. (2002). Redox-mediated decolorization of synthetic dyes by fungal laccases. *Appl. Microbiol. Biotechnol.*, 59: 672-678.

Claus, H., Tenzer, S., Sobe, M., Schlander, M., König, H., Fröhlich, J., (2014) Effect of carboxymethyl cellulose on tartrate salt, protein and colour stability of red wine. *Austr. J. Grape Wine Res.*, DOI: 10.1111/ajgw.12070.

Collado, I. G., Sánchez, A. J., Hanson, J. R., (2007). Fungal terpene metabolites: biosynthetic relationships and the control of the phytopathogenic fungus *Botrytis cinerea. Nat. Prod. Rep.*, 2007, 24, 674-686.

Cotoras, M., Folch, C., Mendoza, L., (2004). Characterization of the antifungal activity on *Botrytis cinerea* of the natural diterpenoids kaurenoic acid and 3β-hydroxy-kaurenoic acid. *J. Agric. Food Chem.,* 52, 2821-2826.

Crupi, P., Pichierri, A., Basile, T., Antonacci, D., (2013). Postharvest stilbenes and flavonoids enrichment of table grape cv Redglobe (*Vitis vinifera* L.) as affected by interactive UV-C exposure and storage conditions. *Food Chem.*, 141, 802-808.

Cuadrado, M. U., Pérez-Juan, P. M., de Castro, M. D. L., Gómez-Nieto, M. A., (2005). A fully automated method for in real time determination of laccase activity in wines. *Anal. Chim. Acta*, 553, 99-104.

Cueva, C., Mingo, S., Muňoz-González, I., Bustos, I., Requena, T., Del Campo, R., Martín-Álvarez, P. J., Bartolomé, B., Moreno-Arribas, M. V., (2012). Antibacterial activity of wine phenolic compounds and oenological extracts against potential respiratory pathogens. *Lett. Appl. Microbiol.*, 54, 557-563.

De Beer, D., Joubert, E., Gelderblom, W. C. A., Manley, M., (2002). Phenolic compounds: a review of their possible role as in vivo antioxidants of wine. *South Afr. J. Enol. Vitic.*, 23, 48-61.

De Campos, L. M. A. S., Leimann, F. V., Pedrosa, C., Ferreira, S. R. S., (2008). Free radical scavenging of grape pomace extracts from Cabernet sauvignon (*Vitis vinifera). Biores. Technol.,* 99, 8413-8420.

De Leonardis, A., Lustrato, G., Mavcciola, V., Ranalli, G., (2010). Application of chemical and physical agents in model systems to controlling phenoloxidase enzymes. *Eur. Food Res. Technol.*, 231, 603-610.

Dewey, F. M., Hill, M., DeScenzo, R., (2008). Quantification of *Botrytis* and laccase in winegrapes. *Am. J. Enol. Vitic.*, 59, 47-54.

Dietrich, H., Pour-Nikfardjam, M. S., (2009). Influence of phenolic compounds and tannins on wine-related microorganisms. In: König, H., Unden, F., Fröhlich, J., editors. Biology of Microorganisms on Grapes, in Must and Wine. Berlin Heidelberg: Springer-Verlag, 307-347.

Dittrich, H., H., Großmann, M., (2010). Mikrobiologie des Weines, 4. Edition, Stuttgart: Ulmer Verlag.

Dubernet, M., Ribereau-Gayon, P., Lerner, H. R., Harel, E., Mayer, A. M., (1977). Purification and properties of laccase from *Botrytis cinerea. Phytochem.,* 16, 191-193.

Du Toit, W. J., Marais, J., Pretorius, I. S., Du Toit, M., (2006). Oxygen in must and wine: a review. *South Afr. J. Enol. Vitic.*, 27, 76-94.

Eder, R., Wendelin, S, (2002). Phenolzusammensetzung und antioxidative Kapazität von Trauben und Weinen. ALVA Jahrestagung 2002, Klosterneuburg, 293-296.

Espín J. C., Wichers H. J., (2000). Study of the oxidation of resveratrol catalyzed by polyphenol oxidase. Effect of polyphenoloxidase, laccase and peroxidase on the antiradical capacity of resveratrol. *J. Food Biochem.*, 24, 225-250.

Favaron, F., Luchetta, M., Odorizzi, S., Pais da Cunha, A. T., Sella, L., (2009). The role of grape polyphenols on trans-resveratrol activity against *Botrytis cinerea* and of fungal laccase on the solubility of putative grape PR proteins. J. *Plant Pathol.*, 91, 579-588.

Fernandes, F., Ramalhosa, E., Pires, P., Verdial, J., Valentao, P., Andrade, P. Bento, A., Pereira, J. A., (2013). *Vitis vinifera* leaves towards bioactivity. *Ind. Crops Products.,* 43, 434-440.

Figuiredo, A. R., Campos, F., de Freitas, V., Hogg, T., Couto, J. A., (2008). Effect of phenolic aldehyds and flavonoids on growth and inactivation of *Oenococcus oeni* and *Lactobacillus hilgardii. Food Microbiol.,* 25, 105-112.

Fortina, M. G., Acquati, A., Rossi, P., Manachini, P. L., Di Gennaro, C., (1996). Production of laccase by *Botrytis cinerea* and fermentation studies with strain F226. *J. Ind. Microbiol.*, 17, 69-72.

Fowler, Z. L., Baron, C. M., Panepinto, J. C., Koffas, M. A. G., (2011). Melanization of flavonoids by fungal and bacterial laccases. *Yeast,* 28, 181-188.

Gamella, A., Campuzano, S., Reviejo, A. J., Pingarrón, J. M., (2006). Electrochemical estimation of the polyphenol index in wines using a laccase biosensor. *J. Agric. Food Chem.,* 54, 7960-7967.

García-Ruiz, A., Bartolomé, B., Martínez-Rodríguez, A., Pueyo, E., Martín-Álvarez, P. J., Moreno-Arribas, M. V., (2008). Potential of phenolic compounds for controlling lactic acid bacteria growth in wine. *Food Control,* 19, 835-841.

García-Ruiz, A., Bartolomé, B., Cueva, C., Martín-Álvarez, P. J., Moreno-Arribas, M. V., (2009). Inactivation of oenological lactic bacteria (*Lactobacillus hilgardii* and *Pediococcus pentosaceus*) by wine phenolic compounds. *J. Appl. Microbiol.,* 107, 1042-1053.

García-Ruiz, A., Moreno-Arribas, M. V., Martín-Álvarez, P. J., Bartolomé, B., (2011). Comparative study of the inhibitory effects of wine polyphenols on the growth of enological lactic acid bacteria. *Int. J. Food Microbiol.*, 145, 426-431.

García-Ruiz, A., Cueva, C., González-Rompinelli, E. M., Yuste, M., Torres, M., Martín-Álvarez, P. J., Bartolomé, B., Moreno-Arribas, M. V., (2012). Antimicrobial phenolic extracts able to inhibit lactic bacteria growth and wine malolactic fermentation. *Food Control,* 28, 212-219.

Giardina, P., Faraco, V., Pezzella, C., Piscitelli, A., Vanhulle, S., Sannia, G., (2010). Laccases: A never-ending story. *Cell Mol. Life Sci.,* 67, 269-385.

Gigi, O., Marbach, I., Mayer, A. M., (1980). Induction of laccase formation in *Botrytis. Phytochem.,* 19, 2273-2275.

Gigi, O., Marbach, I., Mayer, A. M., (1981). Properties of gallic acid-induced extracellular laccase of *Botrytis cinerea. Phytochem.*, 20, 1211-1213.

Gil-ad, N. L., Bar-Nun, N., Mayer, A. M. (2001**).** The possible function of the glucan sheath of *Botyrtis cinerea*: effects on the distribution of enzyme activities. *FEMS Microbiol. Lett.*, 199, 103-113.

Giordano, M., Rolle, L., Zeppa, G., Gerbi, V., (2009). Chemical and volatile composition of three Italian sweet white Passito wines. *J. Int. Sci. Vign. Vin.*, 43, 1-10.

Gomes, S. A. S. S., Nogueira, J. M. F., Rebelo, M. J. F. (2004). An amperometric biosensor for polyphenolic compounds in red wine. *Biosens. Bioelectr.*, 20, 1211-1216.

González-Candelas, L., Gil, J. V., Lamuela-Raventós, R. M., Ramón, D., (2000). The use of transgenic yeasts expressing a gene encoding a glycosyl-hydrolase as a tool to increase resveratrol content in wine. *Int. J. Food Microbiol.*, 59, 179-183.

Gonzalez-Rompinelli, E. M., Rodriguez-Bencomo, J. J., Garcia-Ruiz, A., Sanchez-Patan, F., Martin-Alvarez, P. J., Bartolome, B., Morena-Arribas, M., (2013). A winery-scale trial of the use of antimicrobial plant phenolic extracts as preservatives during wine ageing in barrels. *Food Control,* 33, 440-447.

Grassin, C., Dubourdieu, D., (1989). Quantitative determination of *Botrytis* laccase in musts and wine by the syringaldazine test. *J. Sci. Food Agric.*, 48, 369-376.

Guzzo, F., Cappello, M. S., Azzolini, M., Tosi, E., Zapparoli, G., (2011). The inhibitory effects of wine phenolics on lysozyme activity against lactic acid bacteria. *Int. J. Food Microbiol.*, 148, 184-190.

Han, X., Shen, T., Lou, H., (2007). Dietary polyphenols and their biological significance. *Int. J. Mol. Sci.*, 8, 950-988.

Hanhineva, K., Törrönen, R., Bondia-Pons, I., Pekkinen, J., Kolehmainen, M., Mykkänen, H., Poutanen, K., (2010). Impact of dietary polyphenols on carbohydrate metabolism. *Int. J. Mol. Sci.* 11, 1365-1402.

He, S., Sun, C., Pan, Y., (2008). Red wine polyphenols for cancer prevention. *Int. J. Mol. Sci.*, 9, 842-853.

Hirschhäuser, S., Fröhlich, J., (2007). Multiplex PCR for species discrimination of *Sclerotiniaceae* by novel laccase introns. *Int. J. Food Microbiol.*, 118, 151-157.

Huang, D., Ou, B., Prior, R. L., (2005) The chemistry behind antioxidant capacity assays. *J. Agric. Food Chem.*, 53, 1841-1856.

Jeon, J. R., Baldrian, P., Murugesan, K., Chang, Y. S., (2012). Laccase-catalysed oxidations of naturally occurring phenols: from in vivo biosynthetic pathways to green synthetic applications. *Microb. Biotechnol.,* 5, 318-332.

Johannes, C., Majcherczyk, A., (2000). Laccase activity tests and laccase inhibitors. *J. Biotechnol.*, 78, 193-199.

Kaschak, E., Göhring, N., König, H., Pfeiffer, P., (2009). Biogenic amines in German wines: analysis and assessment according to the application of different HPLC-process. *Deutsche Lebensmittel-Rundschau,* 6, 375-384.

Kassemeyer, H. H., Berkelmann-Löhnertz, B. B., (2009). Fungi of grapes. In: König, H., Unden, F., Fröhlich, J., editors. Biology of microorganisms on grapes, in must and wine. Berlin Heidelberg: Springer-Verlag, 61-87.

Katalinic, V., Mozina, S. S., Generalic, I., Skroza, D., Ljubenkov, I., Klancnik, A., (2013). Phenolic profile, antioxidant capacity, and antimicrobial activity of leaf extracts from six *Vitis vinifera* L. varieties. *Int. J. Food Properties,* 16, 45-60.

Kittl, R., Mueangtoom, K., Gonaus, C., Khazaneh, S. T., Sygmund, C., Haltrich, D., Ludwig, R., (2012). A chloride tolerant laccase from the plant pathogen ascomycete *Botrytis aclada* expressed at high levels in *Pichia pastoris*. *J. Biotechnol.,* 157, 304-314.

König, H., Kaschak, E., Sebastian, P., Petri, A., Blättel, V., Wirth, K., Claus, H., Pfeiffer, P., (2013). Biogene Amine: analytische, mikrobiologische und gesundheitliche Aspekte. In: König, H., Decker, H., editors. Kulturgut Rebe und Wein. Spektrum Akademischer Verlag, Heidelberg, 207-220.

Koren, K., Kohen, R., Ginsburg, I., (2010). Polyphenols enhance total oxidant-scavenging capacities of human blood by binding to red blood cells. *Exp. Biol. Med.*, (Maywood) 235, 689-699.

Kumar, S. N., Siji, J. V., Nambisan, B., Mohandas, C., (2012). Activity and synergistic interactions of stilbenes and antibiotic combinations against bacteria in vitro. *World J. Microbiol. Biotechnol.*, 28, 3143-3150.

Ky, I., Lorrain, B., Jourdes, M., Pasquier, G., Fermaud, M., Gény, L., Rey, P., Doneche, B., Teissedre, P. L., (2012). Assessment of grey mould (*Botrytis cinerea*) impact on phenolic and sensory quality of Bordeaux grapes, musts and wines for two consecutive vintages. *Austr. J. Grape Wine Res.*, 18, 64-72.

Landete, J. M., (2012). Updated knowledge about polyphenols: functions, bioavailability, metabolism, and health. *Crit. Rev. Food Sci. Nutr.*, 52, 936-948.

Landete, J. M., Rodríguez, H., de las Rivas, B., Muñoz, R., (2007). High-added-value antioxidants obtained from the degradation of wine phenolics by *Lactobacillus plantarum*. *J. Food Prot.*, 70, 2670-2675.

Li, H., Förstermann, U., (2012). Red wine and cardiovascular health. *Circulation Res.,* 111, 959-961.

Li, H., Guo, A., Wang, H., (2008). Mechanisms of oxidative browning in wine. *Food Chem.,* 108, 1-13.

Li, H., Xia, N., Förstermann, U., (2012). Cardiovascular effects and molecular targets of resveratrol. *Nitric Oxide,* 26, 102-110.

Lorrain, B., Ky, I., Pechamat, L., Teissedre, P. L., (2013). Evolution of analysis of polyphenols from grapes, wines and extracts. *Molecules,* 18, 1076-1100.

Makris, D. P., Boskou, G., Andrikopoulos, N. K., (2007). Polyphenolic content and in vitro characteristics of wine industry and other agri-food solid waste extracts. *Int. J. Food Comp. Anal.*, 20, 125-132.

Marbach, I., Harel, E., Mayer, A. M., (1985). Pectin, a second inducer for laccase production by *Botrytis cinerea*. *Phytochem.,* 11, 2559-2561.

Martínez-Ávila, G. C., Aguilera-Carbó, A. E., Rodríguez-Herrera, R., Aguilar, C. N., (2012). Fungal enhancement of the antioxidant properties of grape waste. *Ann. Microbiol.*, 62, 923-930.

Masih, E. I., Alie, I., Paul, B., (2000). Can the grey mould disease of the grape-vine be controlled by yeast? *FEMS Microbiol. Lett.*, 189, 233-237.

Maier, G., Dietrich, H., Wucherpfennig, K., (1990). Winemaking without $S0_2$ – with the aid of enzymes? *Weinwirtschaft-Technik,* 126, 18-22.

Mayer, A. M., (2006). Polyphenol oxidases in plants and fungi: going place? A review. *Phytochemistry,* 67, 2318-2331.

Mayer, A. M., Staples, R. C., (2002). Laccase: new functions for an old enzyme. *Phytochem.*, 60, 551-565.

Mills, D. A., Johannsen, E. A., Cocolin, L., (2002). Yeast diversity and persistence in *Botrytis*-affected wine fermentations. *Appl. Environ. Microbiol.*, 68, 4884-4893.

Minussi, R. C., Pastore, G. M., Durán, N., (2002). Potential applications of laccase in the food industry. *Trends Food Sci. Technol.,* 13, 205-216.

Minussi, R. C., Rossi, M., Bologna, L., Rotilio, D., Pastore, G. M., Durán, N., (2007). Phenols removal of musts: strategy for wine stabilization by laccase. *J. Mol. Cat. B: Enzymatic,* 45, 102-107.

Moreno-Arribas, M. V., Polo M. C., (2010). Wine Chemistry and Biochemistry. New York: Springer Science and Business Media.

Moţ, A. C., Coman, C., Miron, C., Damian, G., Sarbu, C., Silaghi-Dumitrescu, R., (2014). An assay for pro-oxidant reactivity based on phenoxyl radicals by laccase. *Food Chem.*, 143, 214-222.

Mukherjee, S., Basak, B., Bhunia, B., Dey, A., Mondal, B., (2013). Potential use of polyphenol oxidases (PPO) in the bioremediation of phenolic contaminants containing industrial wastewater. *Rev. Environ. Sci. Biotechnol.*, 12, 61-73.

Negro, C., Tommasi, L., Miceli, A. 2003. Phenolic compounds and antioxidant from red grape marc extracts. *Biores. Technol.,* 87, 41-44.

Nisiotou, A., Spiropoulos, A. E., Nychas, G. J. E., (2007). Yeast community structures and dynamics in healthy and *Botrytis*-affected grape must fermentations. *Appl. Environ. Microbiol.,* 73, 6705-6713.

Norrie, P. A., (2009). Resveratrol-enhanced wine. *Int. J. Wine Res.*, 1, 187-188.

Oliveira, C. M., Ferreira, A. C. S., De Freitas, V., Silva, A. M. S., (2011). Oxidation mechanisms occurring in wines. *Food Res. Int.,* 44, 1115-1126.

Oszmianski, J., Sapis, J. C., Macheix, J. J., (1985). Changes in grape seed phenols as affected by enzymatic and chemical oxidation in vitro. *J. Food Sci.*, 50, 1505-1506.

Papadopoulou, C., Soulti, K., Roussis, I. G., (2005). Potential antimicrobial activity of red and white wine phenolic extracts against strains of *Staphylococcus aureus*, *Escherichia coli* and *Candida albicans*. *Food Technol. Biotechnol.*, 43, 41-46.

Pastorkova, E., Zakova, T., Landa, P., Novakova, J., Vadlejch, J., Kokoska, L., (2013). Growth inhibitory effect of grape phenolics against spoilage yeasts and acetic acid bacteria. *Int. J. Food Microbiol.*, 161, 209-213.

Pezet, R., (1998). Purification and characterization of a 32-kDa laccase-like stilbene oxidase produced by *Botrytis cinerea* Pers.:Fr. *FEMS Microbiol. Lett.*, 167, 203-208.

Pezet, R., Pont, V., Hoang-Van, K., (1991). Evidence of oxidative detoxification of pterostilbene and resveratrol by a laccase-like stilbene oxidase produced by *Botrytis cinerea*. *Physiol. Mol. Plant Pathol.*, 39, 441-450.

Ponzoni, C., Beneventi, E., Cramarossa, M. R., Raimondi, S., Trevisi, G., Pagnoni, U. M., Riva, S., Forti, L., (2007). Laccase-catalyzed dimerization of hydroxystilbenes. *Adv. Synth. Cat.*, 349, 1497-1506.

Postolache, E., Popescu, C., Ciubucă, A., Râpeanu, G., Bulancea, M., (2012). Dynamics of oxidative enzyme activity during the white grapes winemaking. *J. Environ. Prot. Ecol.*, 13, 1608-1615.

Rabosto, X., Carrau, M., Paz, A., Boido, E., Dellacassa, E., Carrau, F. M., (2006). Grapes and vineyard soils as sources of microorganisms for biological control of *Botrytis cinerea. Am. J. Enol. Vitic.*, 57, 332-338.

Reguant, C., Bordons, A., Arola, L., Rozès, N., (2000). Influence of phenolic compounds on the physiology of *Oenococcus oeni* from wine. *J. Appl. Microbiol.*, 88, 1065-1071.

Reiss, R., Ihssen, J., Richter, M., Eichhorn, E., Schilling, B., Thony-Meyer, L., (2013) Laccase versus laccase-like multi-copper oxidase: a comparative study of similar enzymes with diverse substrate spectra. *PLOS ONE,* 8, DOI 10.1371/ journal.pone.0065633.

Sabel, A., Martens, S., Petri, A., König, H., Claus, H., (2013). *Wickerhamomyces anomalus* AS1: a new strain with potential to improve wine aroma. *Ann. Microbiol.*, DOI 10.1007/s13213-013-0678-x.

Schouten, A., Wagemakers, L., Stefanato, F. L., van der Kaaij, R. M., van Kan, J. A. L., (2002). Resveratrol acts as a natural profungicide and induces self-intoxication by a specific laccase. *Mol. Microbiol.*, 43, 883-894.

Scola, G., Conte, D., Spada, P. W. D. S., Dani, C., Vanderlinde, R., Funchal, C., Salvador, M., (2010). Flavan-3-ol compounds from wine wastes with in vitro and in vivo antioxidant activity. *Nutrients,* 2, 1048-1059.

Servili, M., de Stefano, G., Piacquadio, P., Sciancalepore, V., (2000). A novel method for removing phenols from grape must. *Am. J. Enol. Vitic.*, 51, 357-361.

Sivertsen, H. K., Dewey, F. M., Heymann, H., (2005) Relationship between sensory descriptive analysis and levels of *Botrytis* antigens in dessert wines. *Am. J. Enol. Vitic.*, 56, 330-335.

Slomczynski, D., Nakas, J. P., Tanenbaum, S. W., (1995). Production and characterization of laccase from *Botrytis cinerea* 61-34. *Appl. Environ. Microbiol.*, 61, 907-912.

Soleas, G. J., Diamandis, E. P., Goldberg, D. M., (1997). Resveratrol: a molecule whose time has come? and gone? *Clin. Biochem.,* 30, 91-113.

Sponholz, W. R., (2000). Suberase: eine biotechnologische Möglichkeit Korken zu reinigen. Schweiz. *Zeitschr. Obst- und Weinbau,* 24, 621-625.

Stevenson, D. E., Hurst, R. D., (2007) Polyphenolic phytochemicals – just antoxidants or much more? *Cell. Mol. Life Sci.*, 64, 2900-2916.

Strong, P. J., Claus, H., (2011). Laccase: a review of its past and its future in bioremediation. *Crit. Rev. Environ. Sci. Technol.,* 41, 373-434.

Stuart, J. A., Robb, E. L., (2013). Bioactive polyphenols from wine grapes. Springer briefs in cell biology. New York, Heidelberg, Dordrecht. London: Springer.

Theobald, S., Pfeiffer, P., Zuber, U., König, H., (2008). Influence of epigallocatchin gallate and phenolic compounds from green tea on the growth of *Oenococcus oeni. J. Appl. Microbiol.*, 104, 566-672.

Thilakarathna, S.H., Rupasinghe, H.P., (2013). Flavonoid bioavailability and attempts for bioavailability enhancement. *Nutrients*, 5, 3367-3387.

Tian, R. R., Pan, Q. H., Zhan, J. C., Li, J. M., Wan, S. B., Zhang, Q. H., Huang, W. D., (2009). Comparison of phenolic acids and flavan-3-ols during wine fermentation of grapes with different harvest times. *Molecules,* 14, 827-838.

Todaro, A., Palmeri, R., Barbagallo, R. N., Pifferi, P. G., Spagna, G., (2008). Increase of trans-resveratrol in typical Sicilian wine using β-glucosidase from various sources. *Food Chem.,* 107, 1570-1575.

Tosi, E., Feárizzi, B., Azzolini, M., Finato, F., Simonato, B., Zapparoli, G., (2012). Effects of noble rot on must composition and aroma profile of Aroma wine produced by the traditional grape withering protocol. *Food Chem.,* 130, 370-375.

Utkhede, R. S., Mathur, S., (2002). Biological control of stem canker of greenhouse tomatoes caused by *Botrytis cinerea. Can. J. Microbiol.*, 48, 550-554.

Uyama, H., Kobayashi, S., (2006). Enzymatic synthesis and properties of polymers from polyphenols. *Adv. Polymer Sci.*, 194, 51-67.

Viterbo, A., Yagen, B., Mayer, A. M., (1993a) Cucurbitacins, attack enzymes and laccase in *Botrytis cinerea. Phytochem.,* 32, 61-65.

Viterbo, A., Yagen, B., Mayer, A. M., (1993b). Induction of laccase formation in *Botrytis* and its inhibition by cucurbitacin – is gallic acid the true inducer? *Phytochem.,* 34, 47-49.

Vivas, N., Lonvaud-Funel, A., Glories, Y., (1997). Effect of phenolic acids and anthocyanins on growth, viability and malolactic activity of lactic acid bacterium. *Food Microbiol.*, 14, 291-300.

Walker, J. R. L., (1975). The biology of plant phenolics. Edward Arnold (Publishers) Limited, London.

Wigand, P., Tenzer, S., Schild, H., Decker, H., (2009). Analysis of protein composition of red wine in comparison with rosé and white wines by electrophoresis and high-pressure chromatography-mass spectrometry (HPLC-MS). *J. Agric. Food Chem.*, 57, 4328-4333.

Witayakran, S., Ragauskas, A. J., (2009). Synthetic applications of laccase in green chemistry. *Adv. Synth. Chem.*, 351, 1187-1209.

Zinnai, A., Venturi, F., Sanmartin, C., Quartacci, M. F., Andrich, G., (2013). Chemical and laccase catalysed oxidation of gallic acid: determination of kinetic parameters. *Res. J. Biotechnol.*, 8, 62-65.

Zivkovic, K., König, H., Claus, H., (2011). Wirkung von Bentonit auf die Laccase-Aktivität in Most und Wein. *Deutsche Lebensmittel-Rundschau,* 107, 575-582.

Zocca, F., Lomolino, G., Spettoli, P., Land, A., (2007). Deriphat 2-DE to visualize polyphenol oxidase in Moscato and Prosecco grapes. *Electrophoresis,* 28, 3992-3997.

Zouari, N., Romette, J. L., Thomas, D., (1987). Purification and properties of two laccase isoenzymes produced by *Botrytis cinerea. Appl. Biochem. Biotechnol.*, 15, 213-225.

In: Wine
Editor: Youssef El Rayess

ISBN: 978-1-63321-048-6

Chapter 7

Trends on Analytical Methods for Resveratrol, a Major Bioactive Compound in Wines

Maria João Cabrita*, Nuno Martins, Bruno Soares, Ana Maria Costa Freitas and Raquel Garcia
ICAAM - Instituto de Ciências Agrárias e Ambientais Mediterrânicas, Universidade de Évora, Núcleo da Mitra, Évora, Portugal

Abstract

Resveratrol (3,5,4-*trans*-trihydroxystilbene) is a member of the stilbene group of phenolic compounds, comprising two aromatic rings linked by an ethylene bridge and is considered one of the most promising bioactive compounds due to its bioactivity. Wine is one of the main source of this bioactive compound in Mediterranean diet. Several authors have dedicated their research to the relationship between diet and health, and concerning wine and health studies, ''French Paradox'' was the starting point.

Resveratrol is found in seeds and skins of grapes and in wines. The red wines are richer than the white ones due to differences in winemaking technology. The amount of resveratrol found in wines varies widely, since a huge number of factors affect its biosynthesis (grape variety, geographic region, climatic conditions, agronomic factors) and its extraction from the solid part of the grapes into the wines (technological practices).

Numerous analytical methods are described in literature to determine resveratrol content in wines, but apart few methods reporting direct injection, a previous sample preparation step is required before the injection into a chromatographic system. Liquid chromatography coupled with a fluorescence detector, a diode array detector or a mass spectrometry detector are the most used separation and identification tools, since the use of gas chromatography also requires a derivative step. Recently, the development of molecular imprinting technology based on a mechanism of molecular recognition, and the further applications of the generated molecularly imprinted polymers (MIPs) as selective

* Corresponding author: ICAAM - Instituto de Ciências Agrárias e Ambientais Mediterrânicas, Universidade de Évora, Núcleo da Mitra, Ap. 94, 7002-554 Évora, Portugal. Email: mjbc@uevora.pt.

sorbents in Solid Phase Extraction (MISPE) or magnetic MIPs (MMIPs) have provided a new versatile tool for the highly selective quantification of resveratrol in wines.

In this chapter, the biosynthesis of resveratrol in grapes, the factors affecting resveratrol content in wines and the current analytical methods for resveratrol determination in wines, will be revisited.

However, the focus of this review will be on the new trends in analytical methods for resveratrol in wines, namely the use of molecularly imprinted polymers (MIP) as a recent and innovative sample preparation methodology.

Keywords: Resveratrol, analytical methods, molecularly imprinted polymers

INTRODUCTION

Phytoalexines are a group of compounds belonging to the stilbene family, whose skeleton is based on the 1,2-diphenylethylene structure (Figure 1). Resveratrol (3,5,4'-trihydroxystilbene) and piceid (3,5,4'-trihydroxystilbene-3-β-D-glucoside) are the major stilbenes found in wines existing in the *trans* and *cis* form, while another stilbene, astrigin (3,5,3',4'-tetrahydroxystilbene-3-β-D-glucoside) seems to exist only in the *trans* form.

Grapes and wine are considered the most important human dietary sources of stilbenes. Resveratrol, a stilbene derivative, is the one gaining more attention over the last years due to its numerous health-promoting properties (Guerrero et al., 2010). The increasing interest in the relationship between diet and health is mainly related to bioactive compounds, being resveratrol one of the most biologically active polyphenols contained in wine, therefore, one of the most popular among scientific community.

The "French Paradox" was the starting point concerning wine and health issues. The strikingly low incidences of coronary heart diseases in France, despite intake of a high-fat diet, have been attributed to the consumption of red wine containing high levels of polyphenolic compounds (Renaud & De Lorgeril, 1992). Although the "French Paradox" was afterwards revisited with the possible involvement of ethanol in the protective effect against cardiovascular diseases (Belleville, 2002), the importance of bioactive compounds such as resveratrol remains unquestionable. Nowadays, it is well known that alcohol favorably modifies the lipid pattern by decreasing total plasma cholesterol, in particular low-density lipoproteins (LDL), and by increasing high-density lipoprotein cholesterol. Cardiovascular risk reduction seems to be linked largely to the effect of non-alcoholic components, mainly resveratrol and other polyphenols, on the vascular wall and blood cells (Gresele et al., 2011).

The health promoting properties of resveratrol were recently reviewed by Férnandez-Mar et al. (2012). These authors have extensively revisited the antioxidant, cardioprotective, anticancer, antidiabetic, neuroprotective and anti-aging activities attributed to resveratrol, but it is important to point out that, despite a large body of evidence, human studies are still lacking and both preventive and therapeutic value of resveratrol remains to be fully elucidated.

Resveratrol in wines was first reported in 1992 by Siemann and Creasy, either for red and white wines. Ever since, research on resveratrol content in different wines all over the world has increased, but actually the focus is clearly on the benefits of this compound to human health.

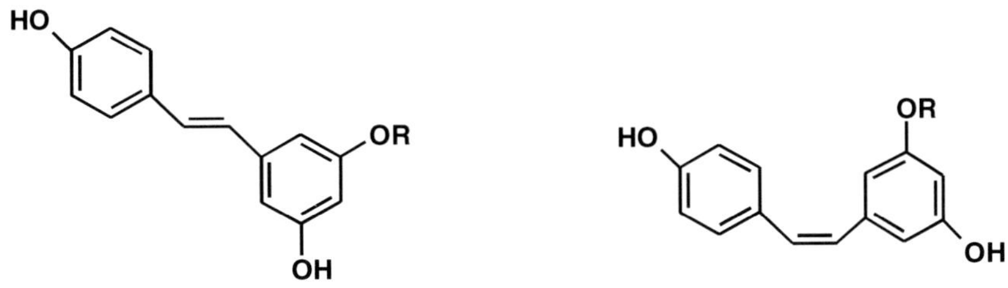

trans-resveratrol (R=H) **trans-resveratrol (R=H)**

trans-piceid (R=Glc) **trans-piceid (R=Glc)**

Figure 1. Structures of resveratrol and piceid.

BIOSYNTHESIS OF RESVERATROL

Resveratrol also called as 3,5,4'- trihydroxystilbene is the most well-known member of a family of substances called stilbenes and is present in a great variety of plant species used for human consumption, such as peanuts, berries and grapes. As discussed previously, the most relevant source of resveratrol for humans is probably grapes and this bioactive polyphenol is also found in wines however the highest amount of this compound is found in red wines. In grapes, resveratrol is synthesized almost entirely in the skin and the synthesis peak occurres before the grapes reach maturity (Stervbo et al., 2007).

The precursors of resveratrol are 4-coumaroyl-CoA and three molecules of malonyl-CoA which are present in all plants. Malonyl-CoA is produced by the elongation of acetyl-CoA units and 4-coumaroyl-CoA is obtained from phenylalanine which, in plants, is synthesized from sugars via the shikimate pathway. As depicted in figure 2, the biosynthetic pathway of resveratrol initiates with the conversion of phenylalanine to cinnamic acid by the loss of its amino group which occurs through an oxidative deamination reaction which is catalyzed by the phenylalanine ammonia lyase. The cinnamic acid is enzymatically hydroxylated to *p*-coumaric acid which, in turn, by means of a specific CoA ligase generates *p*-coumaroyl-CoA. Further, the condensation of *p*-coumaroyl-CoA with three molecules of malonyl-CoA by the activity of the stilbene synthase leads to the release of four moles of CO_2 for each mole of resveratrol produced (Soleas et al., 1997). The final enzyme that participates in the biosynthesis of resveratrol is activated in response to exogenous stress factors, mainly injury, ultra violet irradiation and chemical signals from pathogen fungi. Thus, resveratrol content reaches the maximum level approximately 24h after stress exposure and decay after 42-72h due to the activation of stilbene oxidase (Stervbo et al., 2007). Since the response to exogenous stress factors has a great impact on the production of resveratrol their levels in wine are predictably influenced by the viticulture and enological practices, among other factors that will be described in detail in the section 3. So, in conclusion, it is difficult to predict the amount of resveratrol in wines because of many factors affecting resveratrol biosynthesis.

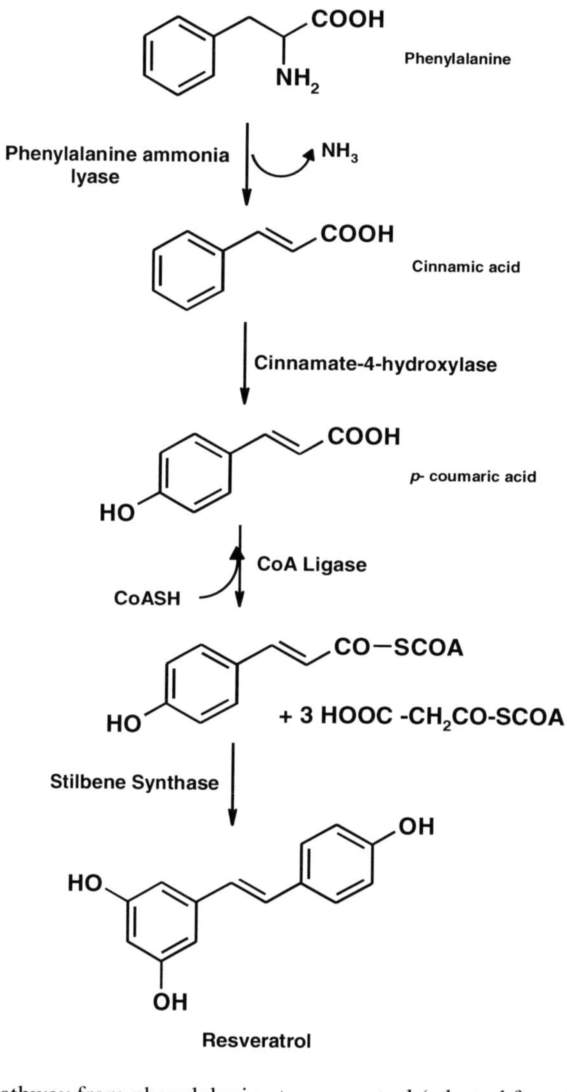

Figure 2. Biosynthetic pathway from phenylalanine to resveratrol (adapted from Soleas et al., 1997).

Factors Affecting Resveratrol Content in Wines

Resveratrol is found in skins and seeds of grapes, and hence, in the grape juice and wine. The amounts of resveratrol in wines vary widely depending on many factors: grape variety, harvest year, geographic region, agronomic factors, climatic conditions, UV light exposure of grape clusters, plant stress conditions and oenological practices (Soleas et al., 1997; Guerrero et al., 2010).

Resveratrol is an antifungic or antimicrobian molecule naturally produced by the plant as self-defense agent, and a natural compound of grape vines (Langcake & Pryce, 1976). It has been reported that grapes attacked by powdery mildew (*Uncinula necator*) have higher *trans*-resveratrol and *trans*-piceid content (Romero-Pérez et al., 2001), but a high pressure of the fungus *Botrytis cinerea* in the vineyard led to a decrease of resveratrol in grape berries (Roldán et al., 2003). This is probably due to the fact, that the *Botrytis* generates an enzyme (stilbene-oxidase) that oxidizes most of the resveratrol to its corresponding metabolites. *Botrytis* causes shriveling of the grapes, and consequently concentrates the contents of the berries, but at the same time the stilbene-oxidase decreases resveratrol and piceid concentrations, in which oxidation is obviously the predominant process.

Once resveratrol is produced in response to exogenous stress factors, it is expected that the levels in wine vary between regions and vintages (Goldberg et al., 1995; Goldberg et al., 1996; Martínez-Ortega et al., 2000; Stervbo et al., 2007). More specifically, cooler and more humid climatic conditions can lead to higher *trans*-resveratrol content (Kolouchova-Hanzlikova et al., 2004), where moist climatic conditions are more favorable to fungi development. When comparing Cabernet Sauvignon wines from different regions, Li et al. (2011) found much more resveratrol in wetter areas than in drier areas of China. These results are in agreement with previously published results which showed that the higher north on northern hemisphere, the higher average levels of *trans*-resveratrol (Stervbo et al., 2007).

A study conducted by Threlfall et al., (1999) confirmed that the resveratrol concentration differs according to variety, whereas UV light exposure of grape clusters can increase the content of resveratrol in some wines. In this regard, Guerrero et al., (2010) conducted a study where postharvest grapes have been treated with ultraviolet light to produce stilbene enriched grapes that were later used in a conventional winemaking process to obtain a red wine enriched in stilbenes, without loss of quality. These authors reported an enrichment of resveratrol both in grapes and in wines, although pointing out that the use of this technology to obtain enriched wine still needs to be optimized.

Resveratrol is present mainly in grape skins, so the enological technique used determines its concentration in the wines. Various factors during the winemaking process affect the levels of resveratrol in the wines like increased temperature, higher levels of SO_2 and/or decreased pH leading to higher resveratrol levels (Mattivi & Nicolini, 1993; Trela & Waterhouse, 1996; Netzel et al., 2003; Gambuti et al., 2004). Finning, a worldwide used technique for clarifying wines can reduce resveratrol content depending on the type of agent used. Polyvinylpyrrolidone (PVPP) and carbon can lead to a decrease in resveratrol content (Threlfall et al., 1999). Wines resulting from longer maturation periods contain a higher content of resveratrol, the concentration of this compound being also higher in red wines as compared to white wines.

A recent study comparing traditional red wine technology with Ganimede technology have demonstrated that the traditional vinification was slightly better than the Ganimede to yield the non-anthocyanin phenolics, namely *trans* and *cis*-resveratrol (Bai et al., 2013). Briefly, in the Ganimede technology, fermentation takes place in a fermenter specially designed to use the CO_2 produced during fermentation to break the floating cap, facilitating the maceration of the skins with the juice. On the other hand, while comparing thermovinification with a more traditional technology, Atanacković et al., (2012) did not found a clear correlation between the winemaking technique and resveratrol content for all varieties under study. Earlier, Sun et al. (2003) have found that resveratrol content in wines made by traditional technologies (fermentation with and without stems) were higher than those in wines made by carbonic maceration, although they found no significant differences between stem and non-stem contact wines, suggesting that skins are indeed the major source of resveratrol. They also found that prolonged maceration at the end of alcoholic fermentation did not affect resveratrol content in wines. It seems that the extraction of resveratrol from grape skins to wine is completed during alcoholic fermentation. López-Alfaro et al. (2013) studied the influence of Pulsed Electric Field (PEF) treatment on resveratrol content of three different grape varieties. PEF technology is based on the application of short duration pulses of high electric field strengths (0.1–50 $kVcm^{-1}$) to products placed between two electrodes at low temperature. Application of PEF in grape skin before the maceration–fermentation step increases the extraction of phenolic compounds during the vinification process, and the authors concluded that this technology could lead to the enrichment of must contents on resveratrol and piceid.

Regardless all the factors affecting resveratrol levels in wines, it seem consensual that variety plays the dominant role and researchers from all over the world have published results regarding the levels of resveratrol in wines from different varieties. Gerogiannaki-Christopoulou et al., (2006) have found for Greek varieties concentrations of *trans*-resveratrol from 0.352 to 1.99 mgL^{-1} in 13 red wines and from 0.005 to 0.57 mgL^{-1} in 18 white wines. Previously, Kallithraka et al., (2001) have reported for 29 Greek red wines values of *trans*-resveratrol slightly higher, ranging from 0.508 to 2.534 mgL^{-1}. In red and rosé wines from Aragon (Spain), *trans*-resveratrol content ranged from 0.32 to 4.44 mgL^{-1} in red wines and from 0.12 to 2.80 mgL^{-1} in rosé wines and *cis*-resveratrol levels ranged from 0.20 to 5.84 $mg.L^{-1}$ in red wines and from 0.02 to 3.17 mgL^{-1} in rosé wines (Abril et al., 2005). For Galician (Spain) red and white wines (24 each) values of resveratrol ranging from 0.65 to 7.95 mgL^{-1} (*trans*-resveratrol, white wines) and 0.06 to 36.13 mgL^{-1} (*trans*-resveratrol, red wines) were reported while maximum levels of *cis*-resveratrol were 0.19 and 0.24 mgL^{-1} for white and red wines respectively (Feijóo, et al., 2008). In nineteen Italian red wines, resveratrol, either the *trans*- or *cis*-form, have been detected ranging from 0.63 to 3.39 mgL^{-1} and from not quantifiable to 4.93 mgL^{-1}, respectively (Buiarelli et al., 2007) Regarding Portuguese varieties Ribeiro de Lima et al., (1999) reported for 50 varietal white wines maximum values of 2.1 and 1.7 mgL^{-1} for *trans* and *cis*-resveratrol, and 5.9 and 9.5 mgL^{-1} for *trans* and *cis*-resveratrol in 34 red varietal wines. Baptista et al., (2001) compared wines from Portugal mainland (9 wines) versus Azores Islands (3 wines) and found values ranging from 0.63-2.87 mgL^{-1} for *trans* and 0.20-0.72 mgL^{-1} for *cis*-resveratrol in Portugal mainland and 2.35-5.21 mgL^{-1} and 0.20-1.70 mgL^{-1} for *trans* and *cis*-resveratrol in Azores Island.

Merlot, Cabernet Sauvignon and Syrah are, probably, the most cultivated red wine varieties. Several authors reported resveratrol content, especially *trans*-resveratrol from wines of these varieties in different countries like Serbia (Atanacković et al., 2012), Italy (Mattivi, 1993), Hungary (Marck et al., 2005), Brazil (Souto et al., 2001), Japan (Sato et al., 1997), Australia (Shao et al., 2003) and United States (Gu et al., 1999). In France, resveratrol content of wines from these varieties are not reported, probably because French wines are blends and not varietal wines.

In spite of all factors that can affect resveratrol content in wines, it should not be exclude the influence of analytical methods used for extraction, separation and even quantification on the differences found among different authors when comparing results. Regardless all factors that can influence resveratrol content of wines, levels of *trans*-resveratrol are always higher than *cis*-resveratrol levels, except for those results reported by Ribeiro de Lima (1999) for red wines. In fact, *cis*-resveratrol found in wines is generally considered to be due to the photochemical isomerization of some *trans*-resveratrol during winemaking process (Jeandet et al., 1995), or it can arise from combined forms existing in grapes by hydrolysis during winemaking process (Mattivi et al., 1995).

CURRENT ANALYTICAL METHODS

The preferred analytical method for resveratrol analysis is high-performance liquid chromatography (HPLC) with diode array UV–Vis detection (Malovaná et al., 2001; Ružić et al., 2011; Atanackovíc et al., 2012), fluorescence (Mercolini et al., 2008; Atanackovíc et al., 2012), electrochemical detection (Kolouchova-Hanzlikova et al., 2004) or mass spectrometry (Wang et al., 2002; Marck et al., 2005; Buiarelli et al., 2007). As an alternative to HPLC, capillary electrophoresis can also be used (Gu et al., 2000; Gao et al., 2002). Methods using gas-chromatography-mass spectrometry (GC/MS), although providing excellent sensitivity and specificity, requires a derivatization step of resveratrol prior to analysis, in order to increase its thermal stability and volatility (Wang et al., 2002)

Using liquid chromatography, separation is achieved by using a gradient with an organic solvent and an acidified aqueous solvent. The most used organic solvent is acetonitrile (Ribeiro de Lima et al., 1999; Vitrac et al., 2005; Vian et al., 2005; de Villiers et al., 2004; Boutegrabet et al., 2011) or methanol (Nave et al., 2007; Ružić et al., 2011; Atanackovic et al., 2012). The most used organic acid to acidified water are acetic acid (Ribeiro de Lima et al., 1999; de Villiers et al., 2004; Vitrac et al., 2005; Vian et al., 2005) or formic acid (Nave et al., 2007; Boutegrabet et al., 2011; Ružić et al., 2011; Atanackovic et al., 2012). The main recent advances in the chromatographic field are mainly related to ultra-high performance liquid chromatography. The key advantages of UHPLC are the increased speed of analysis, higher separation efficiency and resolution, higher sensitivity and much lower solvent consumption as compared to other analytical approaches. This is all enabled by specially designed instruments and sub-2-micron particle packed analytical columns.

But the major constrains of bioactive compounds analysis is sample preparation methods that still remains highly labor-intensive and time-consuming. In respect to resveratrol, several authors reported direct injection of wine (Vian et al., 2005; Ružić et al., 2011; Atanackovic et al., 2012;), however due to the high complexity of the wine matrix a cleanup procedure is often required, being sample preparation methods based on liquid–liquid extraction (Bravo et al., 2008; Malovaná et al., 2001), solid-phase extraction (Kallithraka et al., 2001; Baptista et al., 2001; Rodríguez-Delgado et al., 2002) or solid-supported liquid-liquid extraction (Nave et al., 2007) often reported, although these techniques requires a large amounts of sample, large volumes of potentially toxic solvents and frequently presents lack of selectivity and specificity, so that a large amount of matrix interferences can be eluted simultaneously with the target analytes, thus becoming necessary a clean-up step before quantitation.

So, it is mandatory to develop analytical approaches allowing the detection of analytes with high selectivity and sensitivity. In this context, miniaturized sample preparation analytical techniques, namely, solid-phase microextraction by direct immersion (Viñas et al., 2008), stir bar sorptive extraction (Viñas et al., 2008) and microextraction by packed sorbent (MEPS) (Gonçalves & Câmara, 2011), have gained attention due to their many special features over classical approaches. Among many advantages, the use of little or no solvent, increasing sensitivity of analysis, and user-friendly system, should be pointed out.

More recently, a cleanup procedure based on molecularly imprinted polymers (MIP) solid-phase extraction (MISPE) has emerged offering a possibility to circumvent the drawbacks of traditional techniques referred above. The idea of using a target molecule as a template to direct assembly of its own recognition site was first proposed in 1940 by Pauling. Molecularly imprinting technology is an ingenious way of producing recognition sites for specific analytes in synthetic polymers. Nowadays, there has been a growing interest in the use of such porous materials as adsorbents in chromatographic applications due to their versatility, high specificity and recognition ability as well as other features, such as stability in organic solvents, ease and low cost for preparation (Ferrer et al., 2000; Mahony et al., 2005; Vasapollo et al., 2011). Several polymerization methods have been developed to produce imprinted polymers, such as "bulk" (Mosbach & Ramström, 1996; Yan & Row, 2006), suspension (Mayes & Mosbach, 1996; Ansell & Mosbach, 1997), two-step swelling (Hosoya et al., 1996; Ensing & de Boer, 1999), precipitation, emulsion and core-shell (Pérez-Moral et al., 2000, 2001), film (Malitesta et al., 1999), aerosol polymerization (Vonderbruggen et al., 1996) and polymerization on silica particles (Dai et al., 1997; Katz & Davis, 2000). Each of these mechanisms involves the control of different parameters during the synthesis yielding polymers with different properties, particularly in terms of morphological characteristics.

Molecularly Imprinted Polymers

The synthesis of MIPs leads to the formation of three-dimensional macromolecular structures through a polymerization process which uses a template molecule and a functional monomer for co-polymerization in the presence of a crosslinking agent and an initiator in a porogenic solvent, as depicted in figure 3 (Ramström & Mosbach, 1999; Vasapollo et al., 2011).

Different aspects must be considered for the template selection but seems to be essential the presence of some groups, which must be stable under the synthetic conditions in order to enable the occurrence of the polymerization (Regal et al., 2012). The suitable choice of the functional monomers is also crucial for the success of molecular recognition, and then this selection should be made according to the template molecule, allowing the formation of a stable template-monomer complex (Ramström & Mosbach, 1999; Qian et al., 2010; Scorrano et al., 2011; Regal et al., 2012). The type and amount of cross-linking agent used in the synthesis of MIPs also needs optimization due to their impact on the selectivity of MIPs since its play a crucial role on the stability of the recognition sites and also on the maintenance of the three-dimensional structure and the chemical functional groups after the removal of the template enabling the rebinding of the template molecule (Regal et al., 2012). Another relevant feature is the choice of the porogenic solvent (chemical nature and amount), because it influences polymer morphology that directly affects the performance of MIPs (Vasapollo et al., 2011). The radicalar initiator also plays an important role on the MIPs synthesis, thus a judicious choice is also essential. In general, a lower percentage amount of initiator generate polymers with good quality imprinting cavities while large quantities of initiator leads to the formation of more rigid polymers with well-defined shape and higher specificity of imprinting cavities (Piletska et al., 2009).

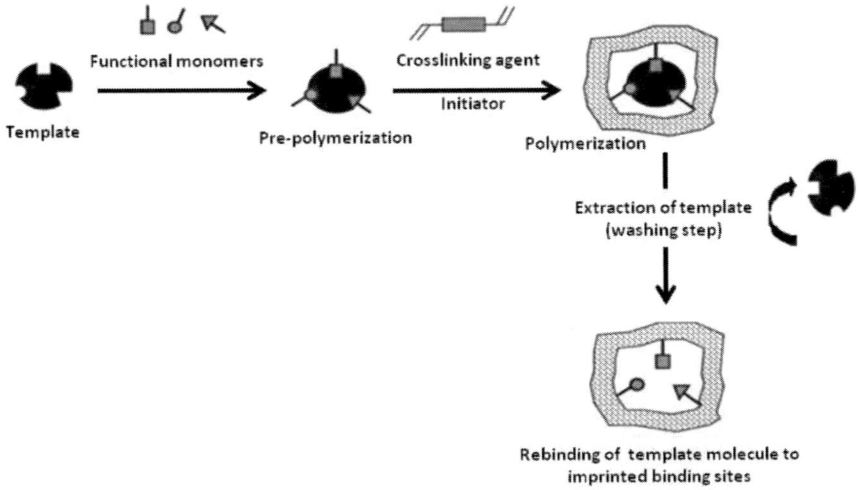

Figure 3. General scheme of the principle of molecularly imprinted polymer technique (adapted from Garcia et al., 2011).

The polymerization takes place in solution, usually initiated by a free radical initiator ensuring that the functional groups of the template molecule are interacting with complementary functional groups of the polymer-forming components (Whitcombe & Vulfson, 2001). Molecular imprinting is classified according to the interaction between monomer and template during the polymerization process. Nowadays the most common methodologies for the preparation of MIPs are based on non-covalent (Lok & Son, 2009), covalent, semi-covalent and metal-mediated interactions (Regal et al., 2012).

After the polymerization is completed, the template molecule is removed by chemical reaction or extraction techniques, through extensive washing steps in order to disrupt the interactions between the template and the monomers, leaving complementary recognition sites in the polymer network (Vasapollo et al., 2011). The size, shape and positions of the functional groups in the recognition sites generated during the polymer synthesis are complementary to those of the original analyte. Thus, molecularly imprinted polymers (MIPs) rebind their original analytes in preference to related molecules. Usually, intermolecular interactions like dipole–dipole, hydrogen bonds, and ionic interactions between the functional groups and template molecule present in the polymer matrix are involved on the molecular recognition phenomenon (Ramstrom & Mosbach, 1999).

MIPS have been broadly used for the imprinting of pesticides, peptides, pharmaceuticals, carbohydrates and other molecules of biological interest (Ramström et al., 2001), however, comparatively little attention has been given to the phenolic compounds, particularly the *trans*-resveratrol which have nonetheless great interest.

As discussed previously, the use of sample preparation methodologies that relies with mechanisms of molecular recognition of the target molecule, as is the case of MIPs, offers a potentially higher degree of sample clean-up than that achieved using the conventional sample preparation methodologies leading to significantly improvements in selectivity allowing to reach lower detection limits, which has a huge impact on the analytical determination.

Aiming to introduce selective methodologies for the selective recognition of *trans*-resveratrol, the pioneered studies of Cao and co-workers concerning the evaluation of new MIPs for the extraction of the target compound from *Polygonum cuspidatum* have explored the possibility of preparing molecularly imprinted materials with predefined selectivity for the *trans*-resveratrol template molecule (Cao et al., 2006). Those investigations have been focused initially on the synthesis of four different molecularly imprinted polymers using resveratrol as the template, methacrylic acid (MAA) or acrylamide (AA) as functional monomers, 2-2'-azobisisobutyronitrile (AIBN) as the initiator by thermo- or photo-induced polymerization, and subsequently on the evaluation of the molecular recognition capability of those materials by means of rebinding and selectivity experiments. Within these studies it was possible to conclude that AA-based MIP synthesized by photo-induced polymerization is the most promising imprinting material since it shows the best properties to be used as selective adsorbent for MISPE applications for the extraction of resveratrol (Cao et al., 2006).

Afterwards, Ma and coworkers intended to isolate *trans*-resveratrol from matrix components in the *Polygonum cuspidatum* extract have studied the possibility of preparing selective MIPs for the recognition of *trans*-resveratrol by non covalent approach using different functional monomers (MAA, AA, 4-vinylpyridine (4-VP)), ethylene glycol dimethacrylate (EDGMA) as crosslinker, AIBN as radical free initiator and ethyl acetate, acetone and a mixture of ethyl acetate:acetone as porogen. Further studies concerning the affinity and selectivity of the imprinted systems were assessed by means of rebinding studies. The results obtained showed that the nature of the functional monomer and the porogen have a significant influence on the molecular recognition behavior of the imprinting material being the MIP prepared using 4-VP as functional monomer and acetone as porogen, the porous material that presents superior recognition properties, enabling the separation of *trans*-resveratrol from matrix components in the *Polygonum cuspidatum* extract. Additionally, these authors have also demonstrated that the resveratrol-MIP allows the selective separation of *trans*-resveratrol from its structural analogs, namely *cis*-resveratrol and bisphenol A and for other constituents of Chinese herbs, like *trans*-polydatin and catechol proving the potential utility of *trans*-resveratrol imprinted polymers for SPE applications (Ma et al., 2007).

Research studies of Schwarz et al. (2011a) aimed to isolate and purify the *trans*-resveratrol bioactive compound from peanut byproduct using molecular imprinting technology. *Trans*-resveratrol-based MIP (MIPRES) were synthesized by self-assembly non-covalent approach using this bioactive stilbene compound as template, 4-VP as functional monomer, EGDMA as cross-linker and AIBN as free radical initiator. The ability of these imprinted materials to recognize the template molecule has been evaluated using static binding isotherm measurements being that evaluation studies also extended to cross-reactivity experiments with several structurally related analogues, namely some naturally occurring polyphenols such as caffeic acid, catechin and *trans*-piceid. The application of MIPRES as sorbent in MISPE procedure for the selective extraction of *trans*-resveratrol from peanut press waste, a byproduct of peanut oil preparation, has been successful leading to a significant cleanup of the peanut meal extract and a consequently enrichment of *trans*-resveratrol, which in some experiments approximate 60-fold enrichment of this bioactive compound from the crude feedstock. These studies also proved that the achieved enrichment can be solely due to the *trans*-resveratrol imprinting effect of MIPRES (Cao et al., 2006). Based on those studies, the MIPRES seems to be very promising as sorbent for SPE extraction purposes since enables the selective pre-concentration of the target *trans*-resveratrol from complex food processing or agricultural by-product matrix (Schwarz et al., 2011a).

Recently, Schwarz and co-workers (2011b) have studied the design of molecularly imprinted polymers for the selective recognition of *trans*-resveratrol using molecular modeling and NMR spectroscopic titration techniques in order to select the most suitable functional monomer and to optimize the molar ratio of the template to functional monomer. These studies enabled the selection of 4-vinylpyridine (4-VP) as functional monomer and a molar ratio 3:1 (4-VP:*trans*-resveratrol)) as an optimal value to achieve a stable pre-polymerization complex. Based on ^{1}H NMR spectroscopy titration studies, interactions between the template and the functional monomer have been studied which occurs whereby the interaction of pyridine moiety and the phenolic OH groups leading to the formation of multilayer clusters via aromatic intermolecular O-H...N hydrogen bonding interactions (Schwarz et al., 2011b). Attending to these results, a *trans*-resveratrol imprinted polymer have been synthesized by the non-covalent self-assembly approach using 4-VP as functional monomer, EGDMA as crosslinker, AIBN as free radical initiator, which shows a highly specific molecular recognition of the template molecule. In this study, MIPs were also prepared for some *trans*-resveratrol congeners, such as some compounds belonging to hydroxylated stilbene family using a similar experimental procedure for comparative purposes. In order to evaluate the molecular recognition of these molecularly imprinted materials, equilibrium binding studies were performed and the influence of polyphenolic hydroxyl number and position has been also assessed (Schwarz et al., 2011b). The data obtained showed that the molecular recognition of the porous materials was greatly influenced by the presence of aromatic OH groups indicating the crucial role of the presence of such groups in the *meta* and or *para* positions. This effect is evident on the interaction of *trans*-resveratrol molecule with the MIP cavity, which occurs through three *meta* and *para* positioned aromatic OH groups, showing a binding capacity of 12.36 $\mu molg^{-1}$ and an imprinting factor of 2.35 showing the high recognition capabilities of the MIP to this template. Competitive static cross-reactivity studies with several *trans*-resveratrol analogues were also carried out and the results have indicated that the MIP synthesized preferentially binds *trans*-resveratrol over its structurally similar analogues attesting their selectivity (Schwarz et al., 2011b).

More recently, Chen and co-workers have explored the suitability of employing a resveratrol-magnetic molecularly imprinted polymer (MMIP) for the extraction and quantification of this bioactive polyphenol from wine samples (Chen et al., 2013). In this approach, a MMIP has been prepared using a multistep procedure based on surface- imprinted polymerization methodology with a super paramagnetic Fe_3O_4 core-shell nanoparticle as supporter using a resveratrol counterpart-rhapontigenin as dummy template molecule, acrylamide (AM) as functional monomer, EGDMA as cross-linker, AIBN as initiator and acetonitrile as porogen. Experimentally, the optimum molar ratio of template/AM/EGDMA was 1:5:30 for the synthesis of MMIPs. Physico-chemical and morphological characterization of the novel MMIPs were carried by scanning electron microscopy (SEM), Fourier Transform Infrared spectroscopy (FTIR), X-Ray diffraction and vibrational sample magnetometer (VSM). In order to evaluate the utility of those magnetic polymers as selective adsorbents for Solid Phase Extraction (SPE) applications, some experiments were performed for the determination of molecular adsorption and the evaluation of their selectivity. The results revealed that the MMIPs showed high adsorption capacity and high selectivity for resveratrol.

Furthermore, this work proves that the use of a dummy template molecule for the synthesis of MMIPs has no effect on the recognition capacity and selectivity of the resveratrol molecule enabling to reduce the occurrence of bleeding phenomenon. As far as we know, this work was the first attempt to synthesize a MMIP using a dummy template for the quantification of resveratrol in such food matrix. The imprinting technology developed in this work implies the use of an external magnetic field to collect the MMIP from the wine samples, with a subsequently desorption step followed by the quantitation of resveratrol contents. The usual packing MISPE procedure is avoided, which makes this new methodology highly suitable and reliable for the enrichment and determination of this bioactive compound from real samples (Chen et al., 2013).

Finally, Hashim et al. (2013) reported an off-line MISPE method for use in the analysis of *trans*-resveratrol in red wine. Briefly, a pre-polymerization complex between the molecular template *trans*-resveratrol and 4-Vinylpiridine was allowed to self-assemble in ACN/EtOH (porogen solvent), prior to the addition of the cross-linking monomer (EGDMA) and the free radical initiator (AIBN). Thermal polymerization at 50◦C for 24 h was followed by an annealing period up to 24 h at 60◦C after which the polymer monoliths were crushed, ground and sieved to a particle size ranging between 63 and 90 μm. The molecular template was removed by repeated washing in an acidic solution of MeOH:AcOH. The NIP was prepared in the same manner but in the absence of the molecular template. MISPE and NISPE cartridges were subsequently assembly and used for wine samples preparation. Elution of the MISPE cartridge increased the yield of *trans*-resveratrol in comparison to the NISPE cartridge, with an imprinting factor (IF = MIP/NIP) of 32. Moreover these authors reported a total recovery of *trans*-resveratrol from the MISPE with spiked wine sample of 99.6%, and reproducible results upon regeneration of the MIP solid phase (Hashim et al., 2013)

Nowadays, the potential of using this kind of polymers for multiclass analysis of different compounds belonging to several classes, but having structural or functional analogies, is being investigated. Euterpio et al. (2013) proposed a method for determination of *trans*-resveratrol and related phenolics, like quercetin in beverages such as wine. They prepared a MIP obtained by bulk polymerization using *trans*-resveratrol as template, 4-VP as functional monomer, EGDMA as cross-linker and AIBN as radical initiator, and also a NIP without adding the template. They performed extractions from spiked wine sample using MISPE cartridges and found excellent recoveries of *trans*-resveratrol and quercitin. Authors also pointed out the advantage of using MISPE technology for the cleanup of the complex matrices allowing to lowering limits of detection in chromatographic separation.

Conclusion

Resveratrol is one of the most important bioactive compounds existing in grapes and wines. Several analytical methodologies can be used to quantify this compound, although liquid chromatography is one of the most used techniques for this purpose. Coupling liquid chromatography with mass detection (MS) allow an unambiguously identification of the compound in complex matrices like wine, but a previously sample preparation method is also required. Over the last years, a great number of new or adapted sample preparation techniques are emerging, some of them presenting a huge potential. One of the most promising techniques is the use of molecularly imprinting polymers for solid-phase extraction.

Molecularly imprinting polymers are an emerging technology in the sample preparation field, and quickly moving on beyond SPE, the most traditional way of using these specially designed polymeric materials. Recent advances in MIP technology allow a greater deployment of tailor-made MIPs for health beneficial target molecules, such as resveratrol. The development of generic protocols for the extraction, detection and enrichment of bioactive compounds from wines can indeed open new frontiers in wine industry.

ACKNOWLEDGMENTS

This work is funded by FEDER funds through the Operational Programme for Competitiveness Factors - COMPETE and national funds through FCT - Foundation for Science and Technology (Strategic Project PEst-C/AGR/UI0115/2011 and Project PTDC/AGR-ALI/117544/2010) and Programa Operacional Regional do Alentejo (InAlentejo) Operation ALENT-07-0262-FEDER-001871/ Laboratório de Biotecnologia Aplicada e Tecnologias Agro-Ambientais

REFERENCES

Abril, M., Negueruela, A. I., Pérez, C., Juan, T. & Estopañán, G. (2005). Preliminary study of resveratrol content in Aragón red and rosé wines. *Food Chem.*, *92*, 729-736.

Ansell, R. J. & Mosbach, K. (1997). Molecularly imprinted polymers by suspension polymerisation in perfluorocarbon liquids, with emphasis on the influence of the porogenic solvent. *J. Chromatogr. A*, *787*, 55–66.

Atanackovic, M., Petrovic, A., Jovic, S., Gojkovic – Bukarica, L., Bursac, M. & Cvejic, J. (2012). Influence of winemaking techniques on the resveratrol content, total phenolic content and antioxidant potential of red wines. *Food Chem.*, *131*, 513–518.

Bai, B., He, F., Yang, L., Chen, F., Reeves, M. J. & Li, J. (2013). Comparative study of phenolic compounds in Cabernet Sauvignon wines made in traditional and Ganimede fermenters. *Food Chem.*, *141*, 3984–3992.

Baptista, J. A. B., Tavares, J. F. P. & Carvalho, R. C. B. (2001). Comparison of polyphenols and aroma in red wines from Portuguese mainland versus Azores Islands. *Food Res. Int.*, *34*, 345-355.

Belleville, J. (2002). The French Paradox: Possible Involvement of Ethanol in the Protective Effect Against Cardiovascular Diseases. *Nutrition*, *18*, 173–177.

Boutegrabet, L., Fekete, A., Hertkorn, N., Papastamoulis, Y., Waffo-Téguo, P., Mérillon, J. M., Jeandet, P., Gougeon, R. D. & Schmitt-Kopplin, P. (2011). Determination of stilbene derivatives in Burgundy red wines by ultra-high-pressure liquid chromatography. *Anal. Bioanal. Chem.*, *401*, 1513–1521.

Bravo, M. N., Feliciano, R., Silva, S., Coelho, A. V., Vilas Boas, L. & Bronze, M. R. (2008). Analysis of *trans*-resveratrol: Comparison of methods and contents in Muscatel fortified wines from Setubal region in Portugal. *J. Food Compos. Anal.*, *21*, 634–643.

Buiarelli, F., Coccioli, F., Jasionowska, R., Merolle, M. & Terracciano, A. (2007). Analysis of some stilbenes in Italian wines by liquid chromatography/tandem mass spectrometry. *Rapid Commun. Mass Sp.*, *21*(18), 2955–2964.

Cao, H., Xiao, J. B. & Xu, M. (2006). Evaluation of new selective molecularly imprinted polymers for the extraction of resveratrol from polygonum cuspidatum. *Macromol. Res.*, *14*, 324-330.

Chen, F. F., Xie, X. Y. & Shi, Y. P. (2013). Preparation of magnetic molecularly imprinted polymer for selective recognition of resveratrol in wine. *J. Chromatogr. A*, *1300*, 112-118.

Dai, S., Shin, Y., Barnes, C. E. & Toth, L. M. (1997). Enhancement of Uranyl Adsorption Capacity and Selectivity on Silica Sol-Gel Glasses via Molecular Imprinting. *Chem. Mater.*, *9*, 2521-2525.

de Viliers, A., Lynen, F., Crouch, A. & Sandra, P. (2004). Development of a solid-phase extraction procedure for simultaneous determination of polyphenols, organic acids and sugars in wines. *Chromatographia*, *59*, 403-409.

Ensing, K. & de Boer, T. (1999). Tailor-made materials for tailor-made applications: application of molecular imprints in chemical analysis. TrAC *Trends Anal. Chem.*, *18*, 138–145.

Euterpio, M. A., Pagano, I., Piccinelli, A. L., Rastrelli, L. & Crescenzi, C. (2013). Development and validation of a method for the determination of (E)-resveratrol and related phenolic compounds in beverages using molecularly imprinted solid phase extraction. *J. Agric. Food Chem.*, *61*, 1640-1645.

Feijoó, O., Moreno, A. & Falqué, E. (2008). Content of *trans*- and *cis*-resveratrol in Galician white and red wines. *J. Food Compos. Anal.*, *21*, 608–613.

Fernández-Mar, M. I., Mateos, R., García-Parrilla, M. C., Puertasa, B. & Cantos-Villar, E. (2012). Bioactive compounds in wine: Resveratrol, hydroxytyrosol and melatonin: A review. *Food Chem.*, *130*, 797–813.

Ferrer, I., Lanza, F., Tolokan, A., Horvath, V., Sellergren, B., Horvai, G. & Barceló, D. (2000). Selective trace enrichment of chlorotriazine pesticides from natural waters and sediment samples using terbuthylazine molecularly imprinted polymers. *Anal. Chem.*, *72*, 3934–3941.

Gambuti, A., Strollo, D., Ugliano, M., Lecce, L. & Moio, L. (2004). *trans*-Resveratrol, quercetin, (+)-catechin, and (-)-epicatechin content in south Italian monovarietal wines: relationship with maceration time and marc pressing during winemaking. *J. Agric. Food Chem.*, *52*, 5747–5751

Gao, L. Y., Chu, Q. C. & Ye, J. N. (2002). Determination of *trans*-resveratrol in wines, herbs and health food by capillary electrophoresis with electrochemical detection. *Food Chem.*, *78*, 255–260.

Garcia, R., Cabrita, M. J. & Freitas, A. M. C. (2011). Application of Molecularly Imprinted Polymers for the Analysis of Pesticide Residues in Food—A Highly Selective and Innovative Approach. *Am. J. Anal. Chem.*, *2*, 16-25.

Gerogiannaki-Christopoulou, M., Athanasopoulos, P., Kyriakidis, N., Gerogiannaki, I. A. & Spanos, M. (2006). *trans*-Resveratrol in wines from the major Greek red and white grape varieties. *Food Control*, *17*, 700–706.

Goldberg, D. M., GarovicKocic, V., Diamandis, E. P. & PaceAsciak, C. R. (1996). Wine: does the colour count? *Clin. Chim. Acta*, *246*, 183–193.

Goldberg, D. M., Ng, E., Karumanchiri, A., Yan, J., Diamandis, E. P. & Soleas, G. J. (1995). Assay of resveratrol glucosides and isomers in wine by direct-injection high-performance liquid-chromatography. *J. Chromatogr. A*, *708*, 89–98.

Gonçalves, J. & Câmara J. S. (2011). New method for determination of (E)- resveratrol in wine based on microextraction using packed sorbent and ultra-performance liquid chromatography. *J. Sep. Sci.*, *34*, 2376–2384.

Gresele, P., Cerletti, C., Guglielmini, G., Pignatelli, P., de Gaetano, G. & Violi, F. (2011). Effects of resveratrol and other wine polyphenols on vascular function: an update. *J. Nutr. Biochem.*, *22*, 201–211.

Gu, X., Chu, Q., O'Dwyer, M. & Zeece M. (2000). Analysis of resveratrol in wine by capillary electrophoresis. *J. Chromatogr. A*, *881*, 471–481.

Gu, X., Creasy, L., Kester, A. & Zeece, M. (1999). Capillary electrophoretic determination of resveratrol in wines. *J. Agric. Food Chem.*, *47*, 3223–3227.

Guerrero, R. F., Puertas, B., Jiménez, M. J., Cacho, J. & Cantos-Villar, E. (2010). Monitoring the process to obtain red wine enriched in resveratrol and piceatannol without quality loss. *Food Chem.*, *122*, 195–202.

Hashim, S. N. N. S, Schwarz, L. J., Boysen, R. I., Yang, Y., Danylec, B. & Hearn, M. T. W. (2013). Rapid solid-phase extraction and analysis of resveratrol and other polyphenols in red wine. *J. Chromatogr. A.*,1313, 284-290.

Hosoya, K., Kageyama, Y., Kimata, K., Araki, T., Tanaka, N. & Frechet, J. M. J. (1996). Preparations and properties of uniform size macroporous polymer beads prepared by two–step swelling and polymerization method utilizing divinyl succinate or divinyl adipate as a crosslinking agent. *J. Polym. Sci. A*, *34*, 2767–2774.

Jeandet, P., Bessis, R., Maume, B. F., Meunier, P., Peyron, D. & Trollat, P. (1995). Effect of Enological Practices on the Resveratrol Isomer Content of Wine. *J. Agric. Food Chem.*, *43*, 316–319.

Kallithraka, S., Arvanitoyannis, I., El-Zajouli, A. & Kefalas, P. (2001). The application of an improved method for *trans*-resveratrol to determine the origin of Greek red wines. *Food Chem.*, *75*, 355–363.

Katz, A. & Davis, M. E. (2000). Molecular imprinting of bulk, microporous sílica. *Nature*, *403*, 286-289

Kolouchova-Hanzlikova, I., Melzoch, K., Fili, V. & Smidrkal, J. (2004). Rapid method for resveratrol determination by HPLC with electrochemical and UV detections in wines. *Food Chem.*, *87*, 151–158.

Langcake, P. & Pryce, R. J. (1976). The production of resveratrol by Vitis vinifera and other members of the Vitacee as a response to infection or injury. *Physiol. Plant Pathol.*, *9*, 77–86.

Li, Z., Pan, Q., Jin, Z., Mu, L. & Duan, C. (2011). Comparison on phenolic compounds in Vitis vinifera cv. Cabernet Sauvignon wines from five wine-growing regions in China. *Food Chem.*, *125*, 77–83.

Lok, C. M. & Son, R. (2009). Application of molecularly imprinted polymers in food sample analysis – a perspective. *Int. Food Res. J.*, *16*, 127-140.

López-Alfaro, I., González-Arenzana, L., López, N., Santamaría, P., López, R. & Garde-Cerdán, T. (2013). Pulsed electric field treatment enhanced stilbene content in Graciano, Tempranillo and Grenache grape varieties. *Food Chem.*, *141*, 3759-3765.

Ma, S., Zhuang, X., Wang, H., Liu, H., Li, J. & Dong, X. (2007). Preparation and Characterization of *Trans-* Resveratrol Imprinted Polymers. *Anal. Lett.*, *40*, 321-333.

Mahony, J. O., Nolan, K., Smyth, M. R. & Mizaikoff, B. (2005). Molecularly imprinted polymer-potential and challenges in analytical chemistry. *Anal. Chim. Acta*, *534*, 31-39.

Malitesta, C., Losito, I. & Zambonin, P. G. (1999). Molecularly imprinted electrosynthesized polymers: New materials for biomimetic sensors. *Anal. Chem.*, *71*, 1366-1370.

Malovaná, S., García Montelongo, F. J., Pérez, J. P. & Rodríguez-Delgado, M. A. (2001). Optimisation of sample preparation for the determination of *trans*-resveratrol and other polyphenolic compounds in wines by high performance liquid chromatography. *Anal. Chim. Acta*, *428*, 245-253.

Mark, L., Nikfardjam, M. S. P., Avar, P. & Ohmacht, R. (2005). A Validated HPLC Method for the Quantitative Analysis of *Trans*-Resveratrol and *Trans*-Piceid in Hungarian Wines. *J. Chromatogr. Sci.*, *43*, 445-449.

Martínez-Ortega, M. V., Carcia-Parrilla, M. C. & Troncoso, A. M. (2000). Resveratrol content in wines and musts from the south of Spain. *Nahrung*, *44*, 253–256.

Mattivi, F. (1993). Resveratrol content in commercial red and rose wines of Trentino (Italy). *Rivista di Viticoltura e di Enologia*, *461*, 37–45.

Mattivi, F. & Nicolini, G. (1993). Influenza della tecnica di vinificazione sul contenuto di resveratrolo dei vini. *L'Enotecnico*, *29*, 81–88.

Mattivi, F., Reniero, F. & Korhammer, S. (1995). Isolation, Characterization, and Evolution in Red Wine Vinification of Resveratrol Monomers. *J. Agric. Food Chem.*, *43*, 1820–1823.

Mayes, A. G. & Mosbach, K. (1996). Molecularly Imprinted Polymer Beads: Suspension Polymerization Using a Liquid Perfluorocarbon as the Dispersing Phase. *Anal. Chem.*, *68*, 3769–3774.

Mercolini, L., Saracino, M.A., Bugamelli, F., Ferranti, A., Malaguti, M. & Hrelia, S. (2008). HPLC-F analysis of malatonin and resveratrol isomers in wine using a SPE procedure. *J. Sep. Sci.*, *31*, 1007-1014.

Mosbach, K. & Ramström, O. (1996). The Emerging Technique of Molecular Imprinting and Its Future Impact on Biotechnology. *Nature Biotechnol.*, *14*, 163 – 170.

Nave, F., Cabrita, M. J. & Costa C. (2007). The Use of Solid-Supported Liquid-Liquid Extraction in the Analysis of Polyphenols in Wine. *J. Chromatogr. A*, *1169*, 23–30.

Netzel, M., Strass, G., Bitsch, I., Könitz, R., Christmann, M., Bitsch, R. (2003). Effect of grape processing on selected antioxidant phenolic in red wine. *J. Food Eng.*, *56*, 223–228.

Pauling, L. (1940). A theory of the structure and process of formation of antibodies. *J. Am Chem Soc.*, *62*, 2643-2657.

Pérez-Moral, N., Whitcombe, M. J. & Vulfson, E. N. (2000). Molecularly imprinted nanoparticles prepared by core-shell emulsion polymerization. *J. Appl. Polym. Sci.*, *77*, 1851-1859.

Pérez-Moral, N., Whitcombe, M. J. & Vulfson, E. N. (2001). Surface imprinting of cholesterol on submicrometer core-shell emulsion particles. *Macromolecules*, *34*, 830-836.

Piletska, E. V., Guerreiro, A. R., Whitcombe, M. J. & Piletsky, S. A. (2009). Influence of the Polymerization Conditions on the Performance of Molecularly Imprinted Polymers. *Macromolecules*, *42*, 4921-4928.

Qian, K, Fang, G. Z., He, J. X., Pan, M. F. & Wang, S. (2010). Preparation and application of a molecularly imprinted polymer for the determination of trace metolcarb in food matrices by high performance liquid chromatography. *J. Sep. Sci.*, *33*, 2079–2085.

Ramström, O. & Mosbach, K. (1999). Synthesis and catalysis by molecularly imprinted materials. *Curr. Opin. Chem. Biol.*, *3*, 759–764.

Ramström, O., Skudar, K., Haines, J., Patel, P. & Brüggemann, O. (2001). Food analyses using molecularly imprinted polymers. *J. Agric. Food Chem.*, *49*, 2105-2114.

Regal, P., Díaz-Bao, M., Barreiro, R., Cepeda, A. & Fente, C. (2012). Application of molecularly imprinted polymers in food analysis: clean-up and chromatographic improvements. *Cent. Eur. J. Chem.*, *10*, 766-784.

Renaud, S & De Lorgeril, M. (1992). Wine, alcohol, platelets, and the French paradox for coronary heart disease. *Lancet*, *339*, 1523-1526

Ribeiro de Lima, M. T., Waffo-Téguo, P., Teissedre, P.L., Pujolas, A., Vercauteren, J., Cabanis, J. C. & Mérillon, J. M. (1999). Determination of Stilbenes (*trans*-Astringin, cis- and *trans*-Piceid, and *cis*- and *trans*-Resveratrol) in Portuguese Wines. *J. Agric. Food Chem.*, *47*, 2666-2670.

Rodríguez-Delgado, M. A., González, G., Pérez-Trujillo, J. P. & García-Montelongo, F. J. (2002). *Trans*-resveratrol in wines from the Canary Islands (Spain). Analysis by high performance liquid chromatography. *Food Chem.*, *76*, 371–375.

Roldán, A., Palacios, V., Caro, I. & Pérez, L. (2003). Resveratrol content of palomino fino grapes: influence of vintage and fungal infection. *J. Agric. Food Chem.*, *51*, 1464–1468.

Romero-Peréz, A. I., Lamuela-Raventos, R. M., Andrés-Lacueva, C. & de la Torre-Boronat, M. C. (2001). Method for the quantitative extraction of resveratrol and piceid isomers in grape berry skins. Effect of powdery mildew on the stilbene content. *J. Agric. Food Chem.*, *49*, 210–215.

Ružić, I., Skerget, M., Knez, Z. & Runje, M. (2011). Phenolic content and antioxidant potential of macerated white wines. *Eur. Food Res. Technol.*, *233*, 465-472.

Sato, M., Suzuki, Y., Okuda, T. & Yokotsuka, K. (1997). Contents of resveratrol, piceid, and their isomers in commercially available wine made from grapes cultivated in Japan. *Biosci. Biotechnol. Biochem.*, *61*, 1800–1805.

Schwarz, L. J., Danylec, B., Yang, Y., Harris, S. J., Boysen, R. I. & Hearn, M. T. W. (2011a). Enrichment of (E)-resveratrol from peanut byproduct with molecularly imprinted polymers. *J. Agric. Food Chem.*, *59*, 3539-3543.

Schwarz, L. J., Danylec, B., Harris, S. J., Boysen, R. I. & Hearn, M. T. W. (2011b). Preparation of molecularly imprinted polymers for the selective recognition of the bioactive polyphenol, (E)-resveratrol. *J. Chromatogr. A*, *1218*, 2189-2195.

Scorrano, S., Mergola, L., del Sole, R. & Vasapollo, G. (2011). Synthesis of molecularly imprinted polymers for amino acid derivates by using different functional monomers. *Int. J. Mol. Sci.*, *12*, 1735–1743

Shao, Y., Marriott, P. & Hugel, H. (2003). Solid–phase microextraction–on–fiber derivatization with comprehensive two dimensional gas chromatography analysis of *trans*–resveratrol in wine. *Chromatographia*, *57*, 349–353

Siemann, E. H. & Creasy, L. L. (1992). Concentration of the Phytoalexin Resveratrol in Wine. *Am. J. Enol. Vitic.*, *43*, 49-52.

Soleas, G. J., Diamandis, E. P. & Goldberg, D. M. (1997). Resveratrol: A molecule whose time has come? And gone? *Clin. Biochemi.*, *30*, 91- 113.

Souto, A. A., Carneiro, M. C., Seferin, M., Senna, M. J. H., Conz, A. & Gobbi, K. (2001). Determination of *trans*–Resveratrol concentrations in Brazilian red wines by HPLC. *J. Food Comp. Anal.*, *14*, 441–445.

Stervbo, U., Vang, O. & Bonnesen, C. (2007). A review of the content of the putative chemopreventive phytoalexin resveratrol in red wine. *Food Chem.*, *101*, 449–457

Sun, B., Ferrão, C. & Spranger, M. I. (2003). Effect of wine style and winemaking technology on resveratrol levels in wines. *Ciência e Técnica Vitivinícola*, *18*, 77-91

Threlfall, R. T., Morris, J. R. & Mauromoustakos, A. (1999). Effect of variety, ultraviolet light exposure, and enological methods on the *trans*-resveratrol level of wine. *Am. J. Enol Vitic.*, *50*, 57–64.

Trela, B. C. & Waterhouse, A. L. (1996). Resveratrol: isomeric molar absorptivities and stability. *J. Agric. Food Chem.*, *44*, 1253–1257.

Vasapollo, G., del Sole, R., Mergola, L., Lazzoi, M. R., Scardino, A., Scorrano, S., Mele, G. (2011). Molecularly Imprinted Polymers: Present and Future Prospective. *Int. J. Mol. Sci.*, *12*, 5908–5945.

Vian, M. A., Tomao, V., Gallet, S., Coulomb, P. O. & Lacombe, J. M. (2005). Simple and rapid method for *cis*- and *trans*-resveratrol and piceid isomers determination in wine by high-performance liquid chromatography using Chromolith. *J. Chromatogr. A*, *1085*, 224-229.

Viñas, P., Campillo, N., Hernández-Pérez, M. & Hernández-Córdoba, M. (2008). A comparison of solid-phase microextraction and stir bar sorptive extraction coupled to liquid chromatography for the rapid analysis of resveratrol isomers in wines, musts and fruit juices. *Anal. Chim. Acta*, *611*, 119-125.

Vitrac, X., Bornet, A., Vanderlinde, R., Valls, J., Richard, T., Delaunay, J. C., Mérillon, J. M. & Teissédre, P. L. (2005). Determination of Stilbenes (δ-viniferin, *trans*-astringin, *trans*-piceid, *cis*- and *trans*-resveratrol, ε-viniferin) in Brazilian Wines. *J. Agric. Food Chem.*, *53*, 5664-5669.

Vonderbruggen, M. A., Wu, K. & Breneman, C. M. (1996). Use of cationic aerosol photopolymerization to from silicone microbeads in the presence of molecular templates. *Chem. Mat.*, *8*, 1106-1111.

Wang, Y., Catana, F., Yang, Y., Roderick, R. & Van Breemen, R. B. (2002). An LC-MS Method for Analyzing Total Resveratrol in Grape Juice, Cranberry Juice, and in Wine. *J. Agric. Food Chem.*, *50*, 431-435.

Whitcombe, M. J. & Vulfson, E. N. (2001). Imprinted Polymers. *Adv. Mat.*, *13*, 467-478.

Yan, H. & Row, K. (2006). Characteristic and Synthetic Approach of Molecularly Imprinted Polymer. *Int. J. Mol. Sci.*, *7*, 155-178.

In: Wine
Editor: Youssef El Rayess

ISBN: 978-1-63321-048-6

Chapter 8

QUALITY MAPPING: AN EXPLORATORY APPROACH FOR EVALUATING THE CHEMICALS DRIVING IN-MOUTH QUALITY IN RED WINES

María-Pilar Sáenz-Navajas[1,2]**, José Miguel Avizcuri**[3]**, Purificación Fernández-Zurbano**[3]**, Vicente Ferreira**[2]* ***and Dominique Valentin**[1,4]*

[1]Centre des Sciences du Goût et de l´Alimentation, Dijon, France
[2]Laboratory for Flavor Analysis and Enology (LAAE),
Aragón Institute of Engineering Research (I3A), University of Zaragoza, Zaragoza, Spain
[3]Instituto de Ciencias de la Vid y el Vin (UR-CSIC-GR), La Rioja, Spain
[4]AGROSUP, Dijon, France

ABSTRACT

Non-volatile molecules have been reported to be important molecules driving wine sensory properties; however the wine industry is far from knowing their exact role in wine quality perception. The present work aims to establish correlations between (1) the non-volatile composition of a set of red wines and their in-mouth sensory attributes and (2) non-volatile composition and the quality scores given by a panel of experts on the same set of wines. A group of 21 wine experts rated the in-mouth quality of 16 Spanish red wines with different non-volatile composition on an unstructured scale anchored with "nil" on the left and "high" on the right. To avoid aroma and color influence in quality assessment, the evaluation was carried out using nose clips and dark glasses. In parallel, a descriptive analysis method was used to evaluate the intensity of in-mouth sensory properties; further non-volatile molecules with known sensory activity were quantified.

Results show an exponential relationship between the in-mouth attribute sourness and both tartaric and succinic acids. Succinic acid and to a lesser extent the alcohol

* Corresponding Author, address: Laboratory for Flavor Analysis and Enology (LAAE), Aragón Institute of Engineering Research (I3A), University of Zaragoza, Zaragoza, Spain; Email: maria@saenz.info.

content present exponential as well as *t*-aconitic acid are exponentially correlated with astringency as are procyanidins B1 and B3 and the total polyphenolic index, which are linearly correlated.

Results show that experts have heterogeneous in-mouth quality concepts. Experts are divided into three main clusters, and their quality judgments were regressed onto the significant factors derived from the PCA calculated on the 22 non-volatile molecules with sensory activity. Seventy-one per cent of expert quality responses were fitted by the elliptical model, while 29% were fitted by the vector model. Experts belonging to cluster 1 (fitted by the vector model: "the more the better") find that high contents in protein-precipitable PAs and in *c*-aconitic acid result in high quality exemplars. Clusters 2 and 3 quality scores were significantly predicted by elliptical models and their anti-ideal and ideal quality points were respectively defined.

Keywords: Quality mapping, in-mouth quality, non-volatiles, sensory

INTRODUCTION

Wine quality is a complex concept that is highly influenced by consumer's level of involvement (Charters and Pettigrew, 2006) or expertise (Ballester et al., 2008) among other factors. More highly involved consumers, which have more knowledge and interest in wine, are able to understand and carry out the process of quality evaluation more easily than less involved consumers (Charters and Pettigrew, 2006). With respect to the level of expertise, wine experts have been shown to have aligned quality concepts probably because they regularly attend formal wine tasting sessions, in which they often have technical information about the wines they taste. This leads to a lower variability and higher consistency in responses compared to less experienced consumers (Urdapilleta et al., 2011), who base their evaluations on their personal consumption experience resulting in personalized and subjective responses (Parr et al., 2011). The fact that quality assessment is based on technical winemaking processes for experts and on individual experience for consumers results in a misalignment in the quality concept between wine professionals and low-experienced consumers (Lattey et al., 2010). Furthermore, among consumers, heterogeneity in responses is usually observed, thus segmentation among less experienced consumers is usually required (Aurifeille et al., 2002). On the contrary, experts have been shown to share quality concepts and to be globally consensual in quality assessment (Lattey et al., 2010; Sáenz-Navajas et al., 2011; Varela and Gambaro, 2006). Yet, some studies showed that although experts may agree on the sensory drivers of quality, ratings of overall quality may be participant-dependent (Gawel and Godden, 2008; Scaman et al., 2001).

Understanding the factors that drive quality perception of experts is important as experts´ opinions influence consumers´ final buying decisions (Marin and Durham, 2007; Marin et al., 2007). This is the reason why there are several works aimed to establish correlations between expert's quality scores and sensory properties of wines. In this context, quality assessments carried out by experts (Lattey et al., 2010; Machado et al., 2011; Sáenz-Navajas et al., 2011; Sáenz-Navajas et al., 2012) or highly-involved wine consumers (Varela and Gambaro, 2006) have been correlated to aroma, taste and in-mouth sensory profiles obtained from trained panels. A consistent correlation between fruity (berry or dried fruit depending on the type of wine) and woody aromas as well as astringency or sourness and quality scores given by experts emerge from these studies.

On the contrary, vegetal, undergrowth or animal aroma characters are usually linked to low quality exemplars since they are correlated with defects derived from the unsuitable management of wine during elaboration. In general, it is observed that experts base their quality judgments primarily on the presence/absence of defective or negative odorants, and secondarily on the presence of fruit-sweet positive aromas (Ferreira et al., 2009). Ultimately, in-mouth (Sáenz-Navajas et al., 2011) and color (Sáenz-Navajas et al., 2011) clues are used for creating the final overall quality judgment (Ferreira et al., 2009). Thus, volatile compounds seem to be the main drivers of wine quality, while non-volatiles seem to play a secondary role. Nevertheless, it is widely known that the conventional oenological parameters measured in wineries for controlling quality are mainly based on non-volatile composition such as: reducing sugars, titratable acidity, total polyphenol content, color intensity, malic and lactic acid or alcohol content. Probably because the methodology employed for evaluating non-volatile composition is less-time and money consuming than the sophisticated and costly methods needed for aroma analysis. To solve this apparent paradox between scientific results and oenology practices it seems very important to understand the real contribution of non-volatile composition to quality perception.

External preference mapping has emerged as an interesting multidimensional statistical tool able to describe the characteristics that contribute to the liking of consumers or groups of consumers as well as to identify which products are most liked or disliked (Schlich, 1995; Schlich and Mcewan, 1992). This tool has been successfully extended to the evaluation and understanding of the concept of quality under the name of *quality mapping* (Delgado and Guinard, 2011). This strategy is promising since it relies less on linear assumptions than more common statistical tools such as General Procruster Analysis or Principal Least Square Regression. Introducing non linearity is essential for better understanding the relationships between sensory and chemical data as these relationships are rarely linear (Chambers and Koppel, 2013).

In this context, the first aim of the present work is to establish relationships between in-mouth sensory attributes perceived in wine and non-volatile composition. Then, quality mapping strategies are used to relate non-volatile chemical composition of red wines to experts´ quality judgments.

Material and Methods

Wines

Sixteen Spanish red wines from different wine making areas, varieties, vintage and with different aging periods in both bottle and oak barrels were selected to cover a wide range of non-volatile chemical composition expected to induce different sensory properties and thus different quality perception. The detailed list of samples, including wine information and basic compositional oenological parameters, is shown in Table 1.

Table 1. The sixteen studied commercial wines and their original oenological parameters

wine code	origin	vintage	grape variety	oak aging (months)	TPI[a]	pH	TA[b]	AV[c]	RS[d]	MA[e]	LA[f]	alcohol (% v/v)
MG_V05	DO Dominio de Valdepusa	2005	cabernet sauvignon	12	83.4	3.65	4.91	0.56	4.35	0.29	0.77	15.2
AY_C05	DO Cariñena	2005	merlot, tempranillo, cabernet sauvignon	10	74.3	3.52	5.86	0.69	3.39	0.33	1.00	14.3
GC_B10	DO Borja	2010	garnacha	4	71.4	3.43	6.14	0.42	3.61	0.25	0.68	14.7
RM_R10	DOCa Rioja	2010	graciano	8	66.4	3.57	5.80	0.41	2.31	0.19	1.45	14.8
CD_C10	DO Cariñena	2010	garnacha, tempranillo, cabernet sauvignon	0	66.4	3.63	5.30	0.53	2.57	0.24	0.90	13.5
CZ_D08	DO Duero	2008	tempranillo	18	62.0	3.65	5.33	0.57	1.71	0.35	2.47	13.4
BO_B10	DO Borja	2010	garnacha, syrah, tempranillo	0	61.0	3.66	5.04	0.47	2.68	0.17	1.07	14.8
CH_R06	DOCa Rioja	2006	tempranillo, viura	0	60.3	3.88	4.45	0.62	1.77	0.20	3.30	14.1
CT_B07	DO Borja	2007	garnacha	15	59.1	3.47	5.66	0.51	4.34	0.30	0.75	13.9
SC_R10	DOCa Rioja	2010	tempranillo, garnacha	0	57.8	3.72	4.84	0.48	2.32	0.18	2.52	13.4
SO_C07	DO Cariñena	2007	garnacha, tempranillo, cabernet sauvignon	18	54.9	3.53	5.66	0.75	3.81	0.18	1.21	13.8
AR_A08	DO Arlanza	2008	tempranillo	12	53.0	3.73	5.57	0.63	1.98	0.24	2.79	13.6
MC_R09	DOCa Rioja	2009	tempranillo, graciano, mazuelo	12	52.3	3.64	4.92	0.52	2.09	0.21	2.11	13.7
NJ_R09	DOCa Rioja	2009	tempranillo, garnacha	18	49.7	3.65	5.35	0.66	1.67	0.18	2.14	13.6
RB_R10	DOCa Rioja	2010	tempranillo, garnacha	18	49.4	3.49	5.37	0.57	2.23	0.23	1.45	14.3
BE_R10	DOCa Rioja	2010	tempranillo, garnacha	0	45.4	3.61	5.09	0.25	1.52	0.18	1.86	13.9

[a] Total Polyphenol Index. Absorbance at 280nm measured in 10-cm cuvettes.

[b] Total titratable acidiy expressed in g L^{-1} of tartatic acid.

[c] Volatile acidity expressed in g L^{-1} of acetic acid.

[d] Reducing sugars expressed in g L^{-1}.

[e] Malic acid expressed in g L^{-1}.

[f] Lactic acid expressed in g L^{-1}.

Quantification of Non-Volatile Molecules

Reagents and Standards

Standards of *t*-aconitic acid, *c*-aconitic acid, syringic acid, protocatechuic acid, protocatechuic acid ethyl ester, *cis-/trans*-gallic acid, *cis-/trans*-caffeic acid, catechin, epicatechin, myricetin, kaempferol, and quercetin were obtained from Sigma-Aldrich (St. Louis, MO, USA). Procyanidin B1, procyanidin B2, *cis-/trans*-caftaric acid, *cis-/trans*-coumaric acid, kaempferol-3-*O*-glucoside, epigallocatechin, epicatechin gallate, epigallocatechin gallate, quercetin-3-*O*-glucoside, quercetin-3-*O*-glucuronide, syringetin-3-*O*-glucoside, isorhamnetin-3-*O*-glucoside, quercetin-3-*O*-galactoside, syringetin-3-*O*-galactoside and isorhamnetin were provided from Extrasynthese (Genay, France). Vanillic acid and trifluoroacetic acid (TFA) were supplied by Fluka (Buchs, Switzerland). UPLC-grade acetonitrile, ethanol, methanol and formic acid were purchased in Scharlau (Spain) and ultrapure water was obtained from a Milli-Q system from Millipore (Molsheim, France).

Conventional Oenological Parameter Determination

Total Polyphenol Index (TPI) was estimated as absorbance at 280 nm (Ribéreau-Gayon 1970). Glycerol, malic and lactic acids were determined by enzymatic methods in accordance with official AOAC analysis methods (AOAC 2002). The analysis of reducing sugars, ethanol content, pH, titratable and volatile acidities were determined by Infrared Spectrometry with Fourier Transformation (IRFT) with a WineScan™ FT 120 (FOSS®, Barcelona), which was previously calibrated with the official OIV methods.

Quantification of Individual Non-Volatile Compounds by UPLC-MS

UPLC analyses were performed using a *Waters Acquity Ultra Performance* LC system (Milford, MA, USA) by direct injection of wine samples, previously filtered with 0.22 μm nylon discs. UPLC separation was achieved using an *acquity BEH* C18 column (100mm×2.1mm, i.d., 1.7 μm particle size, Waters), kept at 40 °C. Mobile phase flow rate was 0.45 mL min^{-1} and the injection volume was 7.5 μL. Solvents were (A) water/formic acid (0.1%), and (B) acetonitrile/formic acid (0.1%). The gradient program employed was as follows: 0–4 min, 99–92% A; 4–11 min, 92–70% A; 11–13.5 min, 70–0% A; 13.5–14.5 min, 0–99% A. The UPLC system was coupled to a microTOF II high-resolution mass spectrometer (Bruker Daltonik, Germany) equipped with an Apollo II ESI (electrospray) source and controlled by Bruker Daltonics Data Analysis software. The ESI source was operated in negative mode recorded in the range of *m/z* 150 and 1500. The optimized conditions of the ESI source were as follows: capillary voltage, 3.5 kV; ESI source temperature, 180 °C; desolvation temperature, 200 °C; cone gas flow, 9 L min^{-1}; the nebulizer gas was set at 3 bar and 25 °C. LC–MS was performed operating in both (continuum) MS mode and in MS/MS mode. The spectra were acquired at the speed of 2 scan/second. Fragmentor voltage for MS/MS acquisition mode was 35e V.

The identity assignation of compounds was carried out by comparison of their retention time (t_R), MS and MS/MS spectra with those of their respective commercially available standards. Quantification was performed by UPLC-MS and using the response (peak area) ratio for each compound. The calibration curves were prepared by calculating the peak area of the different standards and were used to determine linearity and instrumental detection (LOD) and quantification (LOQ) limits. LOD and LOQ were calculated using 3 S/m and 10 S/m (S is the standard deviation of the response; m is the slope of the calibration curve), respectively. Calibration curves were prepared in methanol. The estimated concentrations of coutaric acid have been expressed as caftaric acid equivalents (mg L^{-1} of caftaric acid) and ferulic acid as caffeic acid equivalents (mg L^{-1} of caffeic acid).

Total Polymeric Proanthocyanidin (PA) Analysis

Wine samples were firstly fractionated by preparative chromatography: Gel Permeation Chromatography (GPC) as described elsewhere (Sáenz-Navajas et al., 2010). Therefore, 2 mL of wine was directly injected in the column, and the flow rate was regulated at 1 mL min^{-1} using a peristaltic pump. Firstly, low molecular-weighted compounds such as organic acids and phenolic compounds were eluted with 60 mL of ethanol/water/trifluoroacetic acid (55:45:0.05, v/v/v). This fraction was discarded and a second fraction containing polymeric PAs was collected by elution with 50 mL of acetone/water (60:40, v/v) and then, brought to dryness under vacuum. This fraction was re-dissolved in 7 mL of methanol and further quantified by the vanillin assay according to the method described elsewhere (Sun et al., 1998) for determining the total content in polymeric PAs.

Protein-Precipitable PA Analysis

Quantification of protein-precipitable PAs was estimated using ovalbumin as the precipitation agent and tannic acid solutions as standards in accordance with a previously described method (Llaudy, et al. 2004).

Sensory Analysis

Descriptive Analysis by a Trained Panel

A total of 52 participants were recruited on the basis of their interest and their availability during 9 months. They were not paid for their participation. For attendance reasons the responses of forty-one panelists (17 males and 24 females from 21 to 57 years old, median = 28) were considered for data analyses.

The panelists received sensory training in eight sessions (about one hour per session) over a period of two weeks. This training period included two phases: a general and a product specific training phase. The wines selected for the general training phase presented intense and easily recognizable taste and astringency properties and included red, white and rosé wines of diverse grape varieties and origins. During the specific training phase panelists became familiar with the type of samples of the study (red Spanish wines with different non-volatile composition).

During a typical training session the panelists became familiar with the rating of six attributes evaluated in-mouth: sweetness, sourness, bitterness, astringency, global intensity and persistence. Different reference standards representative of taste and astringency terms were presented as described elsewhere (Sáenz-Navajas et al., 2011). Then, they had to evaluate three to five different wines by rating tastes and astringency on a 10-point scale (0 = "absence", 1 = "very low" and 9 = "very high"), global intensity on a 9-point scale (1 = "very low" and 9 = "very high") and global persistence on a nine-point scale (1 = "very short" and 9

= "very long"). The session ended with a discussion during which the panel leader compared the taste intensity scores given by panelists to describe each wine.

Trained panelists described wines following the procedure described in Sáenz-Navajas et al. (2011). Fifteen-mL wine samples were presented in dark approved wine glasses (ISO NORM 3591, 1977) labeled with 3-digit random codes and covered by plastic Petri dishes according to a random arrangement. Each panelist completed two sessions (ca. 45 min each) for the analysis of 20 samples (16 samples + 4 replicates: two in each session for evaluating the reproducibility of the panel) involving 10 wines per session. Panelists were asked to rate the sweetness, sourness, bitterness, astringency, global intensity, and global persistence of the samples using the above mentioned structured scales for each wine. Trained panelists rated samples using a sip and spit protocol described by Colonna, et al. (2004).

All wines were served at room temperature and were evaluated in individual booths. Panelists were not informed about the nature of the samples to be evaluated.

Quality Evaluation by Experts

The panel of experts was composed of 21 established winemakers from DOCa Rioja (Spain), twelve females and nine males ranging from 28 to 57 years of age (median = 35).

Each participant completed one session (ca. 20 min each) in individual booths. Experts evaluated the in-mouth wine quality of the 16 samples. They were asked to wear nose clips to avoid aroma interaction. Similarly, to avoid color interactions, samples were presented randomly in coded dark approved wine glasses (ISO NORM 3591, 1977). Samples were covered with a Petri dish and presented at room temperature, once in the proposed order to minimize any bias introduced by the order of presentation. Afterwards, they could taste the samples as many times as they wanted and in any order. Unstructured 15-cm-long scales anchored with ''nil'' and ''high'' were used to score quality.

Data Analysis

Sensory Descriptive Analysis

One Principal Component Analysis (PCA) for each of the six attributes evaluated in mouth was run in order to assess judges' agreement. For that, a table with the wines in rows and the judges in columns was employed. The responses of panelists that were spread on the PCA, and thus were not in accordance with the rest of the panel, were kept out of the data analysis.

For assessing the discrimination ability of attributes a one-way analysis of variance (ANOVA), in which wine was the fix factor, was performed on each in-mouth descriptor derived from the trained panel.

Linkage between Sensory and Chemical Composition

Scatter plots were firstly used to visually determine linear, exponential or other possible associations between sensory attributes and non-volatile molecules with known sensory activity (Sáenz-Navajas, et al. 2012). The regression coefficients (R^2) were calculated with EXCELL and only regressions with $R^2 > 0.25$ are reported.

Cluster Analysis Based on Quality Scores

Hierarchical cluster analysis (HCA) was used to cluster experts using a Euclidian distance dissimilarity matrix created from expert quality scores. HCA was applied to the transposed matrix of expert scores, in which the experts are the rows and the wines are the columns. The lines of this matrix have been previously centered. Ward's method was used in the agglomeration.

Linkage between Chemical Composition and Quality Evaluation: External Quality Mapping

A first PCA was performed on the mean scores of two replicates of 22 chemical variables, which are expected to be sensory-active. According to the Kaiser law dimensions, an eigenvalue higher than one was retained.

Then, external preference mapping was conducted by regressing the centroid of each cluster derived from quality evaluation onto the significant PCs obtained from chemical variables (independent variables) using XLSTAT. Regressions adjusted to a vectorial model are called "the more, the better", meaning that the ideal quality wine is located as far as possible towards the arrow projected on the PCA. For the elliptical model either an anti-ideal (-) or an ideal (+) quality point is obtained. This means that a maximum or minimum level of chemical response exists, contrary to what is observed in the vectorial model: "more is better".

RESULTS AND DISCUSSION

Performance of Trained Panelists

A PCA run on the in-mouth attributes revealed that judges' projections were grouped in the loading plot (data not shown) for sourness, bitterness and astringency, global intensity and persistence. This indicates that the panel agrees in the interpretation of these terms. On the contrary, for sweetness judges were spread over the loading plot.

These results could be explained by the fact that either the assessors do not interpret similarly the sweet attribute or sensory differences among wines are too small for being perceived by most panelists. This last hypothesis is supported by the fact that the studied samples are all dry red wines with a content in reducing sugars lower than 5 g L^{-1}. Similarly, it is not probable that glycerol (a compound usually linked to sweetness perception in wines) contributes to the sweet taste in most of the studied wines, since only one sample (NJ_R09) presents a concentration higher than its sensory threshold, which is reported to be 7.5 g L^{-1} in dry red wines (Hufnagel and Hofmann 2008). Hence, sweetness was not considered in subsequent analysis.

In the five one-way ANOVAs calculated with the sensory attributes evaluated in-mouth (sourness, bitterness, astringency, global intensity and persistence), the effect of wine was significant in all cases ($P < 0.001$), which suggests that these attributes are useful in characterizing sensory differences among this set of wines.

Understanding Global Intensity and Persistence

Simple linear regressions have been calculated between the in-mouth attributes evaluated by the trained panel. Results show that global intensity and persistence are significantly correlated ($R^2 = 0.81$; $P < 0.001$) as has been also observed in other sets of red wines (Sáenz-Navajas et al., 2010; 2011). Besides, both terms are mainly related to the bitter taste (R^2 = 0.33 and 0.36, respectively; $P < 0.05$), differently from previous results (Sáenz-Navajas et al., 2011) that showed that they were especially related to astringency. This result could be derived from the fact that the trained panel reported in (Sáenz-Navajas et al., 2011) did not have aligned concepts for bitter perception, which could be the result of the lack of sensitivity to this taste. This is not surprising as roughly 25% of the human population inherit a trait that renders them unable to detect bitterness (Kim and Drayna, 2005). Thus, the concepts of persistence and global intensity were mainly based in astringency, differently from the panel of the present study which was able to establish differences among the 16 wines in terms of bitterness and thus mainly related these two attributes to the bitter taste.

Sensory Activity

To evaluate the link between chemical composition and sensory differences perceived by the trained panel, two-dimensional regressions on sensory descriptive and chemical data were conducted.

Sourness

Linear regression coefficients show that the sour taste presents exponential relationships with tartaric ($R^2 = 0.36$; $P < 0.05$) and succinic ($R^2 = 0.59$; $P < 0.01$) acids. These results are in relative good accordance with previous literature reporting that tartaric acid measured by means of titratable acidity (Sáenz-Navajas et al., 2010) as well as succinic acid (Biasoto et al., 2010) concentrations are good predictors of the perceived sourness in wines since significant linear correlations are found. However, the present results demonstrate that the predictions are considerably improved by considering an exponential rather than a linear trend as was recently suggested for volatile composition and aroma perception (Chambers and Koppel, 2013). As an example, Figure 1 displays the exponential trend observed between sourness and succinic acid concentration in 12 samples of the 16 wines studied. The presence of four outliers can be the consequence of interactions (perceptual and physicochemical) between succinic acid and other non-volatile or even between volatile compounds as the trained panel evaluated the in-mouth properties in the presence of aroma. These sensory interactions have been reported in real (Sáenz-Navajas et al., 2010) and reconstituted red wines (Sáenz-Navajas et al., 2010) and could also explain the absence of relationships between the sour scores given by the trained panel and non-volatile molecules with known sour taste and present at concentrations well above their sensory threshold such as malic and lactic acids or between variables such as pH, volatile and titratable acidity values (see Table 2). Another possible explanation to the fact that these variables do not present any relationship with the sour taste in the studied wines is that their ranges are not wide enough to be able to create sensory differences among the samples.

Interestingly, although citric, *c*- and *t*-aconitic acids have been described to be sour, their sensory impact in the studied wines seems to be limited, probably because their presence is below their sensory threshold for the sour taste as it is shown in Table 2.

Table 2. Sensory thresholds and reference for sour compounds, concentration ranges expressed in mg L^{-1}, median and the difference between the maximum (max) and minimum (min) concentration, regression equation and its coefficient (R^2) in the studied wines

	threshold[c]	concentration range	median	max-min	equation[d]	R^2
pH		3.4-3.9	3.6	0.5		
titratable acidity (g L^{-1})[a]		4.4.-6.1	5.3	1.7		
volatile acidity (g L^{-1})[b]		0.3-0.7	0.5	0.5		
tartaric acid (g L^{-1})	0.044 (Hufnagel and Hofmann, 2008)	0.7-1.2	0.9	0.5	$2.74x^2-3.84x+3.81$	0.36
citric acid (mg L^{-1})	0.35 (Warmke et al., 1996)	0.0-69	17	69		
succinic acid (mg L^{-1})	0.47 (Warmke et al., 1996)	112-287	215	174	$2*10^{-5}x^2-0.002x+2.36$	0.59
malic acid (g L^{-1})	0.076 (Warmke et al., 1996)	0.2-0.4	0.2	0.2		
lactic acid (g L^{-1})	1.4 (Warmke et al., 1996)	0.7-3.3	1.5	2.6		
c-aconitic acid (mg L^{-1})	87 (Hufnagel and Hofmann, 2008)	2.5-3.9	3.1	1.4		
t-aconitic acid (mg L^{-1})	87 (Hufnagel and Hofmann, 2008)	5.1-10.3	8.7	5.2		

[a] Expressed in tartaric acid.
[a] Expressed in acetic acid.
[c] Calculated for individual compounds in water in the reference (see below) in parenthesis.
[d] The independent variable (x) is the concentration of the corresponding compound and the dependent variable is the score given by the trained panel to the sour taste.

Bitterness

There are 15 bitter compounds analyzed, including monomers and dimmers of flavanols, flavonols and ethyl esters of hydroxicinamic acids. These compounds are at normal concentration ranges reported in Spanish wines (Sáenz-Navajas et al., 2012; Sáenz-Navajas et al., 2010), however they are not likely to contribute to bitterness since their content is well below sensory threshold for bitterness (Table 3).

Interestingly, both succinic acid ($R^2 = 0.62$; $P < 0.01$) and to a lesser extent the alcohol content ($R^2 = 0.55$; $P < 0.01$) present exponential and linear relationships, respectively, with the perceived bitter taste (Table 3). The bitter character of succinic acid has been already reported (Whiting, 1976), however it is the first time that its relevance in the bitterness of wines is suggested. The linkage between alcohol content and bitterness seems reasonable as alcohol has been already proved to enhance perceived bitterness at typical wine levels in both model systems (Fontoin et al., 2008) and real wine samples (Noble, 1994).

Table 3. Sensory thresholds and reference for bitter compounds (nd: not detected), concentration ranges expressed in mg L^{-1}, median and the difference between the maximum (max) and minimum (min) concentration, regression equation and its coefficient (R^2) in the studied wines

	Threshold (mg L^{-1})[b]	range	median	max-min	equation[c]	R^2
flavanols						
catechin	290 (Scharbert et al., 2004)	8.72-24.6	15.1	15.84		
epicatechin	270 (Scharbert et al., 2004)	2.64-22.5	6.4	19.82		
procyanidin B1	231 (Hufnagel and Hofmann, 2008)	16.5-36.7	27.4	20.26		
procyanidin B2	280	2.92-23.8	8.4	20.91		
procyanidin B3	289 (Hufnagel and Hofmann, 2008)	0.86-1.81	1.4	0.95		
flavonols						
myricetin	10 (Dadic and Belleau, 1973)	1.15-8.37	5.59	7.22		
quercetin	10 (Dadic and Belleau, 1973)	0.15-9.55	4.40	9.40		
kaenmpferol	20 (Dadic and Belleau, 1973)	<0.03-1.25	0.38	1.22		
phenolic acid ethyl esters						
caffeic acid ethyl ester	229 (Hufnagel and Hofmann, 2008)	122-0.04	0.30	1.18		
gallic acid ethyl ester	438 (Hufnagel and Hofmann, 2008)	177-40.9	77.5	136		
p-coumaric acid ethyl ester	137 (Hufnagel and Hofmann, 2008)	1.76-0.10	0.50	1.67		
vanillic acid ethyl ester	125 (Hufnagel and Hofmann, 2008)	nd				
ferulic acid ethyl ester	158 (Hufnagel and Hofmann, 2008)	nd				
protocatechuic acid ethyl ester	182 (Hufnagel and Hofmann, 2008)	1.25-0.05	0.20	1.20		
syringic acid ethyl ester	130 (Hufnagel and Hofmann, 2008)	nd				
miscellaneous						
alcohol content (% v/v)		13.4-15.2	13.9	1.86	$0.29x-1.2$	0.55
succinic acid		112-287	215	174.73	$3*10^{-5}x^2-0.01x+3.14$	0.62

[a]Average degree of polymerization calculated for oligomers of flavanols (from trimers).
[b]Calculated for individual compounds in water in the reference (see below) in parenthesis.
[c]The independent variable (x) is the concentration of the corresponding compound and the dependent variable is the score given by the trained panel to the bitter taste.

Table 4. Sensory thresholds and reference for astringent compounds (nd: not detected), concentration ranges expressed in mg L^{-1}, median and the difference between the maximum (max) and minimum (min) concentration, regression equation and its coefficient (R^2) in the studied wines

	Sensory threshold[b]	range	median	max-min	equation[c]	R^2
FLAVANOLS						
catechin	119 (Scharbert et al., 2004)	8.72-24.6	15.1	15.8		
catechin gallate	110 (Scharbert et al., 2004)	<0.03-0.33	0.0	0.30		
epicatechin	270 (Scharbert et al., 2004)	2.64-22.5	6.4	19.8		
epicatechin gallate	110 (Scharbert et al., 2004)	<0.04-4.29	0.5	4.25		
epigallocatechin	169 (Scharbert et al., 2004)	0.31-2.18	0.9	1.87		
procyanidin B1	139 (Hufnagel and Hofmann, 2008)	16.5-36.7	27.4	20.2	$0.058x + 1.32$	0.63
procyanidin B2	110 (Hufnagel and Hofmann, 2008)	2.92-23.8	8.4	20.9		
procyanidin B3	139 (Hufnagel and Hofmann, 2008)	0.86-1.81	1.4	0.95	$0.94x + 1.62$	0.60
IPTs		45.4-83.4	59.7	38.0	$0.030x + 1.21$	0.78
Protein-precipitable PAs		0.20-0.62	0.4	0.42	$4.95x^2 - 1.49x + 2.72$	0.62
Polymeric PAs		161-723	281	561	$-1*10^{-6}x^2 + 0.003x + 2.21$	0.47
PHENOLIC ACIDS						
gallic acid	45 (Hufnagel and Hofmann, 2008)	35.50-146.8	60.5	111		
protocatechuic acid	32 (Hufnagel and Hofmann, 2008)	1.32-16.6	2.5	15.3		
caffeic acid	13 (Hufnagel and Hofmann, 2008)	3.28-13.7	5.4	10.4		
caftaric acid	5 (Hufnagel and Hofmann, 2008)	42.6-148	69.9	105.5		
coumaric acid	23 (Hufnagel and Hofmann, 2008)	1.09-4.42	2.1	3.33		
coutaric acid	25 (Okamura and Watanabe, 1981)	5.69-44.30	17.0	38.6		
vainillic acid	53 (Hufnagel and Hofmann, 2008)	1.45-11.5	4.9	10.1		
ferulic acid	13 (Hufnagel and Hofmann, 2008)	0.72-3.75	1.3	3.03		
syringic acid	52 (Hufnagel and Hofmann, 2008)	2.75-11.4	4.9	8.60		
PHENOLIC ACID ETHYL ESTERS						
protocatechuic ethyl ester	9 (Hufnagel and Hofmann, 2008)	<0.06-1.25	0.2	1.20		
caffeic ethyl ester	69 (Hufnagel and Hofmann, 2008)	0.04-1.22	0.3	1.18		
ferulic ethyl ester	15 (Hufnagel and Hofmann, 2008)	nd				

	Sensory threshold[b]	range	median	max-min	equation[c]	R^2
gallic ethyl ester	37 (Hufnagel and Hofmann, 2008)	40.9-177	77.5	136		
vanillic ethyl ester	25 (Hufnagel and Hofmann, 2008)	nd				
coumaric ethyl ester	27 (Hufnagel and Hofmann, 2008)	0.10-1.76	0.50	1.67		
syringic ethyl ester	4 (Hufnagel and Hofmann, 2008)	nd				
flavonols						
quercetin-3-O-galactoside	0.2 (Hufnagel and Hofmann, 2008)	<0.02-0.51	0.2	0.49		
quercetin-3-O-glucoside	1.0 (Hufnagel and Hofmann, 2008)	<0.02-2.43	0.1	2.41		
isoharmnetin-3-O-glucoside	1.1 (Hufnagel and Hofmann, 2008)	<0.03-0.75	0.0	0.72		
myricetin	10 (Dadic and Belleau, 1973)	1.15-8.37	5.6	7.22		
	Sensory threshold[b]	range	median	max-min	equation[c]	R^2
quercetin	10 (Dadic and Belleau, 1973)	0.15-9.55	4.4	9.40		
kaempferol	20 (Dadic and Belleau, 1973)	<0.03-1.25	0.4	1.22		
MISCELLANEOUS						
alcohol content % (v/v)		13.4-15.2	13.9	1.86		
c-aconitic acid	0.1 (Hufnagel and Hofmann, 2008)	2.50-3.89	3.1	1.39		
t-aconitic acid	0.1 (Hufnagel and Hofmann, 2008)	5.10-10.3	8.7	5.20	$0.058x + 1.32$	0.63

[a]Average degree of polymerization calculated for oligomers of flavanols (from trimers).

[b]Calculated for individual compounds in water in the reference (see below) in parenthesis.

[c]The independent variable (x) is the concentration of the corresponding compound and the dependent variable is the score given by the trained panel to astringency.

Astringency

A significant positive linear trend (R^2 = 0.78; P < 0.001) is observed between the total polyphenol content (measured by the TPI) and the astringency scored by the trained panel (Table 4). Among this total polyphenolic composition both protein-precipitable PAs (R^2 = 0.62; P < 0.01) and polymeric PAs (higher than timers) (R^2 = 0.47; P < 0.05) present exponential trends with astringency (Table 4). Non-polymerized flavanols such as monomers and dimmers are not expected to be important drivers of the perceived astringency in this set of wines as all of them are below their sensory threshold (Table 4), however procyanidins B1 (R^2 = 0.63; P < 0.01) and B3 (R^2 = 0.60; P < 0.01) show a positive correlation with astringency. This could be an indirect link derived from the formation of more polymerized flavanols (i.e., protein-precipitable PAs and polymeric PAs), which takes place from its oligomers. This suggests that wines with higher concentrations in dimmers of flavanols such as B1 or B3 would lead to a higher formation of polymerized PAs, which have been demonstrated to be the main drivers of sensory astringency in wines (Sáenz-Navajas et al., 2012).

Among phenolic acids only the derivatives formed by esterification such as caftaric and coutaric acids as well as gallic acid ethyl ester, the flavonols quercetin-3-*O*-glucoside and quercetin-3-*O*-galactoside, and both *c*- and *t*-aconitic acids are at concentrations above their sensory threshold (Table 4), however no systematic relationship with astringency was found.

Apart from the polyphenolic composition, which is widely accepted to be the main driver of astringency in wines, the content in ethanol has been reported to be also an important factor increasing astringency in model solutions (Fontoin et al., 2008), however in real wine samples this effect has not been found (Fischer et al., 1994), what is in accordance with the data derived from the present study. Similarly, *t*-aconitic acid has been repeatedly found to be an important driver of astringency in red wines (Hufnagel and Hofmann, 2008; Sáenz-Navajas et al., 2012) as it is observed with the present set of wines since it presents an exponential trend (R^2 = 0.35; P < 0.05) with astringency sensation evaluated by the trained panel.

Quality Assessment by Experts

A cluster analysis calculated on the 21 experts´ scores has yielded three main clusters as shown in Figure 2. The first cluster is constituted by 7 experts/judges (J), the second and the third clusters are both formed by six and eight experts, respectively. The fact that expert quality judgments based on exclusively in-mouth properties derives in three main clusters demonstrates the heterogeneity among experts in the in-mouth quality construct. Previous research has shown that for experts the in-mouth properties seem to be less relevant than aroma stimuli, when evaluating wine quality (Sáenz-Navajas et al., 2011). This could have led to develop dissimilar criteria when evaluating exclusively in-mouth sensations. This is further supported by the results of an independent study (data not shown) carried out with the same wines and the same panel of experts, where quality was firstly evaluated based exclusively on orthonasal aroma and then using all stimuli (i.e., as experts usually do in normal wine tasting). Results showed that under both conditions, experts shared the same quality concept and the global quality assessment was highly correlated to the orthonasal-based quality ($R^2 = 0.60$; $P < 0.01$), while in-mouth-based quality is far less correlated ($R^2 = 0.40$; $P < 0.05$).

Correlation between Chemical and Quality Variables: External Quality Mapping

Non-volatile molecules with known sensory activity in wine were quantified (Sáenz-Navajas et al., 2012). Chemical compounds that are below their sensory threshold in all wines are not expected to have a significant sensory impact on the studied wines. These compounds are eliminated from further data analyses. Exceptionally, procyanidins B1 and B3 are considered as they present significant linear correlations with astringency, which has been found to be related to wine quality (Sáenz-Navajas et al., 2011). Similarly, reducing sugars are also included since there is evidence that they are involved in quality perception even when their content is below sensory threshold for sweet taste (Sáenz-Navajas et al., 2012, Sáenz-Navajas, et al., 2010). The compounds above their sensory threshold such as glycerol, tartaric, malic and lactic acids, procyanidins B1 and B2, quercetin-*O*-3-glucoside, quercetin-*O*-3-galactoside, coutaric, caftaric and *t*-aconitic acid, gallic acid ethyl ester are considered for PCA. Besides, variables showing significant correlations with bitterness such as alcohol content or with astringency such as IPTs, polymeric PAs or protein-precipitable PAs are also taken into account. Conventional oenological variables pH, titratable and volatile acidities, which are widely used in wineries for quality controls, are also considered in the PCA.

To sum up a total of 22 chemical variables are submitted to PCA. According to the Kaiser criterion the first six principal components (PC), which account for more than 84% of the original variance, are considered. Table 5 shows the correlations between the chemical parameters and the first six PCs.

The average quality scores obtained for the three clusters of consumers are further regressed onto the six PCs obtained from the PCA calculated across wines with the chemical variables. Figures 3a and 3b show the quality maps where clusters are projected on PC1-PC2 and PC1-PC3, respectively. Figures 4a and 4b show the projections of the wines onto the PC1-PC2 and PC1-PC3 subspaces, respectively. These maps allow identifying the wines that are considered by experts to be of high quality as well as the chemical drivers of quality.

The arrow representing cluster 1 (formed by 33% of the panel) in Figures 3a and 3b indicates that the quality evaluated by the participants of this cluster is adjusted to a vector model (linear relationship). The coefficients of the model are shown in Table 6. This model is called "the more, the better" suggesting that for experts fitted by this model, the ideal quality wine is located as far as possible towards the arrow. This means that the chemical variables correlated with this vector have to be maximized for achieving high quality wines. The correlation matrix calculated between the six significant PCs of the PCA and the quality scores of clusters shows a significant negative correlation between PC5 and cluster 1 ($r = -0.60$). This suggests that chemicals weighting heavily on this PC may be more critical to quality perception for cluster 1. Hence, high contents in protein-precipitable PAs and in *c*-aconitic acid result in higher quality scores. Both compounds have been reported to be important drivers of astringency in wines (Sáenz-Navajas et al., 2012). Thus, this result is well in accordance with previous studies carried out with Spanish (Sáenz-Navajas et al., 2011), Australian (Lattey et al., 2010) and Uruguayan (Varela and Gambaro, 2006) experts showing that perceived astringency of wines is positively correlated with quality perception. A positive correlation is observed between PC3 (explains 13.5% of the original variance) and cluster 1 (r = 0.42), which is positively correlated with IPTs, caftaric acid and quercetin-3-*O*-galactoside. Beside, a minor negative correlation is observed between PC6 and cluster 1 (r = -0.32). This PC is positively correlated with succinic acid, which suggests that high concentrations of this acid are related to low quality ratings for this cluster of experts. However, given the low correlation coefficient and the hardly 7% of total variance explained by this PC6, its role in quality perception seems to be limited.

Table 5. Correlation coefficients between chemical variables and the six first factors (PCs) obtained from the PCA. Significant correlations are marked in bold

	PC1	PC2	PC3	PC4	PC5	PC6
IPTs	-0.448	-0.508	0.531	-0.242	-0.043	0.336
volatile acidity	0.002	-0.533	-0.345	-0.390	0.346	-0.033
reducing sugars	-0.730	-0.273	0.336	-0.349	-0.030	-0.159
alcohol content	-0.651	0.155	0.377	-0.161	-0.192	0.458
tartaric acid	-0.233	0.235	0.189	0.795	0.220	0.034
succinic acid	-0.534	0.416	-0.340	-0.057	0.118	0.603
glycerol	-0.474	0.722	-0.116	-0.013	-0.224	0.121
protein-precipitable PAs	-0.181	-0.428	0.220	-0.210	0.520	0.492
polymeric PAs	0.560	-0.567	0.013	-0.119	-0.299	-0.168
procyanidin B1	0.501	-0.525	0.319	0.453	0.167	0.128
procyanidin B3	0.295	-0.386	0.432	0.636	0.097	0.118
t-aconitic acid	0.595	0.414	0.363	-0.082	0.089	0.078
titratable acidity	-0.707	-0.234	-0.150	0.434	-0.055	-0.256
pH	0.806	0.056	0.060	-0.247	0.072	0.411
malic acid	-0.353	-0.701	-0.256	-0.058	-0.384	0.186
lactic acid	0.905	-0.012	-0.280	-0.005	0.066	0.229
caftaric acid	-0.030	-0.152	0.646	-0.190	-0.549	-0.035
coutaric acid	0.820	-0.225	-0.029	-0.137	-0.316	0.059
c-aconitic acid	0.166	0.198	0.537	-0.201	0.610	-0.307
quercetin-3-*O*-galactoside	0.058	0.346	0.861	0.199	-0.198	0.003
quercetin-3-*O*-glucoside	-0.310	-0.006	0.270	-0.677	0.272	-0.322
galic acid ethyl ester	-0.630	-0.545	-0.099	0.392	0.189	-0.007

The groups of experts that constitute cluster 2 (33% of the panel) and cluster 3 (38%) are significantly fitted by the elliptical model ($P < 0.05$ in both cases) with an anti-ideal (-) and an ideal (+) quality point, respectively (the coefficients of both models are shown in Table 6). This means that a maximum or a minimum level of chemical composition exists, contrary to what is observed in the vector model: "more is better". Thus, experts belonging to cluster 2 have an anti-ideal point on high negative values for PC1. This anti-ideal quality wine is high in reducing sugars, alcohol content, titratable acidity and gallic acid ethyl ester, while low in pH-values, *t*-aconitic, coutaric and lactic acids. Besides, the anti-ideal point is on the origin of the map for PC2 and PC3, which can be seen as wines with medium chemical composition driven by both PCs. This means that the anti-ideal quality wine would have average concentrations in volatile acidity, glycerol, polymeric PAs, malic acid as well as in IPTs, caftaric acid and quercetin-3-O-galactoside.

Table 6. Coefficients of the quality models obtained for the three clusters of experts. The centroid of the quality scores calculated for each cluster is the predicted variable (Y) and the independent variables are the first three PCs derived from the PCA calculated on the 22 non-volatile molecules with sensory activity. The equation is: $Y = a + b_1PC1 + b_2PC2 + b_3PC3 + c_1PC1^2 + c_2PC2^2 + c_3PC3^2$

	Intercept (a)	b_1	b_2	b_3	c_1	c_2	c_3
Cluster 1 (n = 7)	0.000	0.049	-0.103	0.235	0.000	0.000	0.000
Cluster 2 (n = 6)	-1.201	0.015	0.444	0.025	0.115	0.028	0.135
Cluster 3 (n = 8)	1.056	-0.137	0.002	-0.051	-0.034	-0.135	-0.123

Table 7. Compositional data for the wines closest (CT_B07, SO_C07) to the ideal point defined for cluster 3, maximum, average and minimum values in the studied set of wines

	CT_B07	SO_C07	Max	average	min
reducing sugars (g L^{-1})	4.34	3.81	4.35	2.65	1.52
glycerol (g L^{-1})	6.53	6.83	7.58	6.70	5.03
alcohol content (%, v/v)	13.9	13.8	15.2	14.1	13.4
pH	3.47	3.53	3.88	3.61	3.43
titratable acidity (g L^{-1})	5.66	5.66	6.14	5.33	4.45
volatile acidity (g L^{-1})	0.51	0.75	0.75	0.54	0.25
malic acid (g L^{-1})	0.30	0.18	0.35	0.23	0.17
lactic acid (g L^{-1})	0.75	1.21	3.30	1.65	0.68
IPTs (Abs)	59.1	54.9	83.3	60.4	45.4
polymeric PAs (g L^{-1})	304	162	723	308	161
procyanidin B1 (mg L^{-1})	21.4	22.1	36.7	26.6	16.5
caftaric acid (mg L^{-1})	126	70.8	148	82.2	42.6
coutaric acid (mg L^{-1})	15.2	16.4	44.3	18.9	5.7
quercetin-3-*O*-galactoside (mg L^{-1})	0.20	0.11	0.51	0.24	0.02

For cluster 3 an ideal point is found near the wines SO_C and CT_B. Interestingly, both wines are among the most aged wines (vintage 2007). Table 7 shows the chemical composition of these two close-to-ideal quality wines (together with the maximum, minimum and average values calculated on the 16 studied red wines). Both wines have in common that they present higher than average content in reducing sugars and titratable acidity. On the contrary, these quality wines have lower than average values for: alcohol content, pH, lactic acid, IPTs, polymeric PAs, procyanidin B1, coutaric acid and quercetin-3-*O*-galacatoside.

CONCLUSION

Results show that experts have heterogeneous quality concepts when exclusively in-mouth properties are evaluated. External quality mapping after segmentation of consumers by hierarchical clustering is presented as a powerful tool able to disclose the main chemical drivers of quality.

Three clusters of experts responding to different in-mouth quality criteria are reported. Interestingly, experts were best fit by the elliptical model (71% of the panel) and to a lesser extent by the vector (29%) model. The significant elliptical models obtained for most of the experts (two clusters out of three) suggest that it exists a maximum or minimum level of chemical chemical composition better than "the more the better" suggested by the traditionally used vector model. To our knowledge, this is the first time that quality mapping techniques are used to evaluate quality perception of wines by a panel of experts and it is presented as a promising statistical tool able to combine both sensory and chemical disciplines.

ACKNOWLEDGMENTS

The research reported in this article was financially supported by the Spanish Ministry of Education (AGL2010-22355-CO2-01/02) and *Instituto de Estudios Riojanos* (La Rioja Government, Spain). M.P.S.N. acknowledges the Spanish Education Ministery (M.E.C.) for her postdoctoral fellowship and J.M.A. the Navarra Government for his predoctoral grant. Authors want also to thank experts and panelists for their participation in the sensory sessions with interest and diligence.

REFERENCES

AOAC (2002). Offical Methods of Analysis. Method 993.05: L-Malic acid/total malic acid ratio in apple juice.

Aurifeille, J.M., Quester, P.G., Lockshin, L., Spawton, T., (2002). Global vs international involvement-based segmentation - A cross-national exploratory study. *Int. Market. Rev.* 19, 369-386.

Ballester, J., Patris, B., Symoneaux, R., Valentin, D., (2008). Conceptual vs. perceptual wine spaces: Does expertise matter? *Food Qual. Pref.* 19, 267-276.

Biasoto, A.C.T., Catharino, R.R., Sanvido, G.B., Eberlin, M.N., da Silva, M.A.A.P., (2010). Flavour characterization of red wines by descriptive analysis and ESI mass spectrometry. *Food Qual. Pref.* 21, 755-762.

Colonna, A.E., Adams, D.O., Noble, A.C., (2004). Comparison of procedures for reducing astringency carry-over effects in evaluation of red wines. *Aust. J. Grape Wine Res.* 10, 26-31.

Chambers, E., Koppel, K., (2013). Associations of Volatile Compounds with Sensory Aroma and Flavor: The Complex Nature of Flavor. *Molecules* 18, 4887-4905.

Charters, S., Pettigrew, S., (2006). Product involvement and the evaluation of wine quality. *Qual. Market Res.* 9, 181-193.

Dadic, M., Belleau, G., (1973). Polyphenols and beer flavor. Proceeding of the American Society of Brewing Chemistry 4, 107-114.

Delgado, C., Guinard, J.X., (2011). How do consumer hedonic ratings for extra virgin olive oil relate to quality ratings by experts and descriptive analysis ratings? *Food Qual. Pref.* 22, 213-225.

Ferreira, V., San Juan, F., Escudero, A., Cullere, L., Fernandez-Zurbano, P., Sáenz-Navajas, M.P., Cacho, J., (2009). Modeling quality of premium Spanish red wines from gas chromatography-olfactometry data. *J. Agric. Food Chem.* 57, 7490-7498.

Fischer, U., Boulton, R.B., Noble, A.C., (1994). Physiological factors contributing to the variability of sensory assessments: Relationships between saliva flow rate and temporal perception or gustatory stimuli. *Food Qual. Pref.* 5, 55-64.

Fontoin, H., Saucier, C., Teissedre, P.-L., Glories, Y., (2008). Effect of pH, ethanol and acidity on astringency and bitterness of grape seed tannin oligomers in model wine solution. *Food Qual. Pref.* 19, 286-291.

Gawel, R., Godden, P.W., (2008). Evaluation of the consistency of wine quality assessments from expert wine tasters. *Aust. J. Grape Wine Res.* 14, 1-8.

Hufnagel, J.C., Hofmann, T., (2008). Orosensory-directed identification of astringent mouthfeel and bitter-tasting compounds in red wine. *J. Agric. Food Chem.* 56, 1376-1386.

Hufnagel, J.C., Hofmann, T., (2008). Quantitative reconstruction of the nonvolatile sensometabolome of a red wine. *J. Agric. Food Chem.* 56, 9190-9199.

Kim, U.K., Drayna, D., (2005). Genetics of individual differences in bitter taste perception: lessons from the PTC gene. *Clin. Genet.* 67, 275-280.

Lattey, K.A., Bramley, B.R., Francis, I.L., (2010). Consumer acceptability, sensory properties and expert quality judgements of Australian Cabernet Sauvignon and Shiraz wines. *Aust. J. Grape Wine Res.* 16, 189-202.

Llaudy, M.C., Canals, R., Canals, J.M., Rozes, N., Arola, L., Zamora, F., (2004). New method for evaluating astringency in red wine. *J. Agric. Food Chem.* 52, 742-746.

Machado, B., Graça, A., Hirson, G., Heymann, H., (2011). Revealing the secret preferences for top-rated dry red wines through sensometrics. 16th International Oenology Symposium Bolzano, Italy, 145-149.

Marin, A.B., Durham, C.A., (2007). Effects of wine bottle closure type on consumer purchase intent and price expectation. *Am. J. Enol. Vitic.* 58, 192-201.

Marin, A.B., Jorgensen, E.M., Kennedy, J.A., Ferrier, J., (2007). Effects of bottle closure type on consumer perceptions of wine quality. *Am. J. Enol. Vitic.* 58, 182-191.

Noble, A.C., (1994). Biterness in wine. *Physiol. Behav.* 56, 1251-1255.

Okamura, S., Watanabe, M., (1981). Determination of phenolic cinnamates in white wine and their effect on wine quality. *Agric. Biol. Chem.* 45, 2063-2070.

Parr, W.V., Mouret, M., Blackmore, S., Pelquest-Hunt, T., Urdapilleta, I., (2011). Representation of complexity in wine: Influence of expertise. *Food Qual. Pref.* 22, 647-660.

Ribéreau-Gayon, P., (1970). Le dosage des composes phénoliques totaux dans les vins rouges. *Chimie Analytique* 52, 627-631.

Sáenz-Navajas, M.P., Avizcuri, J.M., Ferreira, V., Fernández-Zurbano, P., (2012). Insights on the chemical basis of the astringency of Spanish red wines. *Food Chem.* 134, 1484-1493.

Sáenz-Navajas, M.P., Echavarri, F., Ferreira, V., Fernandez-Zurbano, P., (2011). Pigment composition and color parameters of commercial Spanish red wine samples: linkage to quality perception. *Eur. Food Res. Technol.* 232, 877-887.

Sáenz-Navajas, M.P., Fernández-Zurbano, P., Ferreira, V., (2012). Contribution of nonvolatile composition to wine flavor. *Food Rev. Int.* 28, 389-411.

Sáenz-Navajas, M.P., Fernandez-Zurbano, P., Ferreira, V., Dizy, M., (2010). Characterization of taste-active fractions in red wine combining HPLC fractionation, sensory analysis and ultra performance liquid chromatography coupled with mass spectrometry detection. *Anal. Chim. Acta* 673, 151-159.

Sáenz-Navajas, M.P., Fernandez-Zurbano, P., Martin-Lopez, C., Ferreira, C., (2011). Sensory properties of premium Spanish red wines and their implication in wine quality perception. *Aust. J. Grape Wine Res.* 17, 9-19.

Sáenz-Navajas, M.P., Fernandez-Zurbano, P., Tao, Y.S., Dizy, M., Ferreira, V., (2010). Relationship between nonvolatile composition and sensory properties of premium Spanish red wines and their correlation to quality perception. *J. Agric. Food Chem.* 58, 12407-12416.

Sáenz-Navajas, M.P., Ferreira, V., Campo, E., Fernandez-Zurbano, P., Valentin, D., (2010). An assessment of the effects of wine volatiles on the perception of taste and astringency in wine. *Food Chem.* 121, 1139-1149.

Sáenz-Navajas, M.P., Gonzalez-Hernadez, M., Campo, E., Fernandez-Zurbano, P., Ferreira, V., (2012). Orthonasal aroma characteristics of Spanish red wines from different price categories and their relationship to expert quality judgements. *Aust. J. Grape Wine Res.* 18, 268-279.

Scaman, C.H., Dou, J., Cliff, M.A., Yuksel, D., King, M.C., (2001). Evaluation of wine competition judge performance using principal component similarity analysis. *J. Sens. Stud.*16, 287-300.

Scharbert, S., Holzmann, N., Hofmann, T., (2004). Identification of the astringent taste compounds in black tea infusions by combining instrumental analysis and human bioresponse. *J. Agric. Food Chem.* 52, 3498-3508.

Schlich, P., (1995). Preference mapping: Relating consumer preferences to sensory or instrumental measurements. In: Bioflavour 95: Analysis - Precursor Studies - Biotechnology, Eds. P. Etievant and P. Schreier pp. 135-150.

Schlich, P., Mcewan, J.A., (1992). Preference Mapping - a Statistical Tool for the Food-Industry. *Sci. Aliment.* 12, 339-355.

Sun, B.S., Ricardo-da-Silva, J.M., Spranger, and I., (1998). Critical factors of vanillin assay for catechins and proanthocyanidins. *J. Agric. Food Chem. v*46, 4267-4274.

Urdapilleta, I., Parr, W.V., Dacremont, C., Green, J., (2011). Semantic and perceptive organisation of Sauvignon blanc wine characteristics Influence of expertise. *Food Qual. Pref.* 22, 119-128.

Varela, P., Gambaro, A., (2006). Sensory descriptive analysis of Uruguayan Tannat wine: Correlation to quality assessment. *J. Sens. Stud.* 21, 203-217.

Warmke, R., Belitz, H.D., Grosch, W., (1996). Evaluation of taste compounds of Swiss cheese (Emmentaler). Zeitschrift Fur Lebensmittel-Untersuchung Und-Forschung 203, 230-235. doi: 10.1007/bf01192869.

Whiting, G.C., (1976). Organic-acid metabolism of yeasts during fermentation of alcoholic beverages - Review. *J. Inst. Brew.* 82, 84-92.

In: Wine
Editor: Youssef El Rayess

ISBN: 978-1-63321-048-6

Chapter 9

Possibilities of the Use of Phenolic Compounds when Assessing the Authenticity of Wine Samples

Pavel Pavloušek**, Michal Kumšta and Jan Kupsa
Mendel University Brno, Faculty of Horticulture Lednice, Department of Viticulture and Enology, Lednice na Moravě, Czech Republic

Abstract

Phenolic compounds can be used for the assessment of wine samples authenticity on the base of their geographic origin, variety and/or vintage. The research that is being carried out in the Czech Republic is focused on the application of phenolic compounds estimation when assessing the geographic origin either of a wine or of a grape variety. In this paper, we present results obtained in HPLC assays when assessing levels of phenolic substances in wine samples within the period of 2010-2013. Possibilities of the use of phenolic compounds when assessing the authenticity of wine samples are illustrated with results of the following three experiments: (i) study on relationships existing between varietal samples of white wine (Riesling) and the region; (ii) study on relationships existing between white wine and terroir or variety and (iii) study on relationships existing between analyzed samples of red wine and terroir. Obtained experimental results proved that phenolic compounds are very suitable for the assessment of wine authenticity just on the base of the variety and/or terroir.

Keywords: Chemometric analysis, Geographical origin, HPLC, Phenolic compounds, Wine

Introduction

Wine is one of the most significant beverages in a worldwide scale and is a very important food commodity. Grapes and wine are a very complex mixture of several hundred various chemical compounds which can also be described as primary and secondary metabolites. These metabolites can be influenced by terroir of vineyard, wine-growing region and the winemaking techniques.

Geographical origin can be an important attribute of commercial plant-derived foods as it gives an indication of quality, phytochemical composition, the amount of phytochemical variation excepted and appeals to certain values of consumers. Additionally, where regions produce crops with specific terroir, being able to classify geographical origin is a way of protecting the reputation and value of specific products as is specially recognized in the European Union under "Protected Designation of Origin" and "Protected Geographical Indication" classifications (McGhie and Rowan, 2012).

According to the definition of the OIV (Organisation Internationale de la Vigne et du Vin), the terroir is a concept which refers to an area in which collective knowledge of the interactions between environment and applied vitivinicultural practices develops, providing distinctive characteristics for products originating from this area. Terroir includes specific soil, topography, climate, landscape characteristics and biodiversity features. Analytical chemistry plays constantly a more significant role in winemaking. Chemical analysis of wine is therefore a basic tool for provision of wine safety and origin.

Phenolic compounds belong to the most significant group of secondary metabolites in grapes and wine (Ali et al., 2010) and can be successfully used for the wine authenticity assessment, as these compounds are characteristic for the type of wine and can provide information on its geographical origin (Andreu-Navarro et al., 2011).

Phenolic compounds occurring in grapes and wine can be divided into non-flavonoids and flavonoids. Non-flavonoids include hydroxycinnamates, hydroxybenzoic acids and stilbenes (Rentzsch et al., 2009; Cheynier et al., 2010). Phenolic acids (hydroxybenzoic and hydroxycinnamic acids) are important phenolic compounds in white wines and are dominant antioxidants. Hydroxycinnamic acids influence the color and the flavor of wine by a significant way, they are thus the main phenolic compounds in white wines where they can serve as oxidation substrates and precursors of browning. Hydroxybenzoic acids are a minor group of phenolic compounds present in wine. Among the significant flavonoids in grapes and wine belong flavan-3-ols, flavonols, and, in the case of the varieties for red wine production, also anthocyanins. Flavan-3-ols are the abundant class of flavonoids and are found mainly in the skin and seeds of grape berries (Waterhouse, 2002).

As far as the color of grapes is concerned, anthocyanins are the most important compounds. The color may be considered as one of the most important parameters of red wine. Anthocyanins can be defined as water-soluble vacuolar pigments that belong to the parent class of molecules called flavonoids. From the chemical point of views, anthocyanins are found in glycosides forms. The most important anthocyanins are pelargonidin, cyanidin, peonidin, delphinidin, petunidin and malvidin. Anthocyanins can also conjugate with hydroxycinnamic acids (e.g., *p*-coumaric, caffeic acid) and some other organic acids (e. g. acetic acid).

The composition of anthocyanins and their contents in grapes and wine may be influenced by environmental and genetic factors. Climatic factors (e.g., air temperature, solar radiation and water supply) influence significantly the content of anthocyanins in grapes (Castellarin et al., 2007, Tarara et al., 2008).

Phenolic compounds can be used as chemometric markers of geographical origin. Phenolic compounds in wines with different geographical origin indicate that the accumulation of phenolic compounds in grapes and wines is significantly influenced by terroir (Rastija et al., 2009; Jaitz et al., 2010; Li et al., 2011).

The aim of this study is to characterize monovarietal white wines and red wines from different wine-growing regions and terroirs in the Czech Republic on the basis of the concentration of phenolic compounds, i.e., hydroxybenzoic acids, hydroxycinnamates, flavan-3-ols, stilbenes and anthocyanins. The purpose is to find markers which enable to differentiate wines on the basis of their geographical origin, i.e., individual wine-growing regions and terroirs.

METHODS

Experiment 1

34 commercial monovarietal Riesling wines were sampled from five wine-growing regions in the Czech Republic. All wines were produced over the years 2006 – 2008. Four wine samples originated from the wine-growing region of Mikulov, 9 from Slovácko, 3 from Velké Pavlovice, 12 from Znojmo, and 6 from Litoměřice region. All wine samples originate from grapes of identical quality in late harvest (Spätlese) category.

In all wine samples of the Riesling variety following phenolic compounds were analyzed: hydroxybenzoic acids (gallic acid, protocatechuic acid, *p*-hydroxybenzoic acid, vanillic acid, syringic acid), hydroxycinnamic acid (caffeic acid, caftaric acid, *p*-coumaric acid, *p*-coutaric acid, ferulic acid, fertaric acid,), stilbenes (*trans*-resveratrol, *cis*-resveratrol, *trans*-piceid, *cis*-piceid) and flavan-3-ols ((+)-catechin, (-)-epicatechin).

Experiment 2

35 commercial monovarietal Riesling wines were sampled from 9 individual terroirs in the Czech Republic. All wines were produced over the years 2006–2008: 12 wines from 2006, 12 from 2007 and 11 from 2008. The number of wines from the individual terroirs are: Archlebov (9 wines), Mikulov (3), Podmolí (3), Popice (4), Přítluky (3), Roudnice nad Labem (3), Syrovín (3), Velké Žernoseky (3), Znojmo (4). The quality of grapes during the harvest varied in the categories "Spätlese" and "Auslese" after "german wine classification".

In all wine samples of the variety Riesling following phenolic compounds were analysed: hydroxybenzoic acids (gallic acid, protocatechuic acid, *p*-hydroxybenzoic acid, vanillic acid, syringic acid), hydroxycinnamates (caffeic acid, caftaric acid, *p*-coumaric acid, *p*-coutaric acid, ferulic acid, fertaric acid, ethyl caffeate, *p*-coumaric acid ethylester, ferulic acid ethylester), stilbenes (*trans*-resveratrol, *cis*-resveratrol, *trans*-piceid, *cis*-piceid) and flavan-3-ols ((+)-catechin, (-)-epicatechin).

Experiment 3

17 monovarietal red wines were sampled from 3 different terroir in region Dolní Kounice: terroir Karlov (10), terroir Šibeniční Hora (5), terroir Na Nivách (2) and from 3 different grapevine varieties: Blaufränkisch (7), Blauer Portugieser (8) and Saint Laurent (2). Wines are collected from different vintages (1999-2009). All wine samples are originated from grapes of identical quality in late harvest (Spätlese) category.

To evaluate the anthocyanin fingerprinting of wines from region Dolní Kounice, 15 anthocyanins were utilized: delphinidin-3-*O*-glucoside (DpGl), cyanidin-3-*O*-glucoside (CyGl), petunidin-3-*O*-glucoside (PtGl), peonidin-3-*O*-glucoside (PnGl), malvidin-3-*O*-glucoside (MvGl), delphinidin-3-*O*-acetylglucoside (DpGlAc), cyanidin-3-*O*-acetylglucoside (CyGlAc), petunidin-3-*O*-acetylglucoside (PtGlAc), peonidin-3-*O*-acetylglucoside (PnGlAc), malvidin-3-*O*-acetylglucoside (MvGlAc), delphinidin-3-*O*-*p*-coumarylglucoside (DpGlCm), cyanidin-3-*O*-*p*-coumarylglucoside (CyGlCm), petunidin-3-*O*-*p*-coumarylglucoside (PtGlCm), peonidin-3-*O*-*p*-coumarylglucoside (PnGlCm), malvidin-3-*O*-*p*-coumaryl glucoside (MvGlCm).

Methodology

HPLC Analysis

Concentrations of the individual phenolic compounds were determined by a method with direct injection of a sample as described below. The samples of wine were centrifuged at 3.000g for 6 min and diluted with ratio 1:1 by 100 mM $HClO_4$. The chromatographic system Shimadzu LC-10A (Shimadzu, Kyoto, Japan) consisted of two pumps LC-10ADvp (Shimadzu, Kyoto, Japan), a column thermostat with manual injection valve, a DAD detector SPD-M10Avp (Shimadzu, Kyoto, Japan) and a personal computer running the chromatographic software LC solution (Shimadzu, Kyoto, Japan). The chromatographic separations were performed on a column Alltech Alltima C18 (3 μm, 3 x 150 mm, equipped with a guard column (3x 7.5 mm ID) filled with the same sorbent. The temperature of separations was fixed at 60°C. The mobile phases were the following: A = 15 mM $HClO_4$ and B = 15 mM $HClO_4$, 10% MeOH, 50 % ACN. The flow rate is of 0.6 ml/min. The total length of the analysis was 43 minutes and the regeneration time was 4 minutes. Data were recorded in the range of wavelength 200-520 nm. The detection wavelength was 200 nm for (+)-catechin and (-)-epicatechin, 260 nm for vanillic, protocatechuic and *p*-hydroxybenzoic acid, 275 nm for gallic and syringic acid, 285 nm for *cis*-piceid and *cis*-resveratrol, 310 nm for *p*-coumaric acid and its derivatives, *trans*-piceid and *trans*-resveratrol, 325 nm for caffeic acid, 520 nm for anthocyanins.

Reagents and Standard Solutions

Acetonitrile (ACN) and methanol (MeOH) were HPLC supergradient purity (Sigma–Aldrich, Prague, Czech Republic). Vanillic acid, protocatechuic acid, *p*-hydroxybenzoic acid, gallic acid, syringic acid, *p*-coumaric acid, *trans*-resveratrol, *trans*-piceid, caffeic acid, ferulic acid, (+)-catechin, (-)-epicatechin and perchloric acid were purchased from Sigma-Aldrich (St. Louis, USA). Malvidin3-glucoside were identified on the basis of standards (purchased from Sigma–Aldrich, Prague, Czech Republic). Other used chemicals were at least analytical grade and were obtained from local suppliers (Lachema-Penta, Brno, Czech Republic). A stock standard solution was prepared by accurately weighing about 10 mg of each phenol in a 25 ml volumetric flask. The standard was dissolved in 10 ml of acetonitrile and bringing up to volume with distilled water. *Cis*-resveratrol was obtained by exposing the *trans*-resveratrol standard solution to direct UV light for 10 minutes. The source of UV light was fluorescent tube Philips Ultraviolet TUV 30W/G30 T (Philips, Rosemont, USA). The sample was placed directly below the tube in a sealed quartz cell. Concentration of *cis*-resveratrol was expressed as a decrease in the concentration of *trans*-resveratrol (71 % conversion).

Statistical Data Treatment

The obtained data were processed according to the wine-growing regions and expressed by mean values and standard deviations. The use of one-way analysis of variance (ANOVA) and a LSD test at $p<0.05$ with the aim to find the influence of a region and vintage was further step. Then the multivariate chemometric methods were used, followed by canonical variate analysis (CVA) as a supervised learning technique for differentiation of wines to groups on the basis of the wine-growing regions and finding markers which show a significant discrimination value. Afterwards, the contents of individual phenolic compounds were compared by a Pearson correlation with the aim to find mutual dependencies. All statistical analyses were performed with the use of a statistical program UNISTAT (Unistat, Brno, Czech Republic).

1. Evaluation of Phenolic Conpounds in Riesling Wines from Different Regions

1.1. Profile of Phenolic Compounds in Riesling Wines

The profile of phenolic compounds studied in Riesling wines from different regions in the Czech Republic is shown in Table 1.

Table 1. Mean values, standard deviations and analysis of variance of hydroxybenzoic acids, hydroxycinnamic acids, stilbenes and flavan-3-ols in Riesling wines from Czech Republic

Chemical parameters (mg/l)	Region					Effect	
	Mikulov (n=4)	Slovácko (n=9)	Velké Pavlovice (n=3)	Znojmo (n=12)	Litoměřice (n=6)	Region F	Year F
Hydroxybenzoic acid							
gallic acid	2.84±3.01	2.32±2.50	2.86±1.31	2.05±1.38	2.07±0.64	0.22 n.s.	3.47 *
protocatechuic acid	1.63±0.37a	2.73±0.49b	3.32±0.43b	2.72±0.95b	2,44±0.52ab	2.97*	1.34 n.s.
p-hydroxybenzoic acid	0.39±0.14	0.68±0.18	0.75±0.14	0.63±0.36	0.87±0.41	1.66 n.s.	3.32 *
vanillic acid	0.84±0.51b	0.79±0.49b	0.54±0.30ab	0.71±0.36b	0.06±0.02a	4.07***	4.63 **
syringic acid	0.23±0.08ab	0.29±0.07b	0.23±0.07ab	0.16±0.05a	0.25±0.12b	4.14***	0.56 n.s.
Hydroxycinnamic acid							
caffeic acid	3.15±1.47a	9.21±9.27b	5.74±4.42ab	2.21±0.55a	2.11±0.62a	3.03*	0.68 n.s.
caftaric acid	15.19±7.52a	9.66±13.63b	18.12±7.89ab	23.84±3.78ab	40.03±12.05c	5.42***	1.20 n.s.
p-coumaric acid	1.66±0.89ab	2.58±1.27b	2.79±1.73b	1.44±0.70a	1.60±0.38ab	2.62*	1.15 n.s.
p- coutaric acid	1.54±0.43a	3.37±1.44bc	2.08±1.58ab	3.12±0.79bc	4.18±0.85c	4.56***	1.35 n.s.
ferulic acid	0.55±0.11	0.74±0.29	0.72±0.24	0.48±0.17	0.60±0.15	2.40 n.s.	2.32 n.s.
fertaric acid	2.83±1.59	4.00±0.49	3.48±0.39	3.35±0.56	3.95±0.72	2.42 n.s.	1.15 n.s.
Stilbenes							
trans-resveratrol	0.12±0.04a	0.21±0.06a	0.25±0.15ab	0.37±0.25b	0.34±0.20b	2.70*	0.94 n.s.
cis-resveratrol	0.08±0.05b	0.22±0.09a	0.22±0.22a	0.14±0.06ab	0.27±0.23a	2.70 *	0.19 n.s.
trans-piceid	0.06±0.03	0.12±0.08	0.11±0.14	0.12±0.08	0.14±0.08	0.54 n.s.	0.78 n.s.
cis-piceid	0.13±0.09	0.26±0.18	0.28±0.37	0.15±0.06	0.23±0.16	1.07 n.s.	5.90 **
Flavan-3-ols							
(+)-catechin	2.94±1.50ab	2.04±0.74a	2.65±1.34ab	4.52±1.72b	3.21±1.51ab	4.19***	1.02 n.s.
(-)-epicatechin	0.57±0.35a	0.42±0.16a	0.70±0.61ab	2.03±1.04c	1.75±0.91bc	7.34***	0.98 n.s.

* Significant for $p<0.05$, ** Significant for $p<0.01$, *** Significant for $p<0.001$. Data followed by different letters in the same row are significantly different by LSD test at $p<0.05$. *n.s.* not significant.

Gallic acid and protocatechuic acid are hydroxybenzoic acids with a close relation to grapes as they occur mainly in the pulp of berries. These acids have therefore a close relation to the place of grapevine cultivation – terroir. The level of hydroxybenzoic acids in wine shows a great variability in dependency on grapevine variety and viticultural conditions (Rentzsch et al., 2009), which was also confirmed in this study.

The highest content of gallic acid and protocatechuic acid was also found in Riesling wines from different regions in the Czech Republic. The highest mean content was found for protocatechuic acid (2.60 mg/L), followed by gallic acid (2.29 mg/L). In the case of protocatechuic acid, the influence of the region on its content was proved.

The content of hydroxybenzoic acids found in Riesling wines from the Czech Republic is relatively significant, as in German Riesling wines of 1989-1998 vintages only non-detectable amounts of main hydroxybenzoic acids (gallic acid, protocatechuic acid) were found (Pour Nikfardjam et al., 2007). A similar content of main hydroxybenzoic acids was also found in white wines from South Africa where the content of gallic acid varied in the range of 2.0-4.7 mg/L, however, the content of protocatechuic acid was lower (De Villiers et al., 2005). Protocatechuic acid also showed to be the main hydroxybenzoic acid for Riesling wines from the Czech Republic. The third most significant hydroxybenzoic acid in Riesling wines was p-hydroxybenzoic acid (0.67 mg/L). Vanillic acid with the mean content of 0.62 mg/L and syringic acid with the content of 0.23 mg/L were of minor representation in the studied wines. The content of these hydroxybenzoic acids was significantly influenced by the region.

Hydroxycinnamic acids are the main phenolic compounds in white wines. Hydroxycinnamic acids in wine originate mostly from hydroxycinnamic tartaric esters present in grapes. Among the main hydroxycinnamic acids mainly belong p-coumaric acid, caffeic acid and ferulic acid that are mostly present in grapes (Cheynier et al., 2010).

Among hydroxycinnamic acids, caftaric acid has the most significant representation which is almost 50% of total hydroxycinnamic acids in wine (Rentzsch et al., 2009). Caftaric acid with mean content of 26.71 mg/L also showed to be the most represented hydroxycinnamic acid in Riesling wines from the Czech Republic. The content of caftaric acid in German Riesling wines was lower (11.2 mg/L) (Dietrich et al., 2004). Caftaric acid is however considered to be one of the most significant phenolic compounds in Riesling wines (Ritter et al., 1994). The content of caftaric acid in all studied wines is very significantly influenced by the region and not influenced by the vintage year. The highest content was found in wines from Litoměřice region (40.03 mg/L) that significantly differs from the other regions. On the other side, Mikulov region has the lowest content (15.19 mg/L) of caftaric acid. Caffeic acid is also significantly represented in Riesling wines from the Czech Republic. Mean content of caffeic acid in all studied wines is 4.47 mg/L. The content of caffeic acid in the Czech Republic was influenced by the region and not influenced by the vintage year. The significantly highest content of caffeic acid was found in Slovácko region (9.21 mg/L) that exceeded all values for Riesling wines published before.

Another important hydroxycinnamic acid in Riesling wines is p-coutaric acid with the mean content of 3.10 mg/l. The content of p-coutaric acid is very significantly influenced by the region of wine origin and is not influenced by the vintage year. This hydroxycinnamic acid significantly separated the regions of Litoměřice (4.18 mg/L) and of Mikulov (1.54 mg/L). In German wines from Rheingau region (Ritter et al., 1994), a similar content of p-coutaric acid as in Mikulov region was found. Higher content of p-coutaric acid in Riesling wines was found by Nagel et al. (1979). On the basis of ANOVA evaluation, it is possible to assume that p-coutaric acid can be a suitable marker of geographical origin of Riesling wines.

In Riesling wines from the Czech Republic, relatively high content of fertaric acid (3.58 mg/L) was found. The content of fertaric acid was however not influenced by region as it was in the case of ferulic acid (0.60 mg/L). The content of ferulic acid was lower in comparison with Croatian Riesling wines (Komes et al., 2007) and also with wines from Ontario region where ferulic acid was given as a phenolic compound of the highest content (Soleas et al., 1997). The last hydroxycinamic acid with a relation to the region is p-coumaric acid. Mean content of p-coumaric acid was 1.91 mg/L. This value is comparable also with German Riesling wines (Pour Nikfardjam et al., 2007).

On the basis of analysis of the hydroxycinnamic acids profile and results of ANOVA and LSD test, it can be supposed that, among hydroxycinnamic acids, it is possible to find markers of authenticity of the geographical origin what will be further verified by chemometric statistical methods.

Stilbenes are synthesized by several species within the family of Vitaceae including Vitis vinifera. Resveratrol exists in two isomer forms (cis- and trans-). 3-O-β-D-glucosides of cis- and trans-resveratrol are called piceids (Rentzsch et al., 2009).

The most significant compound of the group of stilbenes with a proved relation to a region was trans-resveratrol. Mean content of trans-resveratrol in the studied Riesling wines was 0.28 mg/L. The highest content of trans-resveratrol was found in wines from Znojmo (0.37 mg/L) and Litoměřice (0.34 mg/L) regions. High content of trans-resveratrol in wines from Litoměřice region in Riesling wines fully corresponded with the findings which stated that the content of resveratrol in grapes is most probably reflected the climatic conditions during the period of vegetation and fungal pressure which is higher in cold and humid conditions in the northern part of the Czech Republic (Melzoch et al., 2001). The values of trans-resveratrol in wines from the Czech Republic also correspond with the values for wines from Germany. The concentrations of trans-resveratrol in white German wines were ranging from 0 to 0.7 mg/L with the mean value of 0.3 mg/L (Pour Nikfardjam et al., 1999).

The major flavan-3-ol monomers found in grapes and wine include (+)-catechin, (-)-epicatechin and (-)-epicatechin-3-O-gallate. In Riesling wines evaluated in this study, higher content of (+)-catechin than of (-)-epicatechin was found. The mean content of catechin was 3.28 mg/L. A similar content of (+)-catechin was found in Brasilian Riesling wine (2.96 mg/L) (Ballus et al., 2012) or also in German Riesling wines (2.5 mg/L) (Ritter et al., 1994). The content of (+)-catechin was highly significantly influenced by the region and not influenced by the vintage year. The highest content of (+)-catechin was found in wines from Znojmo region (4.52 mg/L). The mean content of (-)-epicatechin in the studied wines was 1.26 mg/l. The content of (-)-epicatechin found in Czech wines was comparable with German Riesling wines (0.9 mg/L) (Pour Nikfardjam et al., 2007) but significantly lower in comparison with Riesling wines from Rheingau region (6.1 mg/L) (Ritter et al., 1994) and also with Croatian Riesling wines (6.73 mg/L) (Komes et al., 2007). The content of (-)-epicatechin was also very significantly influenced by the region of wine origin. The highest content was found, similarly as for (+)-catechin, in Znojmo region (2.03 mg/L).

On the basis of the results it can be assumed that the studied flavan-3-ols which are very significantly influenced by the geographical origin and not influenced by the year of vintage can thus represent suitable markers of the geographical origin of wine.

1.2. Chemometric Analysis for Evaluation of Geographical Origin

Phenolic profile of grapes and wines has a distinctive relation to a variety and is genetically controlled. On the other side, environmental effects such as geology, soil type and climatic factors can also influence the phenolic profile. For wine authenticity assessment in dependence on the geographical origin, it is therefore ideal to use monovarietal wines, similarly as in this study.

For finding markers of the geographical origin, the application of chemotaxonomic statistical analysis methods is suitable. Canonical discriminant analysis is a dimension-reduction technique related to PCA and canonical correlation. Given a nominal classification of variables together with several interval variables, canonical discriminant analysis derives canonical variables that explain the inter-class variation in the same way that principal components summarize total variation (Moreno-Rojas et al., 2010). Canonical variate analysis is often used for analysis of wines on the basis of their geographical origin, e.g., wines from South Italy (Galgano et al., 2008) or Italian red wines (Galgano et al., 2011). From the evaluation with CVA, some compounds were excluded because they didn't show a significant effect of terroir to their content with ANOVA, like in studies Galgano et al. (2008), Galgano et al. (2011). Therefore, the following phenolic compounds: p-hydroxybenzoic acid, ferulic acid, fertaric acid, trans-piceid, cis-piceid were excluded.

The resulted explained variance represents 69.38% for CV1 and 18.13% for CV2 which constitutes 87.51% of total variation. The differences between the regions were significant (Wilk´s lambda=0.0034 at the value of p=0.00). When the Wilks´ lambda value is near zero, it indicates a good discrimination function of the individual CVs. The evaluation of standardized canonical discriminant coefficients for the individual CVs indicates which phenolic compounds are suitable for differentiation of Riesling wines on the basis of their geographical origin.

The canonical function 1 (CV1) loading of discriminant analysis indicated that trans-resveratrol and (-)-epicatechin in a positive way, and p-coutaric acid in a negative way were the most important discriminating variables among the terroir using canonical components. The canonical function 2 (CV2) loading of discriminant analysis indicated that cis-resveratrol, (-)-epicatechin and p-coutaric acid in a positive way, and (+)-catechin in a negative way were the most important discriminating variables among the terroir using canonical components. This analysis is successful in designating function that separates the Riesling wines on the basis of their geographical origin.

While evaluating Croatian wines, it was found that the samples from Central and Southern Dalmatia have the highest content of trans-resveratrol, whereas the wines from Istria have the lower content. Trans-resveratrol was proved to be one of the markers of geographical origin (Rastija et al., 2009), which was also confirmed by this study with Riesling wines.

Trans-resveratrol and cis-resveratrol also participated in discrimination of Italian red wines on the basis of their geographical origin (Galgano et al., 2011). For the discrimination of Greek red wines according to their geographical origin, compounds from the group of flavan-3-ols such as (+)-catechin and (-)-epicatechin (Makris et al., 2006) were also used. It was also possible to use trans-resveratrol and (+)-catechin for discrimination of monovarietal Cabernet Sauvignon wines on the basis of their geographical origin (Li et al., 2011). These results confirm that the compounds from the group of stilbenes are very often used as markers for differentiation of the geographical origin of white wines (Rastija et al., 2009) as well as red wines (Galgano et al., 2011; Makris et al., 2006; Li et al., 2011), which was also confirmed in this study with monovarietal Riesling wines from the Czech Republic. Similarly as in this study, (-)-epicatechin was found as a marker of the geographical origin for the Chardonnay variety in Australia (Bellomarino et al., 2009).

In the Italian and Spanish wines, hydroxybenzoic acids were used as markers of the geographical origin. This fact was not confirmed in this study. Furthermore, gallic acid was stated as a discriminant parameter of the geographical origin, e.g., for Italian and Spanish wines (Andreu-Navarro et al., 2011). The importance of gallic acid as a discriminant parameter was not confirmed in this study.

The composition of phenolic compounds in wines can be significantly influenced by the vintage year (Ali et al., 2011). None of the studied compounds, which were found as discriminant parameters of the geographical origin in this study, was however influenced by the vintage year, which emphasizes their suitability as markers of geographical origin. The composition of phenolic compounds and metabolic fingerprint of wines can be used for their classification on the basis of the geographical origin (Li et al., 2011), which was also confirmed by this study.

A graphic presentation of the studied wines defined by the first two canonical variates (CVs) is shown in Figure 1, where can be seen a good differentiation of wines according to the individual regions. CV 1 separates well the regions a (Mikulov), d (Znojmo) and e (Litoměřice) which are mainly characterized on the basis of trans-resveratrol and (-)-epicatechin from the regions b (Slovácko) and c (Velké Pavlovice) which are mainly characterized by coutaric acid. CV 2 then separates very well the region a (Mikulov) and c (Velké Pavlovice) which are mainly characterized by (+)-catechin from the region e (Litoměřice) which is mainly characterized by cis-resveratrol and (-)-epicatechin. The use of canonical variate analysis allowed to classify correctly (100%) all the studied wines.

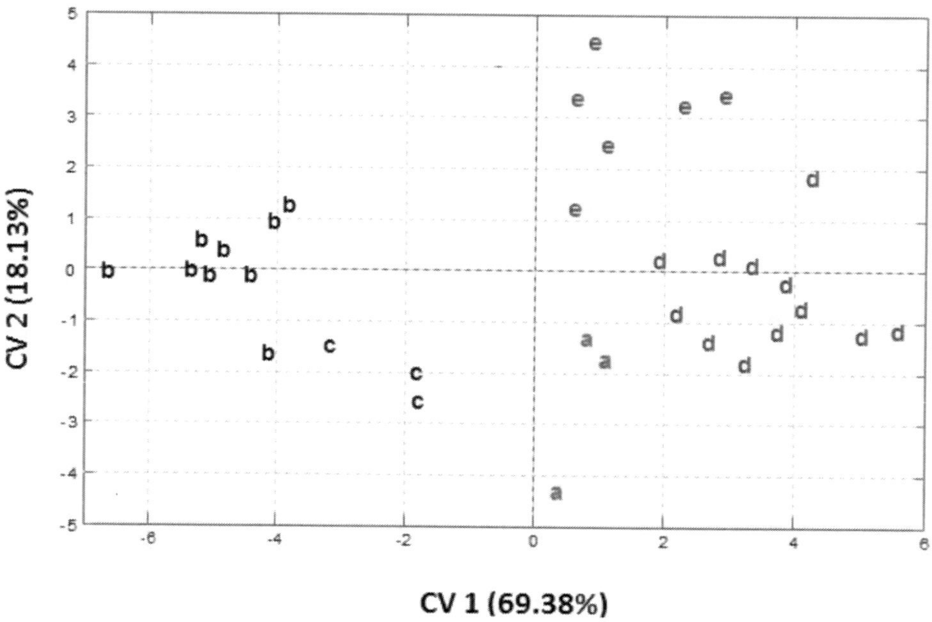

Figure 1. Canonical variate analysis for first (CV1) and second (CV2) canonical variate of wines from different terroirs: a (Mikulov), b (Slovácko), c (Velké Pavlovice), d (Znojmo), e (Litoměřice).

From the view of using phenolic compounds as possible markers of the geographical origin, their relation to grapes and thus to vineyard terroir is important. From the group of hydroxycinnamic acids, p-coutaric acid was found as a marker of the geographical origin for its close relation to grapes and terroir. P-coutaric acid (trans-p-coumaryoltartaric acid) as one of hydroxycinnamic acids which originates within the phenylpropanoid pathway, occurs naturally in grapes.

From the group of stilbenes, trans-resveratrol and cis-resveratrol were proved as markers of the geographical origin in this study. Trans-resveratrol, trans - and cis-piceid were identified in grapes (Bavaresco et al., 2007). Cis-resveratrol is not present in grapes, but only in wine (Waterhouse, 2002). It results from the fact that cis-resveratrol is formed by photochemical isomerization of trans- form to the cis-form during the winemaking process (Ali et al., 2010).

Significant markers of the geographical origin in monovarietal Riesling wines are (+)-catechin and (-)-epicatechin in this study. Flavan-3-ols, the group to which these compounds belong, represent the most significant group of flavonoids in grapes.

There are many environmental factors which influence the formation, transport and accumulation of these compounds in grapes (Braidot, et al., 2008). Among the main factors of terroir expression belong soil, climate and variety and mutual effects of these factors (Van Leeuwen, 2010). This is why just these parameters, very closely related to the origin of grapes and terroir, were chosen as markers of the geographical origin of wines in this study.

1.3. Conclusion

Assessment of a wide spectrum of monovarietal Riesling wines in relation to the geographical origin of wine is one of the first studies of that type in the Czech Republic. Analysis of the phenolic profile of 34 monovarietal Riesling wines by the use of chemometric methods proved that it is possible to use phenolic compounds to discriminate wines on the basis of their geographical origin. Analysis of phenolic compounds using the HPLC method and, subsequently the CVA method were able to differentiate 100% of the studied wines on the basis of wine-growing regions in the Czech Republic. Among the compounds which are applicable for wine discrimination on the basis of the geographical origin, belong hydroxycinnamic acid (p-coutaric acid), flavan-3-ols ((+)-catechin, (-)-epicatechin) and stibenes (trans-resveratrol and cis-resveratrol).

2. Evaluation of Phenolic Compounds in Riesling Wines from Different Terroirs

2.1. Profile of Phenolic Compounds in Riesling Wines

The main objective of our research was to verify the possibility how to use phenolic substances when specifying individual localities (i.e., terroirs). The Czech Republic introduces the Roman (i.e., French) appellation system of wine classification. This means that it is necessary to classify in detail individual terroirs and also to verify if they contribute to the uniqueness of wines originating from individual localities. As already mentioned above, the profile of phenolic substances is not only dependent on the genotype of the variety but also on climatic and pedological conditions of the vineyard. Evaluated profiles of phenolic substances contained in wine originating from individual localities (terroirs) are presented in Tables 2 and 3.

In this study, the highest content was found for protocatechuic acid (mean 2.54 mg/l) followed by gallic acid (mean 2.39 mg/l). ANOVA proved that the content of all studied hydroxybenzoic acids is significantly influenced by terroir. The most significant influence was found in the case of protocatechuic acid and *p*-hydroxybenzoic acid. The content of *p*-hydroxybenzoic acid and also vanillic acid is influenced by the year of vintage

Table 2. Mean values, standard deviations and analysis of variance of hydroxybenzoic acids and hydroxycinnamic acids in Riesling wines

Chemical parameters (mg/l)	Terroir									Effect	
	Archlebov (n=9)	Mikulov (n=3)	Podmolí (n=3)	Popice (n=4)	Přítluky (n=3)	Roudnice nad Labem (n=3)	Syrovín (n=3)	Velké Žernoseky (n=3)	Znojmo (n=4)	Terroir F	Year F
Hydroxybenzoic acid											
gallic acid	2.09±2.11 AB	5.17±1.19 C	1.61±0.57 AB	0.43±0.15 A	2.86±1.31 BC	2.64±0.57 AB	3.30±2.51 BC	1.94±0.42 AB	1.51±0.45 AB	2.73**	0.92 n.s.
protocatechuic acid	2.73±0.58 CDE	1.39±0.13A	3.11±0.65 DE	2.34±0.53 BCD	3.32±0.43E	3.15±0.46E	3.02±0.45 DE	2.07±0.47 ABC	1.75±0.52 AB	5.81***	1.86 n.s.
p-hydroxybenzoic acid	0.69±0.19B	0.30±0.05A	0.83±0.50B	0.61±0.17AB	0.75±0.14B	1.18±0.15C	0.77±0.03B	0.88±0.17BC	0.36±0.17A	5.38***	3.07*
vanillic acid	0.70±0.46B	1.03±0.43B	0.92±0.33B	0.72±0.47B	0.54±0.30AB	0.08±0.02A	0.84±0.57B	0.07±0.03A	0.45±0.38AB	2.27*	4.03*
syringic acid	0.27±0.06BC	0.20±0.05AB	0.21±0.07AB	0.23±0.13ABC	0.23±0.07ABC	0.25±0.05ABC	0.33±0.08C	0.15±0.06A	0.16±0.02A	2.27*	1.28 n.s.
Hydroxycinnamic acid											
caffeic acid	3.10±1.42 A	4.01±1.04 AB	2.75±0.42 A	2.36±0.33 A	5.74±4.42 B	2.79±0.43 A	21.46±0.26 C	2.10±0.97 A	1.64±0.07 A	52.16***	0.71 n.s.
caftaric acid	31.30±12.9 B	11.64±5.06 A	25.14±4.22 AB	23.87±3.02AB	18.12±7.89 A	46.49±7.41 C	17.72±0.02 A	36.73±6.63 BC	24.05±2.26 AB	5.21***	0.95 n.s.
p-coumaric acid	1.95±0.54 AB	2.04±0.79 AB	2.03±0.54 AB	1.57±0.84 AB	2.79±1.73 BC	2.13±0.35 AB	3.65±1.71 C	1.46±0.50 AB	0.81±0.49 A	2.87**	1.00 n.s.
p- coutaric acid	3.83±1.14 C	1.52±0.58 A	3.60±0.84 BC	2.00±0.18 A	2.08±1.58 A	4.19±0.64 C	2.05±0.04 A	3.59±0.56 BC	2.55±0.57 AB	4.78**	0.97 n.s.
ferulic acid	0.71±0.25	0.54±0.13	0.53±0.22	0.80±0.21	0.72±0.24	0.81±0.24	0.88±0.34	0.51±0.11	0.49±0.08	1.52 n.s.	0.27 n.s.
fertaric acid	3.87±0.91 B	1.91±0.80 A	3.55±0.30 B	4.15±0.72B	3.48±0.39 B	4.12±0.65 B	3.42±0.18 B	4.25±0.79 B	3.41±0.52 B	3.19*	0.34 n.s.
ethyl caffeate	1.04±1.02	2.33±0.84	0.62±0.19	0.41±0.20	3.43±4.06	0.42±0.08	1.40±0.15	0.66±0.56	0.23±0.07	2.10 n.s.	0.52 n.s.
p-coumaric acid ethylester	0.37±0.35 ABC	0.92±0.27 D	0.19±0.05 AB	0.11±0.06 A	0.75±0.58 CD	0.09±0.01 A	0.71±0.51 BCD	0.23±0.22 ABC	0.07±0.02 A	3.43**	0.54 n.s.
ferulic acid ethylester	0.03±0.02 A	0.05±0.02 A	0.04±0.02 A	0.05±0.02 A	0.08±0.03 A	0.04±0.01 A	0.34±0.25 B	0.03±0.01 A	0.03±0.01 A	6.07***	1.25 n.s.

*Significant for $p<0.05$, ** Significant for $p<0.01$, *** Significant for $p<0.001$. Data followed by different letters in the same row are significantly different by LSD test at $p<0.05$. *n.s.* not significant.

Table 3. Mean values, standard deviations and analysis of variance of stilbenes and flavan-3-olsin Riesling wines

Chemical parameters (mg/l)	Terroir									Effect	
	Archlebov (n=9)	Mikulov (n=3)	Podmolí (n=3)	Popice (n=4)	Přítluky (n=3)	Roudnice nad Labem (n=3)	Syrovín (n=3)	Velké Žernoseky (n=3)	Znojmo (n=4)	Terroir F	Year F
Stilbenes											
trans-resveratrol	0.18±0.08 A	0.13±0.03 A	0.49±0.15 C	0.17±0.09 A	0.25±0.15 AB	0.36±0.17 BC	0.20±0.01 AB	0.17±0.07 A	0.12±0.09 A	4.55**	0.10 n.s.
cis-resveratrol	0.15±0.11	0.07±0.01	0.21±0.05	0.11±0.05	0.22±0.20	0.25±0.23	0.26±0.02	0.09±0.06	0.08±0.04	1.15 n.s.	0.57 n.s.
trans-piceid	0.13±0.06	0.05±0.03	0.10±0.06	0.13±0.06	0.11±0.14	0.14±0.10	0.08±0.06	0.15±0.11	0.07±0.05	0.63 n.s.	1.39 n.s.
cis-piceid	0.24±0.15	0.12±0.09	0.15±0.10	0.22±0.08	0.28±0.37	0.12±0.05	0.19±0.22	0.21±0.21	0.14±0.07	0.42 n.s.	6.93 **.
Flavan-3-ols											
(+)-catechin	2.35±0.67 AB	3.94±0.69 CD	4.70±0.48 D	1.66±0.27 AB	2.65±1.34 ABC	3.45±0.79 CD	1.48±0.46 A	1.96±0.32 AB	2.72±1.29 BC	6.22***	0.11 n.s.
(-)-epicatechin	0.48±0.22 AB	0.71±0.30 AB	2.08±0.36 D	0.34±0.06 A	0.70±0.61 AB	1.59±0.50 CD	0.39±0.05 AB	1.04±0.23 ABC	1.16±1.02 BC	4.72**	0.30 n.s.

* Significant for $p<0.05$, ** Significant for $p<0.01$, *** Significant for $p<0.001$. Data followed by different letters in the same row are significantly different by LSD test at $p<0.05$. *n.s.* not significant.

From the group of hydroxycinnamic acids, *p*-coutaric acid, caftaric acid and fertaric acid are the most important components of wine (De Luca, 2010). Riesling wines from the Czech Republic have also very significant contents of caffeic acid (mean 5.11 mg/l). Especially high content of caffeic acid was proved in wines from the terroir of Syrovín (mean 21.46 mg/l). Caffeic acid as a hydroxycinnamic acid with the highest content in wine is also given by (Ballus et al., 2012) who found content up to 70 mg/l in the case of white wines.

ANOVA proved the influence of terroir on almost all studied compounds except ferulic acid and ethyl caffeate. The content of none of the studied compounds was influenced by the year of vintage. On the basis of the LSD test, the differences between the individual terroirs were proved. Mainly the terroirs of Syrovín and Roudnice nad Labem showed more considerable differences from the other terroirs. On the basis of caffeic acid content, very significant difference was proved for the terroir of Syrovín. Similarly, considering the content of caftaric acid, the terroir of Roudnice nad Labem was strongly separated, and the similarity exists only with the terroir of Velké Žernoseky. The similarity of both terroirs is logical as these terroirs are geographically very close. On the basis of the results of ANOVA and LSD test, it can be supposed that authenticity markers of the geographical origin could be found among the hydroxycinnamates what will be further verified using chemometric statistical methods.

Stilbenes are a class of non-flavonoid phenolic compounds in grapes and attracted a great interest because of their potential valuable health effects but occur in relative low concentrations when compared to other phenolic compounds. Among the terroirs with the highest content of *trans*-resveratrol belongs the terroir of Podmolí (mean 0.49 mg/l) which, on the basis of an LSD test, shows the similarity only with the terroir of Roudnice nad Labem (0.36 mg/l). Among the terroirs with low content of *trans*-resveratrol belong the terroirs of Archlebov, Mikulov, Popice, Velké Žernoseky, and Znojmo.

High content of *trans*-resveratrol in wines from the terroir Roudnice nad Labem fully corresponded with findings of Melzoch et al. (2001). They stated that the content of resveratrol in grapes most probably reflected the climatic conditions during the period of vegetation and fungal pressure which is higher in cold and humid conditions of the North Bohemia. Similar values of *trans*-resveratrol in Riesling wines from the Czech Republic were also investigated by Faitova et al. (2004) by the analysis of 76 samples of wine. The content of *trans*-resveratrol in samples from Bohemian wine-growing region ranged from 0.033 to 0.421 mg/l with the mean value of 0.117 mg/l. The content of *trans*-resveratrol in samples from Moravian wine-growing region ranged from 0.033 to 0.875 mg/l with the mean value of 0.123 mg/l.

In Riesling wines studied in this work, the total content of (+)-catechin (2.77 mg/l) was found higher than of (-)-epicatechin (0.94 mg/l). ANOVA also proved a significant influence of terroir on the content of (+)-catechin and (-)-epicatechin. In the terroirs with considerably lower content of (+)-catechin ranks the region of Syrovín (1.48 mg/l), and, on the contrary, very high content is shown by the region of Podmolí (4.70 mg/l). In wines from the terroir of Podmolí, high content of (-)-epicatechin (2.08 mg/l) was also found. The content of (+)-catechin and (-)-epicatechin was not influenced by the year of vintage.

2.2. Chemometric Analysis for Evaluation of Terroir Origin

CVA (canonical variate analysis) determined 8 discriminant functions (CVs) which altogether represent 100% of the total variance. First discriminant function marked CV1 represents 87.89% of the total variance, whereas CV2 represents 7.59%. These two functions are the most significant and together represent 95.48% from the total variance. The Wilk´s λ value 0.0000 means that these functions enable satisfactory discrimination.

The canonical function 1 (CV1) loading of discriminant analysis indicated that caffeic acid, *cis*-resveratrol correlating in a positive way and ethyl caffeate and *p*-coumaric acid ethylester correlating in a negative way were the most important discriminating variables among the terroir using canonical component. The canonical function 2 (CV2) loading of discriminant analysis indicated that caftaric acid, *cis*-piceid, (-)-epicatechin and *trans*-resveratrol in a positive way and *p*-coutaric acid, *trans*-piceid and *cis*-resveratrol in a negative way were the most important discriminating variables among the terroir using canonical component. This analysis is successful in designating function that separates Riesling wines according to their geographical origin.

Compounds from the group of stilbenes were also found to be markers of wine authenticity on the basis of its geographical origin. By the assessment of Croatian wines, it was found that wine samples from Central and Southern Dalmatia had the highest content of *trans*-resveratrol, than wines from Istria who had low content. *Trans*-resveratrol was confirmed as one of geographical origin markers (Rastija et al., 2009) that was also confirmed for Riesling wines in this study.

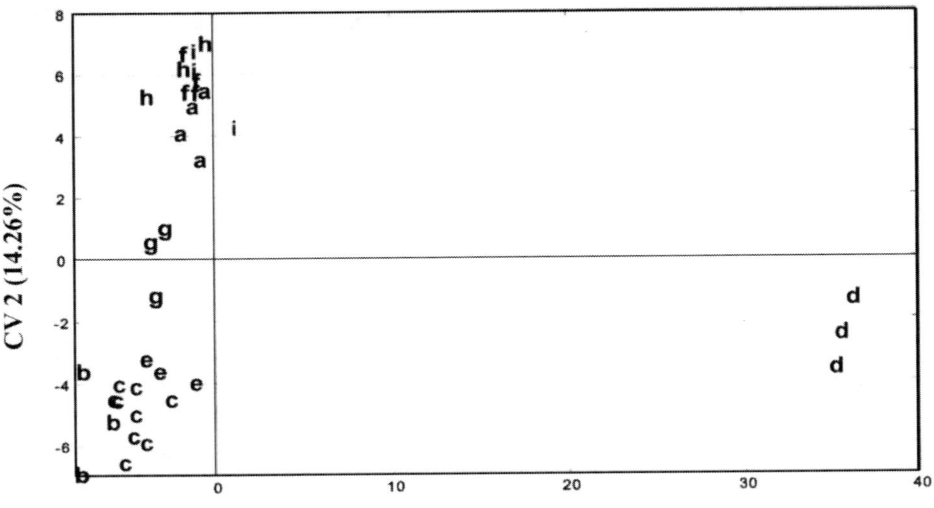

Figure 2. Results of canonical variate analysis for first (CV1) and second (CV2) canonical variate of wines from different terroirs: a - Popice, b-Mikulov, c-Archlebov, d-Syrovín, e-Přítluky, f-Znojmo, g-Podmolí, h-Velké Žernoseky, i-Roudnice nad Labem.

By differentiation of individual terroirs according to CV1 and CV2, correct classification for 100% of evaluated wines on the basis of terroir, i.e., geographical origin, was achieved. The graphical representation of canonical variate analysis is shown in Figure 2. From the graphical presentation, it is clearly evident that CV1 markedly separates the locality d (Syrovín) from all other evaluated localities, the main discriminant factor being caffeic acid. CV 2, on the basis of the discriminant parameters found, separates the localities b (Mikulov), c (Archlebov) and e (Přítluky) from the localities a (Popice), f (Znojmo), g (Podmolí), h (Velké Žernoseky) and I (Roudnice nad Labem) (Figure 2).

2.3. Conclusion

Results of this study with Riesling wines showed that it is possible to classify wines according to their geographical origin on the basis of the profile of phenolic compounds. Phenolic compounds can be therefore considered as authenticity markers of the geographical origin of wines. ANOVA and canonical variate analysis showed a significant influence of terroir on the profile of the most relevant phenolic compounds present in white wines. To the main discriminant parameters, by the use of which it was possible to classify white wines in this study, belong caffeic acid, *p*-coutaric acid, caftaric acid, ethyl caffeate, *p*-coumaric acid ethylester, (-)-epicatechin and all compounds from the group of stilbenes, the most significant marker from them being *trans*-resveratrol. Therefore mainly compounds from the group of hydroxycinnamates, flavan-3-ols and stilbenes can be considered suitable authenticity markers of the geographical origin.

Utilisation of Anthocyanins Profiles to Differentiation of Terroir

The aim of this part was to identify effects of anthocyanin spectrum and composition of red wines from the viewpoints of identification of the terroir. Another objective of this study was to find out and identify those anthocyanin pigments that characterize the terroirs.

Anthocyanins are very important compounds due to the fact that the determined the colour of both grapes and red wine. As far as the red wine is concerned, colour can be regarded as one of the most important parameters of red wine.

Principal component analysis (PCA) is a multivariate method analysing quantitatively the dependent variables. Principal component analysis was performed with all variable values from all anthocyanins. The two first components represent 58.6% of total variance (PC1 = 35.8 %, PC2 = 22.8 %). On the basis of PCA method, the variables for canonical discriminant analysis were selected. The following flavonoid phenolic compounds were thus eliminated from further assessment: PnGl, CyGlAc, MvGlCm. These compounds turned out to be nonsignificant for wine differentiation on the basis of geographical origin using PCA method. In order to increase the discrimination value obtained by PCA method the geographical origin of wines was verified by canonical discriminant analysis (CDA). This method includes only those variables which proved high significance for differentiation of geographical origin and varietal origin using PCA method.

Canonical discriminant analysis (CDA) was used for determination of chemical markers which are also the most important for the wine differentiation on the basis of vineyard site and grape variety in this study.

The resulted explained variance for the differentiation of terroirs represents 98.5% for CV1 and 1.5% for CV2 which is 100.0 % of total variation. The differences between the regions were significant (Wilk´s lambda = 0.0072 at the value of p = 0.01). The Wilks´ lambda value near zero indicates a good discrimination function of the individual CV. It can be seen from the achieved results that the CV1 most considerably correlates with DpGl, MvGl and DpGlCm in positive way and with CyGl, PtGl and DpGlAc in negative way. The CV2 is positively connected with DpGlAc and CyGl and negatively with PtGlAc and CyGlCm.

The CV 1 discriminates very markedly the vineyard site *Karlov* from the remaining two vineyard sites (called Šibeniční vrch and Na nivách). Wines originating from the vineyard site Karlov are well characterised by the presence of anthocyanins DpGl, MvGl and DpGlCm. (i.e., a monoglucoside derivative with p- coumaric acid). However, the canonical function CV2 cannot discriminate wines originating from vineyards sites Šibeniční vrch and Na Nivách (Figure 3). This means that when discriminating wines from different vineyard sites on the base of their profiles of anthocyanin pigments, the most important parameters were monoglucosides DpGl, MvGl and DpGlCm.

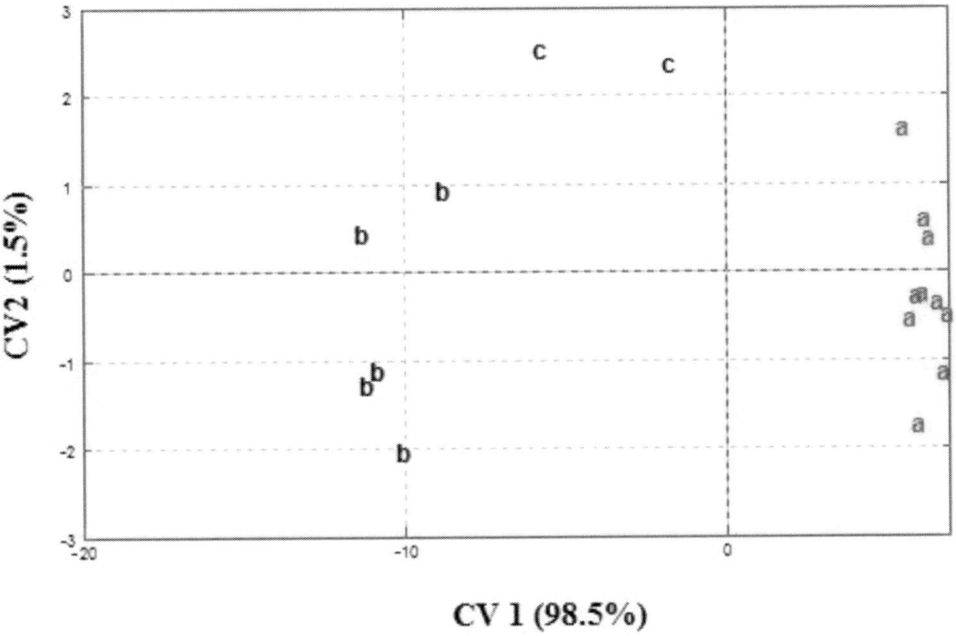

Figure 3. Results of canonical variate analysis for first (CV1) and second (CV2) canonical variate of wines from different terroirs: a – Karlov, b- Šibeniční Hora, c- Na Nivách.

It was found out that for the discrimination of individual vineyard sites the most important compounds were anthocyanins monoglucosides and their acetic acid derivates. Similar results were published also by MAKRIS et al. (2006) who discriminated the geographic origin of Greek red wines using malvidin-3-*O*-acetylglucoside, delphinidin-3-*O*-glucoside, malvidin-3-*O*-glucoside and malvidin-3-*O*-glucoside. Similar anthocyanins markers were used for the discrimination of South African wines and the most important ones were petunidin-3-*O*-glucoside, delphinidin-3-*O*-glucoside, malvidin-3-*O*-glucoside, peonidin-3-*O*-glucoside, malvidin-3-*O*-acetylglucoside and petunidin-3-*O*-*p*-coumarylglucoside (ROSSOUW and MARAIS, 2004).

Results of this study demonstrated that anthocyanin monoglucosides can be used for the discrimination of wines originating from different vineyard sites.

Conclusion

Three different studies performed with Czech wines demonstrated that phenolic compounds can be successfully used for the discrimination of wine on the base of its geographic origin. It was found out that phenolic compounds can be used when discriminating not only larger geographic units (i.e., wine-growing subregions) but also smaller ones (i.e., individual, concrete terroirs).

As far as white wine is concerned, it is possible to use successfully those phenolic compounds that are significantly influenced by environmental conditions, i.e., above all flavan-3-ols ((+)-catechin, (-)-epicatechin) and stilbenes (*trans*-resveratrol and *cis*-resveratrol). As far as the group of hydroxycinnamic acids (that represents an important group of aromatic compounds occurring in white wine) is concerned, it is possible to use also caffeic acid and *p*-coutaric acid. In case of red wines, it is possible to use some anthocyanin monoglucosides for the discrimination of individual terroirs, e.g., DpGl and MvGl.

References

Ali, K., Maltese, F., Hae Choi, Y., Verpoorte, R., (2010). Metabolic constituents of grapevine and grape-derived products. *Phytochem. Rev.* 9, 357-378.

Ali, K., Maltese, F., Töpfer, R., Choi, Y.H., Verpoorte, R., (2011). Metabolic characterization of Palatine German white wines according to sensory attributes, varieties and vintages using NMR spectroscopy and multivariate data analyses. *J. Biomol. NMR* 49, 255-266.

Andreu-Navarro, A., Russo, P., Aguilar-Caballos, M.P., Fernández-Romero, J.M., Gómez-Hens, A., (2011). Usefulness of terbium-sensitised luminescence detection for chemometric classification of wines by their content in phenolic compounds. *Food Chem.* 124, 1753-1759.

Ballus, C.A., Dillenburg Meinhardt, A., Grado de Oliveira, R., Teixerira Godoy, H., (2012). Optimization of capillary zone electrophoresis separation and on-line preconcentration of 16 phenolic compounds from wines produced in South Africa. *Food Res. Int.* 45, 136-144.

Bavaresco, L., Pezzutto, S., Gatti, M., Mattivi, F., (2007). Role of the variety and some environmental factors on grape stilbenes. *Vitis* 46, 57-61.

Bellomarino, S.A., Conlan, X.A., Parker, R.M., Barnett, N.W., Adams, M.J., (2009). Geographical classification of some Australian wines by discriminant analysis using HPLC with UV and chemiluminiscence detection. *Talanta* 80. 833-838.

Braidot, E., Zancani, M., Petrussa, E., Peresson, C., Bertolini, A., Patui, S., Macri, F., Vianello, A., (2008). Transport and accumulation of flavonoids in grapevine (Vitis vinifera L.). *Plant Signal. Behav.* 3, 626-632.

Castellarin, S.D., Pfeiffer, A., Sivilotti, P., Degan, M., Peterlunger, E, di Gaspero, G., (2007). Transcriptional regulation of anthocyanin biosynthesis in ripening fruits of grapevine under seasonal water deficit. *Plant Cell Environ.* 30, 1381–1399.

Cheynier, V., Schneider, R., Salmon, J.M., Fulcrand, H., (2010). Chemistry of Wine. In: Mander, L., Liu, H.W. (eds). Comprehensive Natural Products II. Elsevier, Oxford. pp.1119-1172.

De Luca, V., (2010). Wines, pp.241-255. In: Comprehensive Natural Products II. Mander, L., Liu, H.W. (ed) Elsevier Oxford.

De Villiers, A., Majek, P., Lynen, F., Crouch, A., Lauer, H., Sandra, P., (2005). Classification of South Africa red and white according to grape variety on the non-coloured phenolic content. *Eur. Food Res. Technol.* 221, 520-528.

Dietrich, H., Patz, C.D., Pour-Nikfardjam, M., Hoffmann, D., Greiner, D., Bauer, H., (2004). Chemische Charakterisierung von Weinen der Sorte Riesling der Jahre 1892 bis 1921 aus dem Rheingau. *Mitt. Klosterneuburg* 54, 222-233.

Faitová, K., Hejtmánková, A., Lachman, J., Pivec, V., Dudjak, J., (2004). The Contents of Total Polyphenolic Compounds and *Trans*-Resveratrol in White Riesling Originated in the Czech Republic. *Czech J. Food Sci.* 22, 215-221.

Galgano, F., Favati, F., Caruso, M., Scarpa, T., Palma, A., (2008). Analysis of trace elements in southern Italy wines and their classification according to provenance. *Food Sci. Technol.* 41, 1808-1815.

Galgano, F., Caruso, M., Perretti, G., Favati, F., (2011). Authentication of Italian red wines on the basis of the polyphenols and biogenic amines. *Eur. Food Res. Technol.* 232, 889-897.

Jaitz, L., Siegl, K., Eder, R., Rak, G., Abranko, L., Koellensperger, G., Hann, S., (2010). LC-MS/MS analysis of phenols for classification of red wine according to geographic origin, grape variety and vintage. *Food Chem.* 122, 366-372.

Komes, D., Kovačevič-Ganič, K., Lovric, T., (2007). Study of phenolic phytochemicals of Croatian white wines. *Bull. OIV* 920-922, 593-598.

Li, Z., Pan, Q., Jin, Z., Mu, L., Duan, C., (2011). Comparison on phenolic compounds in Vitis vinifera cv. Cabernet Sauvignon wines from five wine-growing regions in China. *Food Chem.* 125, 77-83.

Makris, D. P., Kallithraka, S., Mamalos, A., (2006). Differentiation of young red wines based on cultivar and geographical origin with application of chemometrics of principal polyphenolic constituents. *Talanta* 70, 1143-1152.

McGhie, T.K. & Rowan, D.D., (2012). Metabolomics for measuring phytochemicals, and assessing human and animal responses to phytochemicals in food science. *Mol. Nutr. Food Res.* 56, 147-158.

Melzoch, K., Hanzlíková, I., Filip, V., Buckiová, D. & Šmidrkal, J., (2001). Resveratrol in Part of Vine and Wine originating from Bohemian and Moravian Vineyard Regions. *Agric. Consp. Scient.* 66, 53-57.

Moreno-Rojas, R., Sánchez-Segarra, P.J., Cámara-Martos, F., Amaro-López, M.A., (2010). Multivariate analysis techniques as tools for categorization of Souther Spanish Cheeses : nutritional composition and mineral content. *Eur. Food Res. Technol.* 231, 841-851.

Nagel, C.W., Baranowski, J.D., Wulf, L.W., Powers, J.R., (1979). The hydroxycinnamic acid tartaric acid ester content of must and grape varieties grown in the Pacific northwest. *Am. J. Enol. Vitic.* 30, 198-201.

Pour Nikfardjam, M.P., Rechner, A., Patz, C-D., Dietrich, H., (1999). Trans-Resveratrol content of german wines. *Die Wein-Wissenschaft* 54, 17-20.

Pour Nikfardjam, M.S., Köhler, H.J., Schmitt, A., Patz, C.D., Dietrich, H., (2007). Polyphenolic composition of German white wines and its use for the identification of cultivar. *Mitt. Klosterneuburg* 57, 146-152.

Rastija, V., Srečnik, G., Medić-Šarić, M., (2009). Polyphenolic composition of Croatian wines with different geographical origins. *Food Chem.* 115, 54–60.

Rentzsch, M., Wilkens, A., Winterhalter, P., (2009). Non-flavonoid Phenolic Compounds. In: Moreno-Arribas, M.V., Polo, M.C. (eds). Wine Chemistry and Biochemistry Springer Science + Bussiness Media. pp.509-527.

Ritter, G., Götz, H., Dietrich, H., (1994). Untersuchung der phenolischen Substanzen in Rheingauer Rieslingweinen. *Wein-Wissenschaft* 49, 71-77.

Rossouw, M., Marais, J., (2004). The Phenolic Composition of South African Pinotage, Shiraz and Cabernet Sauvignon Wines. *S. Afr. J. Enol. Vitic.* 25, 94-104.

Soleas, G. J., Dam, J., Carey, M., Goldberg, D. M., (1997). Toward the fingerprinting of wines: Cultivar-related patterns of polyphenolic constituents in Ontario wines. *J. Agric. Food Chem.* 45, 3871-3880.

Tarara, J., Lee, J., Spayd, E.S., (2008). Berry temperature and solar radiation alter acylation, proportion, and concentration of anthocyanin in Merlot grapes. *Am. J. Enol. Vitic.* 59, 235–247.

Van Leeuwen, C., (2010). Terroir: the effect of the physical environment on vine growth, grape ripening and wine sensory attributes. In: Reynolds, A.G. (Ed). Managing wine quality, Volume 1: Viticulture and wine quality. Woodhead Publishing Limited, Cambridge. pp. 273-315.

Waterhouse, A.L., (2002). Wine Phenolics. *An. N.Y. Acad. Sci.* 957, 21-36.

In: Wine
Editor: Youssef El Rayess

ISBN: 978-1-63321-048-6

Chapter 10

CHEMOMETRIC METHODS FOR THE CLASSIFICATION OF WHITE WINES

José Manuel Camiña*[*1,2], Marianela Savio[1,2], Silvana Mariela Azcarate[1,2], Octavio Furlong[3] and Eduardo Jorge Marchevsky[1,4]

[1]Consejo Nacional de Investigaciones Científicas y Técnicas (CONICET)
[2]Instituto de Ciencias de la Tierra y Ambientales de La Pampa (INCITAP) and Facultad de Ciencias Exactas y Naturales, Universidad Nacional de La Pampa
[3]Instituto de Física Aplicada (INFAP) and Universidad Nacional de San Luis
[4]Instituto de Química de San Luis (INQUISAL) and Facultad de Química, Bioquímica y Farmacia, Universidad Nacional de San Luis, Argentina

ABSTRACT

Wines are products whose cost depends on several quality factors, required by the customer: geographical origin, variety of grape, oak aging, etc. The Controlled Denomination of Origin (CDO) of wines is frequently desired due to several properties, which depend on characteristics of the different places of origin around the world, including weather, grape variety, crop, temperature variation, winery practices, etc.

The control of the properties mentioned above, is usually difficult by traditional methods, since it is necessary to determine several specific variables such as trace elements, organic acids, phenolic compounds, etc., which require expensive equipment, expert operators, long-time analysis and pretreatment of samples, among other undesired aspects. However, the introduction of chemometric tools has simplified the interpretation and analysis of data, allowing the use of a great number of variables (data matrix), filtering only the most important information and leaving out the "noisy" data. These tools represent a fundamental advantage since they allow the implementation of several spectroscopic methods, which were not useful for such complex analysis: UV-Vis, infrared spectroscopy, nuclear magnetic resonance, and fluorescence methods, which could be used thanks to their multivariable advantage.

* Corresponding Author, address: Instituto de Ciencias de la Tierra y Ambientales de La Pampa (INCITAP) and Facultad de Ciencias Exactas y Naturales, Universidad Nacional de La Pampa. Email: jcamina@conicet.gov.ar

Based on the benefits of chemometric methods in comparison to traditional methods, this chapter will involve the implementation of the most recent tools used for wine analysis around the world. The determinations include Controlled Denomination of Origin (CDO), geographical and/or botanical origin, and other important quality properties of wines. Chemometric tools include artificial neural network analysis (ANN), Principal Component Analysis (PCA), Cluster Analysis (CA), Linear Discriminant Analysis (LDA), Support Vector Machine (SVM), Partial Least Square Discriminant Analysis (PLS-DA), Soft Independent Modeling of Class Analogy (SIMCA), among others.

This chapter has been organized in topics based on the application of methods for quality analysis in white wines, according to botanical and geographical classification, as well as other quality analysis, describing for each one, all the chemometric and analytical methods available for the determination of quality properties, for an easy reading and understanding.

Keywords: Chemometric, multivariate, classification, white, wines

LIST OF ABBREVIATIONS OF CHAPTER

AAS = Atomic Absorption Spectroscopy
ANN = Artificial Neural Networks
ANOVA = Analysis of Variance
CA = Cluster Analysis
CCA = Canonical Correlation Analysis
CDA = Canonical Discriminant Analysis
CBO = Certified Brands of Origin
CDO = Controlled Denomination of Origin
CE = Capillary Electrophoresis
CVA = Canonical Variate Analysis
DAD = Diode Array Detector
DFA = Discriminant Function Analysis
DPLS = Discriminant Partial Least Squares Analysis
ESI-MS = Electrospray Ionization Mass Spectrometry
ETAAS = Electrothermal Atomic Absorption Spectroscopy
FA = Factor Analysis
FAAS = Flame Atomic Absorption Spectrometer
FTIR = Fourier Transform Infrared Spectroscopy
FTNIRS = Fourier Transformed Near Infrared Spectrometry
GC-FID = Gas Chromatography with Flame Ionization Detector
GC-MS = Gas Chromatography Mass Spectrometry
GF-AAS = Graphite Furnace Atomic Absorption Spectrometry
^{1}H NMR = Proton Nuclear Magnetic Resonance
HPCE = High-Performance Capillary Electrophoresis
HPLC = High Performance Liquid Chromatography
HS-SPME = Head-Space Solid-Phase Microextraction
ICA = Independent Component Analysis

ICP-OES = Inductively Coupled Plasma Atomic Emission Spectroscopy
ICP-MS = Inductively Coupled Plasma Mass Spectrometry
IMS-UV = Ion Mobility Spectrometry with Ultraviolet Ionization
IRMS = Isotope Ratio Mass Spectrometry
KANN = Kohonen Artificial Neural Networks
k-NN = *k*-Nearest Neighborhood
LDA = Linear Discriminant Analysis
MIRS = Mid-Infrared Spectroscopy
MLP-NN = Multilayer Perceptron Neuronal Network
M_r = Relative Molecular Mass
MS = Mass Spectrometry
MSPD = Matrix Solid-Phase Dispersion
NIRS = Near-Infrared Spectroscopic
NMR = Nuclear Magnetic Resonance
O-PLS = Bidirectional Orthogonal Projections to Latent Structures
OPLS-DA = Orthogonal Projection to Latent Structure Discriminant Analysis
PCA = Principal Components Analysis
PCF = Principal Component Factoring
PCS = Principal Component Similitude
PDO = Protected Designations of Origin
PGI = Protected Geographical Indication
PLS = Partial Least Square
PLS-DA = Partial Least Squares Discriminant Analysis
PLSR = Partial Least Square Regression
PNN = Probabilistic Neural Networks
RB-NN = Radial Basis Neural Networks
SAW = Surface Acoustic Wave
SDA = Stepwise Discriminant Analysis
SHS = Sampling Head Space
SIMCA = Soft Independent Modeling of Class Analogy
SL-SPE = Solid Liquid and Solid Phase Extraction
SNIF-NMR = Site-Specific Natural Isotopic Fractionation by ^{2}H Nuclear Magnetic Resonance
SPME = Solid Phase Micro Extraction
SVM = Support Vector Machines

INTRODUCTION

The origin of the products based on their denomination of origin, has been increasing in the recent years since there are some products whose value is based mainly on their origin. This fact has generated the emergence of new analysis to define the origin of these products. One of the most widely used tools has been the implementation of new techniques based on chemometric methods, which have allowed an important worldwide development in the field of classification.

From this point of view, the origin analysis of wines involves the determination of the controlled determination of origin (CDO), the protected geographical indication (PGI), as well as the protected denomination of origin (PDO). These latters contribute to the certification of origin which is the most important component of the product value. This arises as a consequence of the concept of "terroir", which results in certain characteristics that make a wine from a specific region, unique in its type, without the possibility of finding another equal in the world. The terroir can involve both geographical and cultural aspects, since it could be due to regional characteristics of the soil, weather, as well as the method of production (milling, fermentation, oak aging, etc.), which define a wine as unique. This is more strongly appreciated in the production of Champaign, whose name has been protected to characterize sparkling wines derived from Champaign (France), from those coming from any other place in the world. However, this controlled denomination of origin is also done in the case of wines with terroir, such is the case of wines of La Rioja (Spain), Bordeaux (France), Lambrusco (Italy), Riesling (Germany), etc.

Even though there are some previous reviews related to the analysis of wines, these are mainly centered on the use of specific analytical methodologies rather than the obtained classification. Thus, we have the analysis of aromas and flavors of beverages in general, by the use of electronic noses and tongues, combined with chemometric tools (Smyth and Cozzolino, 2012); the analysis of polyphenolic compounds in wines using chromatographic and electrophoretic methods with electrochemical detection (Sanchez Arribas et al., 2012); the use of electronic noses in diverse alcoholic beverages (Martí et al., 2005); strategies to characterize wines based on their composition profiles (content of organic acids of low molecular weight, volatile species, polyphenols, amino acids, biogenic amines and inorganic species) applying chemometrics to facilitate the extraction of information (Saurina, 2010), and the application of quality control methods to evaluate the wine authenticity performing the differentiation and classification based on the geographical origin by means of a multivariate analysis (Arvanitoyannis et al., 1999)

Therefore, in this chapter, the existent studies on the classification of white wines using chemometric methods for the data analysis, in the period between 1998 and 2013, will be analyzed based on the type of classification performed. Thus, in order to facilitate the comprehension of the text, the sections have been divided based on countries, varietals, qualities of interest, etc, and arranged in chronological order for its quick search and viewing.

Evaluation of the Geographical Origin of White Wines

Argentina

The Torrontés represents the variety of white wine with the highest production in this country. Thus, the geographical classification of Torrontés wines from Argentina has been performed by using UV spectroscopy and multivariate tools such as Principal Component Analysis (PCA), Cluster Analysis (CA), Linear Discriminant Analysis (LDA) and Partial Least Square Discriminant Analysis (PLS-DA). By the use of these tools, the classification of Torrontés wines from four different provinces was achieved: Salta, Río Negro, San Juan and Mendoza (Azcarate et al., 2013a). The study shows that in all cases, the geographical classification has been correctly performed, seeing the separated grouping of the four studied regions, situation that is confirmed by the discriminatory tables, which show a 100% correct classification in each one of the regions. The authors conclude that the proposed method can be of interest for the determination of geographical origin of Torrontés wines, in the Argentinean national market as well as the international one.

Similarly, the same authors (Azcarate et al., 2013b) were able to classify the Sauvignon blanc variety produced in Argentina, by UV spectroscopy and multivariate tools such as PCA, CA, LDA and PLS-DA. The results showed the classification of this variety according to three provinces: San Juan, Mendoza and Río Negro. The classification table obtained by LDA and PLS-DA showed a 100% correct classification for each one of the regions. The authors concluded that due to the low cost analysis and the low sample treatment, the method could be of interest for the determination of geographical origin of Sauvignon blanc in the wine industry.

For its part, the multi-elemental composition of different varieties of wine (Chardonnay-Frambua, Torrontés Riojano-Alicant, Chardonnay) and soils has been studied, both from three provinces of Argentina (San Juan, Córdoba and La Rioja). Eleven elements were determined: K, Na, Ca, Mg, Fe, Cu, Zn, Mn, Ni, Co and Cr, by means of flame atomic absorption spectrometer (FAAS), in 31 wine samples and 137 soil samples. Stepwise discriminant analysis (SDA) was performed to classify the analyzed wines according to the region of origin. The authors obtained a 100% correct classification of the wines from seven parameters (K, Fe, Ca, Cr, Mg, Zn and Mn), and a 96% correct classification for soils (Ca, Cr, K, Fe, Cu, Zn and Mg). They evaluated the correspondence between the wines and the soil profiles by canonic analysis, obtaining a correlation coefficient of 0.85 (p-value < 0.001), which indicates that an 85% of the variability observed between wines is due to the composition of the soil where the grapes were cultivated. They concluded that the multi-elemental determination of wines and soils, along with chemometric tools, contribute to the study of the chemical traceability of wines, and the verification of their geographical origin (Fabani et al., 2010).

Australia

The geographical classification of Chardonnay wines of two regions (Coonawarra and Geelong) was performed in this country by data multivariate analysis techniques. They analyzed the wines by High Performance Liquid Chromatography (HPLC) with UV and acidified potassium permanganate chemiluminescence detection, using PCA and LDA to classify the wines according to the production region (Bellomarino et al., 2009). They achieved the separation with the two detection systems. The identification of the key components that lead to the discrimination of the wines was performed by mass spectrometry (MS). The linear discriminant analyses with UV detection pointed out that coumaric and gallic acids, and epicatequine, were important for the geographical determination of Australian wines, obtaining a 100% correct classification. The acidified potassium permanganate chemiluminescence showed that the main components responsible for the geographical determination were myricetin and cinnamic and tartaric acids, reaching a 75% classification of the correctly discriminated Chardonnay wines (Bellomarino et al., 2009).

Brazil

A study was carried out to evaluate sensorial characteristics in Chardonnay wines coming from four geographical regions (Vale dos Vinhedos, Flores da Cunha, Vacaria and Pinto Bandeira) of the Rio Grande do Sul province (Brazil), by means of headspace solid phase micro extraction (HS-SPME), gas chromatography coupled with mass spectrometry (GC-MS) and gas chromatography with flame ionization detector (GC-FID). Initially, the optimum headspace and chromatographic conditions were evaluated and then a total of 32 volatile compounds were analyzed. The multivariate analysis was performed by CA. The results showed that the wines of the Vale dos Vinhedos region presented the best sensorial characteristics due to the presence of a great amount of aromatic esters; while on the contrary, the wines from the Vacaria region were the ones with the least aromatic qualities. The authors concluded that the wines of the Vale dos Vinhedos can be differentiated form those coming from the other regions thanks to their composition in esters, reason why the method can be useful as classification model and to improve the aromatic quality of Chardonnay wines in the production of Brazilian wines (Welke et al., 2012).

Czech Republic

In this country, the multivariate classification of wines produced in the four most important producing regions of Bohemian, Czech Republic, (Žernosecko, Mělnicko, Mostecko, and Roudnicko), was performed during the years 2000 and 2001. They carried out a multi-elemental analysis of 41 samples of white wines by inductively coupled plasma mass spectrometry (ICP-MS) and inductively coupled plasma optical emission spectrometry (ICP-OES), to evaluate if these methods could provide data to determine the wine region of origin. Šperková and Suchánek (2005) performed the determination of the content of 27 metals: Al, Ba, Mn, Rb, Sr, Zn, As, Co, Cr, Cu, Ni, Pb, V, Ce, Cs, Li, Mo, Sb, Sn, Th, U, Y by ICP-MS, and Ca, Fe, K, Mg, Na by ICP-OES. They used PCA and LDA for the wines differentiation and classification according to the geographical origin. They obtained a 97.4% correct classification of the samples according to their origin from the content of Al, Ba, Ca, Co, K, Li, Mg, Mn, Mo, Rb, Sr, V, and the proportion of the elements Sr/Ba, Sr/Ca, Sr/Mg. They also evaluated the influence of the production year in the multi-elemental determination. The authors concluded that the study indicates that several of the measured parameters do not depend on the production year, and that the analysis of trace elements provides a good perspective for the determination of the region of origin of the Bohemia wines (Šperková and Suchánek, 2005).

Greece

The chemical and sensorial analysis were performed for the determination of geographic origin of 33 Greek wines, from which 10 were white coming from the regions of Mantinia, Santorini, Rodas and Kefalinia; including Assyrtiko, Moschofilero, Athiri and Robola varieties. In the chemical analysis, the concentrations of nine phenolic compounds (catechin, procyanidine B1, procyanidine C1, gallic acid, vainillinic acid, cafeic acid, epicatechin, syringic acid and total phenols), six anthocyanins (delphinidin-3-glucoside, cyanidine-3-glucoside, petunidin-glucoside, peonidin-3-glucoside, malvidin-3-glucoside and acylated malvidin-3-glucoside) and nine metals (Na, K, Ca, Mg, Zn, P, Fe, Cu and Mn) were determined; while in the sensorial analysis the astringency, acidity, sweetness, body, aroma and aftertaste were determined. In the case of white wines, the most important variables were Zn, myricetin and gallic acid. The PCA analysis showed the wines grouped, however, the particular geographic distinction of each of them is not clear, reason why the method was not conclusive for white wines (Kallithraka et al., 2001).

Italy

In this country, the traceability study of Moscato white wines coming from the region of d'Asti, which has Protected Designation of Origin (PDO), was performed. The analysis of rare earth elements (REE) in wines, soils and grapes of vineyards of the region, was performed by ICP-MS. The authors showed that the REE study in soils presents a defined and characteristic profile that would allow finding the geographical origin. However, the geographical classification of these wines was difficult, since in the wine clarification process the use of bentonite is employed as routine practice, which provides significant amounts of REE that interfere with the analysis, reason why more studies are required for its application (Aceto et al., 2013).

Furthermore, the geographical classification of Greco Bianco wines coming from the Italian regions of Calabria and Campania, and from two different vintages, was performed. The analysis was performed on 40 wine samples and included: physicochemical parameters, acetic, lactic, tartaric, malic, citric and succinic acids, proline, alanine, arginine, isoamyl alcohol, glycerol, pH, total acidity, density color, intensity tonality, total polyphenol index, abs (280 nm), volatile acidity, total polysaccharides, total and free SO_2, glucose, fructose and ethanol. These parameters were analyzed by traditional methods, as well as HPLC and Proton Nuclear Magnetic Resonance (^{1}H NMR) analyses; and the classification tools used were PCA and PLS-DA. The scores plot obtained by PCA showed a clear classification of the wines based on the studied regions, which is confirmed by PLS-DA, where an excellent geographical classification of the wine samples can be observed. The variables of greatest impact in the classification were the total acidity and organic acids, whose profiles are different in each region, reason why the authors concluded that the profile of acids can be an adequate tool to determine the geographical origin of Greco Bianco wines coming from Calabria and Campania (Caruso et al., 2012).

Portugal

Martelo-Vidal et al. (2013) have performed the geographical discrimination of different varieties of Albariño white wines coming from the Rias Baixas region of the Condado, Salnés, Rosal and Ribeira do Ulla areas. Other wine varieties were analyzed (Godello, Treixadura and blend white wines) coming from other regions (Valderroas, Riberira Sacra, Monterrei, Ribeiro, Rioja and Table). The analyses were performed by UV-Vis and Near-Infrared Spectroscopy (NIRS), and chemometric tools: PCA, soft independent modeling of class analogy (SIMCA), LDA and support vector machine (SVM). The analysis by PCA showed that the combined analysis UV and NIRS, are in principle the most appropriate, since they enable the discrimination of some geographical regions. This observation was then confirmed with the different chemometric tools used, highlighting a 100% correct classification for the Condado region (UV/VIS/NIR), Salnés (UV/VIS) and Ribeira do Ulla (UV/VIS), by the use of SIMCA; SVM gave a correct classification for wines of Condado and Salnés, while the LDA in general gave lower percentages of correct classification for all the regions.

Serbia

The contents of five metals (Cu, Mn, Fe, Cd and Pb) were determined by flame and graphite furnace atomic absorption spectrometry (FAAS and GF-AAS), in white wines coming from different regions of Serbia. The samples were obtained in the Serbia market: Kosovski Riesling, Rubin Riesling, Džervin Riesling and Italian Riesling, from southern, eastern, central and northern Serbia, respectively. PCA and factor analysis were applied to highlight the relationship between these elements; then in the data reduction, three main factors that control the variability were identified. The authors showed that the application of hierarchical cluster analysis to the studied wines indicates a differentiation of the samples with respect to their origin (Ražić et al., 2007).

Slovenia

In a study done by Košir et al. (2001) on the use of site-specific natural isotopic fractionation by ^{2}H nuclear magnetic resonance (SNIF-NMR) and isotopic ratio by mass spectrometry (IRMS), the geographical classification of white wines coming from coastal regions (Dava and Sava), was performed. The multivariate analysis was carried out by PCA, CA and Kohonen artificial neural networks (KANN), and included the analysis of white and red wines, as well as the sugar artificial enrichment. By the analysis of the main components, it was observed that the geographical classification between the coastal regions was superimposed with the continental regions of Sava and Dava. However, this type of analysis showed a marked difference between natural and sugar enriched wines. Nevertheless, by means of KANN, a better classification of wines based on geographical origin was achieved; while with CA, a good classification based on geographical origin and the sugar enrichment was obtained. From this study, it is concluded that PCA, CA and KANN can achieve a good geographical classification and an adequate discrimination between natural and sugar enriched samples. Thus, these methods could be used to determine the geographical origin of Slovenian wines and to evaluate possible adulterations.

Spain

This country is characterized by having wines with PDO. Therefore, the classification of Spanish wines coming from different PDO was performed in a study through their elemental profile in combination with SVM (Jurado et al., 2012). The authors determined the content of Al, Ba, B, Ca, Cr, Cu, Fe, Mg, Mn, Ni, P, K, Si, Na, Sr, S and Zn in white wines from four production areas of Spain (Condado de Huelva, Penedés, Ribeiro, and Rueda) by ICP-OES. They applied pattern recognition methods to construct the classification models, using PCA to visualize the tendencies of the data, and stepwise discriminant analysis (SDA) to reduce the number of input variables. From this preliminary analysis, they determined that the concentrations of Cr, Mn, Si, Na and Sr were the relevant variables to build a SVM classification model. The application of the SVM model led to the differentiation of the four PDO, with a 100% prediction efficiency. The authors highlight that SVM is based on a structural risk minimization classification model, controlling the complexity of the model and avoiding over- fitting, advantages that offer against other recognition techniques used for the construction of models (Jurado et al., 2012).

Garrido-Delgado et al. (2009) differentiated and identified varieties of white wine (Pedro Ximenez, Palomino Fino, Zalema, Airen, Macabeo, Chardonnay, Verdejo, Sauvignon Blanc and Muscatel) through the most abundant organic acids (succinic, malic, tartaric, citric, acetic and lactic) by electropherogram fingerprints obtained with capillary electrophoresis (CE), and the help of multivariate statistics. The white wines were certified brands of origin (CBO) in southern Spain: Jerez and Montilla–Moriles (15-23% ethanol); Condado de Huelva and Valdepeñas (9-14.5% ethanol)). They used sensitive and specific chemometric models, as PCA and LDA, to classify the wines according to the CBO, allowing a 98% correct classification in the validation samples. The authors concluded that the CE in combination with chemometric techniques can help to prevent frauds in the CBO.

In addition, the HS-SPME and GC procedures were used to determine the composition of the samples volatile fraction of white wines of several Spanish CBO. Four CBO were studied: Rueda, Ribeiro, Penedés, and Condado de Huelva. The present compounds were previously identified by GC–MS, and the quantitative determinations were carried out by GC-FID. The determined compounds responsible for the taste of the beverages were: 2-methyl-1-propanol, 2-phenylethylethanol, 3-methyl-1-butanol, and 2-methyl-1-butanol, ethyl acetate, isoamyl acetate, ethyl hexanoate, hexyl acetate, ethyl sorbate, ethyl octanoate, 2-phenethyl acetate, ethyl decanoate, isoamyl octanoate, ethyl 2-decenoate, ethyl laurate, ethyl myristate, and ethyl palmitate. Some of the important volatile fraction components in the wine bouquet were: ethyl, isoamyl, hexyl, and 2-phenylethyl esters. In addition, higher alcohols, responsible for the fruity, ethereal-fruity, and flowery aromas, characterize these wines and make them adequate compounds for authentication studies. The most important variables for the discrimination were: ethyl acetate, ethyl hexanoate, hexyl acetate, 2-phenylethanol, ethyl decanoate, and ethyl myristate. The Rueda wines presented the highest concentrations of ethyl acetate, isoamyl acetate, hexyl acetate and 2- phenethyl acetate. These wines, along with the Ribiero and Penedés wines, presented high concentrations of compounds such as ethyl hexanoate and ethyl decanoate. According to the results obtained, and by the application of pattern-recognition procedures, PCA, LDA, and multilayer perceptrons neural networks (MLP-NN), the CBO differentiation was achieved. PCA showed that the Rueda and Condado de Huelva separated naturally. It was observed that the recognition abilities were of 100% by means of LDA, but the prediction abilities were in a range of 81.3–95% for the wines of Condado de Huelva, Ribeiro, and Penedés. For classification, a MLP-NN model was made with a 6:5:4 architecture. As a result, the authors concluded that it is possible to discriminate samples of white wines and predict the CBO (Jurado et al., 2008).

Acevedo et al., (2007) combined UV-visible spectroscopy with multivariate statistical techniques to characterize Spanish wines of different denomination of origin: La Mancha, Madrid, Penedés, Rioja and Valdepeñas. The authors of this work proposed SVM to improve the classification rates when using UV-visible spectrophotometric methods. They performed spectrophotometric measurements of 82 samples of white wines in specific regions of the electromagnetic spectrum (200-800 nm). From the obtained UV-visible spectra, they performed the classification by SVM. They obtained a classification rate above 96%. Afterwards, they compared the obtained results with other pattern recognition methods used in chemometric such as SIMCA, PLS-DA, MLP-NN, and k-Nearest Neighborhood (k-NN). The results obtained by SVM were comparable to those obtained for MPL-NN, since both could solve the problem of UV-visible spectral data, although SVM presents an inherent simplicity in comparison with the neural networks. The authors confirmed that this methodology allows the discrimination of wines according to their denomination of origin using UV-visible spectroscopy, in combination with chemometrics, as an affordable technique that could be applied in different industrial sectors.

Moreover, the possibility of developing a method based on a simple and inexpensive instrumentation, to differentiate and classify white wines of controlled designation of origin: La Mancha, Quintanar de la Orden, Fuente de Pedro Naharro and Mota del Cuervo, coming from Spain, has been studied. They used UV-visible spectroscopy to develop a simple and low cost detection method, combined with chemometric analysis for the analysis of the data. The UV-visible spectra of 120 samples were used and a PCA analysis was performed, while the development of classification models was carried out by means of SIMCA, based on the wines origin. The authors concluded that the UV-visible spectral region was of great importance for the differentiation of wines according to origin within the same denomination of origin, obtaining a correct classification of around 90% (Urbano et al., 2006).

Zamúz and Vilanova (2006) studied wines produced from the Albariño white grape of three different areas of Rías Baixas, north-east of Spain (Val do Salnés, Condado do Tea and O Rosal), vintage 2004. By means of GC-MS, they determined the content of volatile components of the produced musts (terpenes, C13-norisoprenoids, alcohols, acetates and ethylic esters). They carried out a PCA analysis to establish the relationship between the composition and the geographical region. From the multivariate analysis, they established that with two main components the 90.7% of the total variance of the data was explain, allowing the grouping of wines according to their respective origins. In addition, they showed through the analysis that: the Albariño wines of O Rosal were characterized by a high content of higher alcohols; the Val do Salnés wines showed high concentrations of free terpenes, acetates and ethylic esters; and the Condado do Tea wines, higher concentrations of C13-norisoprenoids. From the results obtained in this study, the authors found significant differences between the different geographical areas of Rías Baixas AOC (north-east of Spain) in regard to the concentration of most of the aromatic compounds; and that those differences could be established by PCA, which allowed the separation of the three groups of wines as a function of their geographical origin.

By the use of FAAS, 11 elements were determined (K, Na, Ca, Mg, Fe, Cu, Zn, Mn, Sr, Li and Rb) in samples of sweet and dry wines coming from the three Spanish islands: El Hierro, Las Palmas and Lanzarote wines, belonging to the archipelago of the Canarias. The multivariate tools included CA, LDA and SIMCA. CA showed a good discrimination between Lanzarote in comparisson to El Hierro and Las Palmas, but it was not clear between the last two regions. By PCA, the score plot showed a low capacity of geographical classification of the wines, since it presented important superposition between the three islands. However, LDA significantly improved the discriminative capacity, since the predicting percentages went from 85.7% for El Hierro to 100% for Lanzarote. Finally, SIMCA gave a better classification model, with a 100% classification for the three islands. The authors concluded that the metallic content in the islands was different due to the maritime influence and the soil volcanic origin. In regard to the multivariate methods, LDA showed good discriminative capacity but was surpassed by SIMCA due to the degree of sensibility and specificity of the method that correctly discriminated samples of white wine coming from the three islands of the Canarias archipelago (Frías et al., 2003).

Classification Including Several Countries

Andreu-Navarro et al. (2011) performed the analysis of white wines coming from Sicily (Italy) and Córdoba (Spain), using HPLC with fluorometric and UV photometric diode array detections, for the analysis of 18 phenolic compounds: gallic acid, protocatechuic acid, p-hydroxybenzoic acid, salicylic acid, vanillic acid, caffeic acid, syringic acid, catechin, vanillin, p-coumaric acid, syringaldehyde, epicatechin, ferulic acid, rutin, trans- and cis-resveratrol, quercetin and kaempferol. By the use of LDA, a 100% correct classification based on the geographical origin was possible; however, it should be considered that the classification was also based on different types of grapes (white, red and rose). From the white wines point of view, the most important contribution of this work was the analysis of phenolic compounds and the use of chemometric tools.

In Oceania, the combination of UV, visible, NIRS and mid-infrared spectroscopy (MIRS) with multivariate analysis of the data has been studied as an alternative low cost and short time analysis to differentiate, based on the geographical origin, Sauvignon blanc wines of Australia and New Zealand. They analyzed the UV, Vis, NIRS and MIRS spectra of 32 different brands of wines. They used PCA, SIMCA and PLS-DA for the classification of wines according to the geographical origin, using full cross validation (leave-one-out) as validation method. The PLS-DA correctly classified 86% of the wines. The best prediction rates were obtained using MIRS. The concatenation of NIRS and MIRS, was 93% for the New Zealand wines for both predictions, and 86% and 93% for the predictions of Australian wines, respectively. The authors pointed out that the results demonstrate the potential of the combination of spectroscopy with chemometric data analysis, as a fast method to classify Sauvignon blanc wines according to their geographical origin (Cozzolino et al., 2011).

Liu et al. (2008) performed studies using visible and NIRS spectroscopy combined with chemometrics to classify Riesling wines (n=50) of different countries (Australia, New Zealand, France and Germany). PCA, PLS-DA and SDA based on PCA scores were used to classify them according to their country of origin. A full cross validation (leave-one-out) was used as validation method when the classification models were developed. The PLS-DA models correctly classified 97.5%, 80% and 70.5% of the wines from Australia, New Zealand and Europe (France and Germany), respectively. The SDA calibration models correctly classified 86%, 67%, 67% and 87.5% of the wines from Australia, New Zealand, France and Germany, respectively. The results demonstrated that the visible and NIRS spectra contain information that allows to distinguish wines of different countries. However, in order to validate the capacity of Vis-NIRS to classify samples of white wines, the authors have clarified that a larger sample set would be required.

Classification of White Wines According to Varietals

In a study performed by Lampíř (2013), the characterization of seven different varieties of white wines from Czech Republic was carried out: Aurelius, Chardonnay, Müller Thurgau, Moravian Muscat, Hibernal, Malverina and Merzling. The determination was performed by studying the phenolic compounds profile by HPLC chromatography and canonical variate analysis (CVA). Fifteen different phenolic compounds were detected in each wine variety. The multivariate classification gave seven well distinguished groups corresponding to each wine variety. In particular, Moravian Muscat was significantly different from the rest. The author concluded that the use of phenolic compounds profiles is of great interest to perform the discrimination of white wines according to varietals.

Kirsanov et al. (2012) studied different varieties of white wines (Chardonnay, Pinot gris and Sauvignon blanc) by applying the electronic tongue, and PLS model. The studied variables included total acidity, pH, ethanol and polyphenolic compounds. The multivariate results obtained by PLS models were compared to those obtained with classical methodologies of wine analysis. The authors found an important correlation between the proposed method and those used as reference.

Gutiérrez et al. (2011) studied the classification of two white varietals from Croatia (Zelenac and Grasevina) and the Macabeu, Parellada, Chardonnay and Xarel-lo varietals from Cataluña (Spain), using an electronic tongue device with microelectrodes, electrochemical detection and optical detection. The model tools used by the authors were PCA and PLS. The results by means of PCA, presented as score plots, showed a good general classification of all varietals. The study also included the analysis of Spanish white wines mixtures. PCA showed superpositions between the Spanish mixed varietals in relation to their pure varietals. The PLS analysis shows a comparative table between the reference method and the method proposed for the analysis of Macabeu, Parellada, Chardonnay and Xarel-lo monovarietals, with a range of relative error from -4.9 to 9.5%. The authors concluded that the multidetector method using electronic tongue and coupled to the multivariate methods was accurate for the quantification and discrimination of the studied monovarietals and their bi-trivarietals mixtures.

Villagra et al. (2012) performed a varietal discrimination of 25 commercial Chilean white wines, of the varieties Chardonnay and Sauvignon blanc. They investigated the potential use of direct injection, electrospray ionization mass spectrometry (ESI-MS), in negative ion mode, in combination with multivariate analysis tools, as a quality control method. The wine samples were just diluted and directly infused in the mass spectrometer. PCA allowed a good classification of the white wines, in which the first three main components explained 96.82% of the total variance of the data. LDA allowed a 100% correct discrimination of the samples, while the validation by the leave-one-out cross-validation method gave a 76% correct classification. Thus, the authors concluded that some of the analyzed wine samples could be blends of more than one variety and therefore, a stronger mathematical model constructed with spectral information of the pure varietals and mixtures varietals, is required to improve the classification percentages.

Garrido-Delgado et al. (2011) carried out the classification of 4 varieties of white wines with CBO: Jerez, Montilla-Moriles, Valdepeñas and Condado de Huelva, using a new analytical technique: ion mobility spectrometry with ultraviolet ionization (IMS-UV) coupled to a continuous flow system and comparing the results by classic GC chromatography. The chemometric tools used were: PCA, LDA k-NN. By using IMS-UV, a good classification was obtained using LDA, with a well-defined discriminative plot. In addition, the score plots by PCA and LDA showed a good classification of the four varietals using GC. The LDA classification was much clearer, since its discriminative table showed a correct prediction percentage of between 90-95%, for the 4 varietals. The authors concluded that even though GC was a precise method, the use of forefront techniques such as IMS-UV for the classification of the CBO varietals was more economic and faster.

Louw et al. (2009) combined Fourier transform infrared (FTIR) spectroscopy, GC and multivariate analysis, to get the chemical composition, and discriminate between different South African young cultivar wines. They used 173 young white wines to exclude variability caused by oak maturation, blending and bottle aging, of single-cultivar Chardonnay and Sauvignon blanc. Through the volatile composition FTIR spectroscopic profiles, and by the analysis of variance (ANOVA), PCA and LDA, it was possible to differentiate the wine varietals. The Sauvignon blanc wines contained significantly higher mean concentrations of decanoic acid, hexyl acetate, and octanoic acid, in comparison to Chardonnay, which contained significantly higher amounts of ethyl hexanoate. Finally, the authors found that the best discrimination model between white wines, based on FTIR spectra, had a 98.3% correct classification.

Seventy two wine samples of Chardonnay, Riesling, Sauvignon blanc and Viognier varieties, produced in 2006 in different regions of Australia, were evaluated by MIRS combined with chemometrics for their discrimination. The spectral data were reduced to a small number of main components by PCA. Then, the authors performed LDA to classify the wines according to different varietals. They achieved correct classification rates of 96%. From this study, the authors demonstrated that MIRS, combined with multivariate statistical techniques, could be used to discriminate between different varieties of white wines. In addition, they suggested that this technique could be used by the wine industry as a rapid and robust detection tool for varietal discrimination (Bevin et al. 2008).

Tredoux et al. (2008) classified wines from South Africa based on their volatile composition by chemometric methods. They developed a simple method for the analysis of the main volatile and semi-volatile compounds of wines, by thermal desorption and as chromatography-mass spectrometry (TDGC- MS). They analyzed a total of 19 wines of three white varietals (Chardonnay, Chenin and Sauvignon blanc), of vintages from 1996 to 2003. They determined 38 main volatile and semi-volatile compounds of the wine, including alcohols, acids, esters, phenols, aldehydes, ketones and lactones. They used the quantitative data of the wines to study if the main compounds analyzed had the capacity to differentiate the wine samples according to the grape variety. The PCA and CA analysis showed that the larger fraction of the variation in volatile composition between the wine samples could be attributed to differences in the wine age, contact with the wood, and the fermentation practices. They also successfully carried out a DA to classify the wines according to the variety. Through the analysis of the results, the authors concluded that the volatile composition contains great amount of information that can be studied with chemometric methods proposed in the work for the classification of wines according to the variety.

Son et al. (2009) characterized wines from Korea by means of ^{1}H NMR analysis coupled to pattern recognition methods, such as PCA and orthogonal projection to latent structure discriminant analysis (OPLS-DA). They studied the following varieties of wine: Muscat bailey A (*Vitis labrusca*), Campbell early (*V. labrusca* B.), Kyoho (*V. labrusca* L.) and Meoru (*Vitis coignetiae*). The metabolites responsible for the differentiation of the four varieties were identified as 2,3-butanediol, glycerol, malate, citrate, tartrate, succinate, lactate, proline, alanine, choline and trigonelline. The Kyoho wines have higher levels of glycerol, 2,3-butanediol, succinate and alcohol, which suggests a higher content of sugar in the grape, in comparison to other varieties. Higher contents of citrate and trigonelline in Muscat bailey A wines, alanine in Campbell early wines, and proline, malate and choline in Meoru, demonstrated that such metabolites vary with the grape variety.

Chabreyre et al. (2008) used high-performance capillary electrophoresis (HPCE) for the identification of wine varietals from the protein profile of white wines previously dialyzed, combined with the concentration of shikimic acid determined according to the OIV official method, by HPLC and multivariate analysis. Authentic monovarietal wines of diverse geographical origin from France were studied, according to a Plackett-Burman sampling plan elaborated in the laboratory: Chardonnay (24 samples), Chenin (24), Petit Manseng (7), Sauvignon blanc (37), Semillon (24) and Ugni Blanc (9). The authors elaborated a matrix that contains the protein results of the wines (mixtures and monovarietals) and shikimic acid concentration. By means of PLS, they made a model that provides a 20% precision for the varietal determination of most of the studied grapes, which could be promising results for control purposes.

Other authors systematically collected voltammetric responses of white wines, selected from different aging and origins, to test the efficiency in relation to the blind analysis of similar matrices. The main goal of the study was to test electrode materials that could complement, or even substitute, the traditional electrochemical sensors used in electronic tongues. Three different modified electrodes were used, which consist of a conductive polymer: poly(3,4-ethylenedioxythiophene) (PEDOT) and of composite materials of Au and Pt nanoparticles embedded in a PEDOT layer. Twelve wine samples were used: Chardonnay (form France; Veneto and Puglia), Trebbiano (from Emilia, Umbria and Romagna) which were analyzed without previous treatment with the differential pulse voltammetry technique. The modified electrodes were tested separately, coupled between them, and even all together, to check the possibility of collecting even more information about the investigated system. The response of the electrodes system was evaluated with PCA. The classification models were built and validated by PLS-DA. The authors demonstrated that all the modified electrodes allowed performing an excellent varietal classification of the white wines. From them, the best classification model was obtained with the PEDOT-modified electrode, giving results of 100% in terms of sensibility and specificity; making it a convincing candidate as detection unit in electronic tongues (Pigani et al., 2008).

Lozano et al. (2007) performed a comparison between several sampling methods in order to discriminate wine varieties: static headspace with dynamic injection, purge and trap and solid-phase micro-extraction (SPME), used in a thin film semiconductor sensor array based electronic nose. Wines of different grape varieties (Airén and Malvar) were used, elaborated with the majority of the varieties in the origin denomination (OD) “Madrid wines”. Linear techniques as PCA and nonlinear ones as probabilistic neural networks (PNN) have been used for pattern recognition. Even though the highest response was obtained by the static headspace, due to the amount of extracted methanol, the best discrimination was achieved with the other two methods. While using purge and trap and SPME, the separation was clear in different clusters with PCA, and the classification performed with PNN was important (100% and 95.8% for SPME and purge and trap, respectively). The authors consider that the selection of the extraction technique depends on the application: static headspace and purge and trap are adequate for in situ and portable systems, while SPME could be used in the laboratory.

Previously, the authors used an array of surface acoustic wave (SAW) sensors using Zinc oxide films ($Si–SiO_2–ZnO$ structure), in order to discriminate Spanish wines from different grape varieties and elaboration processes. Wines of the Madrid region were elaborated with the varieties Airén and Malvar. Sensor array is composed of eight surface acoustic wave sensors, one of them used as a reference sensor and the others coated with diverse polymers by spray coating technique. Again, linear techniques as PCA and non-linear ones as probabilistic neural networks (PNN) have been used for pattern recognition, obtaining a 90% satisfactory varietal classification (correct predicted number over total number of measurements) (Lozano et al., 2006).

In another study, 36 wine samples of white varieties were studied: Madeira, Boal, Malvazia, Sercial and Verdelho, in order to estimate the fraction of free monoterpenols and C13 norisoprenoids (terpenoid compounds) using HS-SPME technique coupled with GC–MS. The mean aging values (1998-2000) showed that these wines have profiles characteristic of terpenoid compounds. The Malvazia wines exhibited the higher values of total free monoterpenes, contrary to the Verdelho wines that showed the lowest levels of terpenoides, but were the ones that produce the highest concentration of farnesol. The use of multivariate analysis techniques allowed establishing relationships between the compounds and the variables under study. PCA and LDA were applied to the data matrix obtaining 89.9% and 100% separation and classification, respectively, between the four groups as a function of their varietal origin (Câmara et al., 2007).

Before that, these same authors differentiated and characterized monovarietal wines of Madeira according to the white grape varieties (Boal, Malvazia, Sercial and Verdelho), from which they were elaborated. They determined 42 volatile components (higher alcohols, fatty acids, ethyl esters and carbonyl compounds) from 36 wine samples by HS-SPME and GC–MS. They studied the main sources of variability presented in the data set by PCA. They established that the most important contributions for the differentiation of the Boal wines were the benzyl alcohol and hex-3-en-1-ol; ethyl octadecanoate, hex-3-en-1-ol and benzoic acid for Malvazia wines; and 2-metil-1-ol for the Sercial wines. The Verdelho wines were more correlated with 5-(ethoxymethyl)-2-furfural, nonanone and cis-9-ethyldecenoate. They applied SDA, for which they obtained 96.4% of the prediction capacity by the selection of 19 variables that maximized the variance of the initial set of data. PCA and SDA allowed obtaining good differentiation, classification and prediction models of wines from Madeira, according to the grape varieties. The authors concluded that all the studied chemical groups achieved a good separation according to the variety, but the higher alcohols and ethyl esters were the most important groups for the characterization of the Madeira wines (Câmara et al., 2006).

Rocha et al. (2006) described a methodology for a fast distinction of white wines from Portugal. For that, they performed the determination of volatile compounds of 16 white monovarietal wines (Ferñao-Pires and Arinto), of the 1999 vintage, by HS-SPME coupled with GC–MS. The applied PCA to the chemical data matrix obtained, aimed to characterize and distinguish the wine varieties. The first two main components represented the 78% of the total variability present in the data set. A clear distinction was observed between the two varieties of wines along with the first principal component (53% of total variability) for all the extraction times used. These results clearly showed that this methodology has the ability of distinguishing Ferñao-Pires from Arinto, representing a simple and effective tool for the distinction of wine by the fast analysis of its headspace volatile fraction. The authors point out that the total time of analysis was of 6.5 min (5 min in for SPME plus 1.5 min for the analysis by GC-MS), which makes it a fast method for the distinction of these varieties of wine.

Boselli et al. (2006) studied the phenolic composition and the quality of two white wines of CDO of Marche (Italia), of the varieties: Passerina (4 samples) and Verdicchio dei Castelli di Jesi (3 samples), through the multivariate analysis. To characterize the wines, they determined their phenolic fraction by HPLC coupled with ESI-MS, and performed the sensorial evaluation, which was carried out by 16 trained panelists that evaluated a set of sensorial descriptors of the wines (herbaceous, fruity, mandle, exotic fruits, honey, acidity, salty, bitter, burning and astringent). They performed a main component analysis to the obtained set of data, by which they determined that the differences in chemical composition were correlated with the sensorial properties of the wines: the burning and astringent taste was associated to caftaric, ethyl coumarate and 2-S-glutathionylcaftaric acid in the Verdicchio variety; while the flavor to honey, fruits and exotic fruits was related to Passerina. The authors concluded that the fingerprint of the phenolic compounds obtained, showed that Verdicchio and Passerina, two wines produced in the same region but with different grape varieties, are completely different with regard to the chemical composition as well as the sensorial properties.

The Riesling and Chardonnay wine varieties from Australia were analyzed through MS-based electronic nose and chemometric analysis by PCA, LDA and Discriminant PLS (DPLS). By means of MS e-nose, the relative masses abundance for each of the varietals, performed on a total of 150 samples, was obtained. The scores plot by PCA showed a classification with an important superposition zone between the two varietals. Besides, the discriminative table obtained by LDA, showed 72 and 85% of correct classification for Chardonnay and Riesling, respectively; while the best classification was obtained by DPLS, with a 93% of correct classification for both varietals. These results confirm that the use of DPLS as chemometric tool tends, in general, to be more effective than other methods for the classification of white wine varietals (Cozzolino et al., 2005).

Before that, the same authors determined the same Australian varietals, Riesling and Chardonnay by PCA and DPLS, but in this case they used UV and NIRS spectroscopy. Using the second derivate of the UV and NIR spectra, a good classification of both varietals was not obtained by PCA. However, by DPLS analysis, a very good classification of both varietals was obtained, from a 97.8 to 100% correct classification. The authors concluded that the DPLS method is useful for the varietal determination of Riesling and Chardonnay wines from Australia, with a 98% average of accuracy, allowing a correct classification despite of the spectral similitude (Cozzolino et al., 2003).

In a study performed by Coimbra et al. (2002), the classification according to varietals of María Gómez and Bical white wines was performed, by the extraction of polysaccharides using FTIR, PCA and canonical correlation analysis (CCA). The studied varieties were subjected to different treatments of clarification and maceration with or without skin. The studied polysaccharides were mannose, arabinose, galactose, hexuronic acid, galacturonic acid, rhamnose and glucose. By means of PCA and CCA, the effect produced by the treatments in the different models was evaluated, observing a strong incidence of the content of mannose, arabinose, galacturonic acid and glucose in the varietal classification. However, by CAA, it was observed that mannose was the variable with the highest incidence on the classification. The authors concluded that the combination of FTIR with PCA and CCA, allowed evaluating the effect of polysaccharides on the production of wines, as well as their varietal classification, obtaining a good correlation between the content of mannose and polysaccharides in both varietals.

The analysis of Chardonnay and Pinot blanc from the regions of Anderson Valley, Carneros District, Napa Valley, and Russian River Valley, was performed using aroma profile and analysis by PCA. The authors could classify white and red varieties (both base and sparkling wine); while at the same time, they could group the different white varieties, reason why the method has been capable of classifying these wines according to aromatic characteristic, where citrus and apple were distinguished for studied white wines (De la Presa-Owens et al., 1998).

JOINT CLASSIFICATION OF WINES ACCORDING TO GEOGRAPHICAL ORIGIN AND VARIETAL

Weldegergis et al. (2011) investigated the content of aroma of young South African wines, determining the principal volatile compounds of 110 wines of Sauvignon blanc and Chardonnay varieties. The wines were from different regions of South Africa including Paarl, Stellenbosch, Worcester, Robertson, Olifants River and Swartland. In order to investigate the possible correlations between the volatile content and the grape variety and/or the geographical origin, ANOVA, FA, PCA, and LDA were used. Significant differences were observed in the levels of certain volatile substances as a function of the region and the variety, being the latter factor more influential. ANOVA indicated significant differences in the content of volatile compounds between cultivars; some were identified as possible predictors of the white wine varieties. FA showed that the fermentation practices and the wood aging are the main responsible factors of the variation in the volatile content of the studied wines. PCA was valuable to relate the different variables responsible of the differentiation between samples of wine. LDA was used to identify the variables useful in the prediction of the wine cultivar. The integration of LDA with PCA showed the contribution of a small number of variables in the differentiation of South African wines of different varietals. In general terms, the results of the statistical approach were complementary, since they illustrate the differences between the two white cultivars.

Kruzlicova et al. (2009) used artificial neural networks (ANN) to evaluate the possibility of classifying white wines of Slovakia according to variety, location and production year. They performed the determination of 26 volatile compounds by means of GC-MS, in Welsch

Riesling, Grüner Veltliner and Mueller Thurgau wines from five producing regions, located at the west and southwest of Slovakia, and of different vintages years (1996, 1997 and 1998). Before the classification analysis, they evaluated the analytical data by ANN, and the classification results were compared with ANOVA. They determined a variable ranking (volatile components) as a function of their individual importance in the classification. The results found showed that the most important volatile compounds to differentiate white wines were a-terpineol for the varietal classification, diethyl succinate for the regional classification and 1-hexanol and 3-hexen-1-ol for the classification by vintage year. The ANN performance was calculated for the three classification criteria, and obtained a 100% successful classification for the vintage year and the production region, and 93.3% for the variety.

Šnuderl et al. (2009) used data multivariate analysis methods to classify white wines produced in two regions of Slovakia (Bratislava and Hlohovec). They worked with a total of 46 wine samples of Welsch Riesling, Grüner Veltliner and Chardonnay varieties, with vintages corresponding to the years 1999 and 2000. A panel of experts performed a sensory analysis evaluating the different wine properties, such as color, limpidity, bouquet and taste. They also performed a chemical analysis by different analytical methods, from which they determined free and total sulfur dioxide, total acidity, volatile acidity, pH, density, sugar-free extract; citric, malic, tartaric and lactic acids; glucose, fructose and total polyphenols. They statistically evaluated the obtained data by PCA and LDA. They obtained a 100% of correct classifications by variety, location and year of production; and a 93.5% correct classification for the sensory quality. The authors suggested that the chemometric approach applied in this work, could be useful for other classifications with other characteristics of the wine such as amino acids profile, polyphenols profile, volatile aromatic compounds, trace elements profile or proteins fractions.

Anastasiadi et al. (2009) developed a simple and sensible method for the classification of wines from Greece based on the variety, geographical origin and vintage year, by using NMR-based metabolomics and chemometric classification. They collected 26 wine samples from the main varieties of white grapes from Greece (Moschofilero and Assyrtiko), cultivated in the region of Nemea in Peloponnese, and the Santorini Island, of two consecutive vintages (2005 and 2006). From these wines they prepared varietals rich in polyphenols using adsorption resin, XAD-4. From the extracts they obtained the spectra of the samples by ^{1}H RMN spectroscopy, which were segmented, integrated and normalized. The obtained data were subjected to PCA and PLS-DA. The chemometric classification of the wines as a function of the phenolic profile, allowed the authors to discriminate between wines from different wineries from the same production area, and between different years of wines of the same variety.

Du et al. (2007) used a combination of ^{1}H NMR and PCA to classify type, origin or geographical production origin of wines. Twenty one wines produced under the national standard of China (GB/T15037-94) with grapes or grape juice (100%), were analyzed. Samples included: dry and medium dry white wines from Shacheng (continental region), and dry white from Yantai (coastal region). The authors obtained an excellent classification of the type and location of the wine production. Significant statistical separations were obtained ($p<0,001$) for dry and medium dry white wine. The main chemical products that contributed to the separation were identified, and their peaks were assigned in ^{1}H NMR. The components can be used as chemical markers for the wine types and location of production. Taking into account that the majority of the winemaking companies use grapes or grape juices of the area as raw material, the location of production can represent the geographical origin of the wines. Finally, it is concluded that the separation of types is mainly due to the fermentation process for different wines, and the environmental variations such as local weather, soil, groundwater, solar intensity and rainfall patterns.

By means of ^{1}D and ^{2}D ^{1}H and ^{13}C NMR, different white wine varietals were analyzed (Chardonnay, Welsch Riesling, Sauvignon blanc and Riesling) from different regions of Slovenia (West Coast, Sava and Dava). The analysis consisted in the determination of succinic acid, butylene glycol, glycerol and the amino acids arginine, citrulline, lysine, proline, isoleucine, alanine and valine. In addition, the anthocyanins malvidine-3-glucoside, peodinine-3-glucoside, petudine-3-glucoside, delphinidine-3-glucoside and cyanidine-3-glucoside, were analyzed. By using analysis of clusters, the study determined that by the use of the amino acids variables, it was possible to obtain a clear classification of white wines by varietals; while if anthocyanins are used as variables, the discrimination by geographical origin is possible, distinguishing three regions: West Coast, Sava and Dava. Therefore, the method resulted satisfactory to determine varietals and geographical origin of Slovenian wines (Košir et al., 2002).

Hungarian authors performed the classification of white wines based on geographical origin (18 Hungarian regions) and botanical (19 varieties of white wines), but also included the analysis according to 9 vintage years (1993, 1994, 1996, 1997, 1998, 1999, 2000, 2001, and 2002) and winemaking technology (high quality and liquor quality). The studied variables were: complete profile of amino acids and biogenic amines; and the multivariate methods included PCA and LDA. The work showed that the PCA model clearly discriminates the wines of high quality form those of liquor quality. By means of LDA, the classification according to wine-making technology was between 91.5 and 97.0%, which indicates an acceptable level of correct classification. However, the geographical classification was relatively low (70.8%), and the botanical classification was unfeasible, with the exception of the native botanic variety Tokaj. Finally, in relation to the vintage year, the LDA classification showed discrimination between the years 1993 and 2000; while the rest of the vintages could not be classified. The authors concluded that PCA and LDA, in conjunction with the amino acids profile, were useful to determine the type of winemaking, and to a much lesser extent, the geographical and botanical origin, and the vintage year (Héberger et al., 2003).

Classification of White Grapes

Baiano et al. (2011) studied the differences in table grape varietals evaluating phenolic profiles and antioxidant activity, by means of UV-visible spectroscopy and HPLC with diode array detector (DAD). Seeds, skins and pulps of seven varietals cultivated in the region of Apulia (Italy) were analyzed, considering the white grapes (Italia, Baresana, Pizzutello, and Thompson Seedless) and red/black grapes (Red Globe, Michele Palieri, and Crimson Seedless). PCA allowed obtaining a good separation of the Italia and Michele Palieri varietals. According to the clusters analysis, the cultivars were grouped by skin color into two groups, one including Michele Palieri, and the other Italia, Baresana, Pizzutello, and Thompson Seedless. Crimson Seedless and Red Globe were not classified. Through the chemical analysis, they concluded that the seeds have a great contribution to the antioxidant activity, since they have a larger phenolic content than the skin and high amounts of proanthocyanidins; however, the strongest antioxidant activity is found in the pulp juices due to their content in hydroxycinnamic acids.

In a study carried out by Dopico-García et al. (2007), they compared matrix solid-phase dispersion (MSPD) with solid liquid extraction followed by a solid-phase extraction (SL-SPE), to separate extracts of phenolic compounds and organic acids from white grapes. Even though the MSPD appeared to be a simple and fast technique, a complete extraction of all the studied compounds could not be achieved, especially that of organic acids. Therefore, to obtain significant recoveries, it was chosen to work with SL-SPE. In order to study the effects of the variety and origin in the profile of the mentioned compounds, a white wine called "Vinho Verde" was used, which is considered a quality wine produced in a specified region, and which is found in the north-west region of Portugal. Samples of different varieties were collected in different localities and vineyards: Famalicão (one vineyard), Felgueiras (one vineyard), Monção (three vineyards: Quinta de Menanços, Quinta de Moreira and Quinta de Ceivães) and Ponte de Lima (two vineyards: Quinta de Barreiros and Quinta da Facha). Ten varieties were studied: Alvarinho, Asal Branco, Avesso, Batoca, Douradinha, Esganoso Castelo Paiva, Loureiro, Pedernã, Rabigato and Trajadura. The most abundant compounds found were quercetin-3-O-glucoside/quercetin-3-O-rutinoside and epicatechin, and tartaric and malic acids. PCA was applied separately to establish the main sources of the variability present in the data set, for phenolic compounds, organic acids and the global data. The analysis of the results by the use of multivariate techniques showed a high influence of the variety and the origin in the composition. PCA showed that the phenolic compounds represented the highest variability (77.9%) with two principal components (PCs), which allowed the characterization of the samples varieties according to their higher content in flavonoids or epicatechin derivatives. The 100 and 70% of the recognition capability was obtained by the application of SDA, using origin and variety as classification variables, respectively. The authors considered that taking into account the strong effect of the origin on the results, a study should be performed with a larger number of samples of each variety and origin, to improve the differentiation of grapes according to their variety. Therefore, the phenolic compounds and organic acids demonstrated to be useful variables to characterize white "Vinho Verde" grapes.

In another study, the evaluation of white grapes, varietal Parellada, coming from Cava (Spain) and corresponding to different production volumes (from 6000 to 17000 kg/ha), was performed. The authors used PCA to determine the effect of the production and maturation index, by the analysis of proteins and polysaccharides in the grape. The studied variables included: sugars, glucose, fructose, titratable acidity, maturation index, absorbance 280 nm, protein, total polysaccharides, acid polysaccharides, neutral polysaccharides; acids: citric, tartaric, galacturonic, malic, succinic and lactic acids; molecular mass compounds: M_r 206000-193000, M_r 62000-48000, M_r 28000-24000, M_r 7000-5000, M_r 3000-2000 and M_r <2000. The work showed the classification obtained by using these variables, where a clear difference was observed between the productions < 10.500 kg/ha and those >10.500 kg/ha. Based on these results, the authors concluded that a low production rate can affect the content of the grape and the maturation degree, since under those conditions a higher content of polysaccharides is produced; thus, the method can be useful for the quality control of Parellada grapes of the region of Cava (Riu-Aumatell et al., 2002).

Two varieties of white grapes from Portugal were evaluated based on their aromatic characteristics. The grape varieties María Gomes and Bical were analyzed by GC-MS, and then analyzed by PCA. Fifty nine variables of aromas were analyzed, observing a clear distinction between both varietals in the scores plot. María Gomes presented a larger amount and higher concentration of volatile compounds than in the Bical variety. Therefore, the authors suggested that, based on the aromas analysis, and from PCA, it can be concluded that some of the differences found for each variety, specific oenological technologies (wine-making) could be applied for each of them (Rocha et al., 2000).

Sensory Evaluation

The sensory analysis of white wines has resulted in a series of interesting studies that have exceeded in number the classification of white grapes, since this type of study has a great interest in terms of wine quality. In this sense, the use of electronic nose is highlighted as a novel tool of analysis of aromatic properties.

Aguilera et al. (2012) performed the sensory analysis of 26 wines (19 red and 7 white) using a system of electronic nose of 16 electrodes sensible to gaseous components, measuring the resistance of each one in time, and comparing the results by a panel of experts. The studied sensory variables were: herbaceous, flower-like, fruity, spicy vegetal, phenolic, microbiologic, chemical, oxidation, woody-like, alteration, persistence, average intensity, off-flavors and wine quality. The multivariate tools included Independent Component Analysis (ICA), PLS and ANN. By ICA the authors were able to classify five of the seven evaluated white wines, since two of them presented some degree of superposition. By PLS, the correlation degree of the obtained sensory properties was evaluated by electronic nose, in comparison to the panel of experts, with correlation coefficients that were from 0.800 to 0.959, except the woody-like that presented a coefficient of 0.508. By means of ANN, the authors obtained the classification matrix reaching an 84% correct classification for white wines, concluding that the multivariate tools were useful for classifying and predicting sensory parameters of wines.

Gonzáles Álvarez et al. (2011) studied the sensory properties and aromatic fingerprinting obtained by GC–MS, in order to identify the aroma characteristics of monovarietal Godello white wine (Valdeorras, Spain). PCA and PLS multivariate methods were used to correlate the data. PCA showed that the discrimination of Godello wines is based on volatile compounds with higher odor activity values or very close to unity, thus, with sensory characteristics; while PLS showed the relationship between sensory descriptors and volatile compounds in the wines. On the basis of the instrumental analysis, the compounds that in their majority contributed to the flavor of Godello wines were the fruity flavor (from ethyl esters and acetates), spicy (fatty acids derivatives) or floral aroma (originated by the terpenes). In regard to the sensory analysis, the descriptors with higher intensity were the fruity and floral aromas (flowers, apple and citric), along with herbaceous notes.

Vilanova et al. (2010) studied a total of 35 samples of Albariño wines that represent five geographical areas of Rías Baixas (Spain), to characterize and correlate the flavor and the wine components, by descriptive sensory analysis and instrumental analysis (GC-FID). They presented the results by multivariate modeling methods such as PCA and PLS regression. The relationships between the sensory descriptors and volatile compounds of Albariño wines were studied through the Pearson correlation and PLS. From the instrumental analysis, the compounds that had the most contribution to the flavor of Albariño wines were fruity flavor (ethyl esters and acetates) and fruity-like aromas (monoterpenes). They found similar results in the sensory analysis where the major descriptors were also the fruity and floral aromas (citric, flowers, fruit, ripe fruit, apple and tropical). The authors demonstrated that the relationships between the sensory data and volatile compounds can be used to evaluate the sensory properties of Albariño wines.

Lozano et al. (2005a) performed a sensory analysis study by SPME, electronic nose and multivariate analysis for the classification of wines of Airén and Malvar variety, from Madrid. This study is complementary to that carried out by Santos et al. (2004), and diverse solid materials were used to perform the extraction of volatile compounds, with subsequent analysis by electronic nose, using PCA and ANN as multivariate tools. By means of PCA, a correct classification of both varietals was observed using different polymers and different polymers thicknesses; while with ANN, a polar plot with the behavior of each sensor according to each polymer used, was obtained. The authors concluded that using coated silica fiber, the best results were obtained for PCA; while the polyacrylate fiber was the best material to perform the analysis by ANN, reaching a 96.67% correct classification in the studied varieties; thus, the use of SPME can be useful to classify varietals by electronic nose.

Lozano et al. (2005b) performed the analysis of 29 aromas in Airén and Malvan wines, and used electronic nose with 15 sensors, PCA and PNN to carry out a classification analysis of aromas in these white wine varieties. The studied aromas were: geranium, rose, violet, rose citric, pollen rose, apple, banana, pear, fruit, strawberry, blueberry, mulberry, green grass, butter, cheese, soap oil herbaceous, cream milky almond, vinegar, green soap, sulphur, gum, rancid chemical, sherry, wood and stable-horses. These aromas were classified in large groups as: floral, fruity, herbaceous, microbiological and chemical aromas. The scores plot analysis by PCA showed good groupings based on the type of aromas (fruity and floral in separate plots). Meanwhile, the analysis by PNN showed a correct classification of 98-100% for all the groups of aromas, except the chemical aromas, which had a lower classification degree. The authors concluded that the aromas profile is useful to define the identification of the Airén and Malvan varietals.

Besides, the previous authors, by the analysis of volatile components and the use of sensor array and GC-MS, performed a study of Airén and Malvar varietals from Madrid (Spain) in order to classify them according to the aromatic characters. The multivariate analysis was carried out by PCA and discriminant function analysis (DFA), and compared to the data obtained by sensors array, which was constructed with SnO_2 sensors doped with In and Cr at diverse thicknesses. Twenty three volatile compounds were analyzed by GC-MS, out of which 12 were used to obtain the DFA model. With this model, a 100% correct classification was obtained from both varietals. In relation to the use of sensors array, the wines were analyzed after the fermentation and after clarification; in addition, diluted ethanol solutions were measured to perform comparisons. The classification table by DFA showed a 100% concordance in the classification of the studied varietals. Based on the results, the authors concluded that the analysis by sensors array gave results comparable to the ones obtained by GC-MS, but it is necessary to find a correlation between each sensor and the corresponding volatile compounds analyzed (Santos et al., 2004).

Once again, by the use of electronic nose with headspace extraction, 8 aromas in different white wines were determined: pear, apple, roses, citric, green grass, almond, wood and butter. The multivariate tools used were PCA and radial basis neural networks (RB-NN). By PCA, a correct classification of the aromas was achieved, observing 8 well defined representative groups. In addition, by the use of polar plots obtained through RB-NN and cross-validation, and by the use of three principal components, the changes in each sensor that each aroma describes are observed, achieving a 100% discrimination capacity of the aromas through the classification matrix. The authors conclude that PCA is a useful tool for the classification and reduction of variables, while RB-NN was useful as supervised classification tool (Lozano et al., 2004).

ANALYSIS OF THE AGING DEGREE

The aging in wines can represent a desired quality when it is performed in a controlled manner, since it allows obtaining a unique bouquet, difficult to imitate and repeat. Therefore, several authors have in particular studied this property.

Thus, the aging prediction has been studied by different analytical techniques, covering a 20 years period of the "Malmsey" grape variety from Portugal. The authors used HPLC with DAD, GC-MS and UV–Vis spectroscopy and evaluated the ability to predict the aging by multivariate prediction models obtained from the PLS regression, for each set of chemical data. They showed that the PLS models are capable of predicting the wine age with good precision from their volatile and phenolic composition, as well as the UV-Vis absorbance measurement. In addition, they established a compromise between the accuracy of the model and the analysis cost, to decide which methodology to use according to the application required. GC-MS and HPLC-DAD are the most time consuming and complex, but lead to more precise results; however, UV-Vis could also acceptably predict the wine age at a lower cost and time (Pereira et al., 2011).

These authors also studied the "Malmsey" Madeira wine, with the main goal of gaining deepening the understanding of the relevant phenomena that take place during the elaboration process, in particular during aging. In order to develop a solid framework for the analysis of the identity, results of the chemical characterization on how the aroma evolves during aging were presented. A prolonged period of aging of twenty years was considered, from which several samples were analyzed in terms of their aromatic composition. In order to identify the relevant patterns, the multivariate structure of this chemical information was processed by multivariate statistical feature extraction techniques such as PCA and PLS-DA. These were properly combined with two classifiers, representative from the parametric (LDA) and non-parametric (*k*-NN) families of classification methodologies. Selection of the number of features to retain (latent variables) was properly addressed in several alternatives, namely using different cross-validation approaches, such as leave-one-out and k-fold Monte Carlo cross-validation. The obtained results showed that it is advised to use *k*-NN on latent features obtained through PLS-DA, and that quite interesting classification performances can be reached, despite of the natural variability present in the winemaking products. These results also provided solid bases that can be used to construct available frameworks that collaborate with quality monitoring and identity assurance tasks. The performance of the achieved model in this scenario had an error rate of 4%, which can be improved decreasing the classification resolution. This methodology should be applied with more chemical information on the samples, in order to improve the current performance of the classifiers (Pereira et al., 2010).

Paneque et al. (2010) studied the effect of white wines from Montilla-Moriles and Villaviciosa (Spain), based on the content analysis of 10 metals (Ca, Mg, Fe, Cu, Mn and Zn by FAAS; and Na, K, Al and Sr by EAS). The multivariate tools used included CA and LDA; by CA the dendrogram was obtained observing the classification of both varietals based on being young or old. By means of the LDA plot, different classifications were performed according to: denomination of origin (Montilla-Moriles and Villaviciosa), aging degree (young and old) and aging-origin together, obtaining a correct classification between 88 and 100%. The authors concluded that the prediction ability obtained was of 100% for young wines from Montilla–Moriles and Villaviciosa, and close to 90% for aged wines; and that the method is useful to determine the age of wines from those regions.

Fernández-Novales et al. (2009) carried out a study on 23 varieties of white grapes, analyzing the content of reducing sugars by NIRS, and its relation with the aging and winemaking using PLS regression. The authors found a very good correlation in the analysis of sugars with all the varietals. They concluded that the feasibility of NIRS spectroscopy for predicting reducing sugar content during grape ripening, winemaking, and aging, allows the use of this tool in the search for fingerprint spectra to develop a simple, efficient, and low-cost instrument.

Study Of Defects in Wine

Gutierrez et al. (2010) presented the results of a new voltammetric electronic tongue that uses electrodes modified with epoxy-graphite, capable of distinguishing and classifying the different types of wine, as well as some of the typical production defects such as vinegar flavor in contact with air or the use of an excess of sulfite as preservative. They chose the samples considering different denominations of origin from Spain, as well as the type and age: four wines of Penedès, three of Rioja, one of Montsant, one of Ribera del Duero, one of Catalunya and another of Valdepeñas. They identified the wines considering the age stage (young, matured and vintage) and grouped them according to variety. The electronic tongue was developed with five voltammetric electrodes, four of them being bulk-modified with different substances: nanoparticles of copper and platinum, and polyaniline and polypyrrole powder, which made this procedure to be extremely versatile and inexpensive to obtain rich cross-response features. The responses were processed using PCA, being able to make a reasonable distinction between samples. However, because the clusters of data could not be directly identified, the resulting PCA scores were modeled with ANN, allowing a qualitative classification of the white, red and rose wines, and/or the detection of defects, even under the aging effect.

Santos et al. (2010) used an electronic nose on the base of thin film semiconductor sensors, to compare with a trained human sensory panel, the performance in the detection and quantification of the optimum concentration of aromatic compounds, in order to prevent the wine defects. The compounds chosen were: whiskeylactone, nonalactone, decalactone and diacetyl for red wines; and 3-methyl butanol, 4-vinyl guaiacol, butyric acid and damascenone for white wines. PCA and ANN were used to process the data. The authors concluded that the use of electronic noses in the determination of compounds is advantageous compared to the human sensory panel in terms of training time and precision.

Francioso et al. (2007), by voltammetric measurements on platinum microelectrodes using PCA, classified multiple concentrations of defect pairs added to a white wine. These defects can be the result of contaminated starting materials, aging or overdosing of preservatives. Acetaldehyde in the wine, for example, is produced from ethanol by acetobacter bacteria, and the infected grapes can produce high concentrations, which decreases the taste experience. Sulfur dioxide is often added during the fermentation to inhibit wild yeast growth, stabilize the wine and bind to acetaldehyde. L-ascorbic acid is an additive that inhibits non-enzymatic oxidation. The ascorbic acid/acetaldehyde defect exhibited low interaction between them, and the pair-wise concentrations were symmetrically placed in a score plot around a center axis of equimolar concentrations. The ascorbic acid/sulfur dioxide pair showed a central axis for the equimolar concentrations shifted to the sample of higher sulfur dioxide concentration studied. The pair of defects with the strongest interaction was acetaldehyde/sulfur dioxide, which had the equimolar concentrations in the score plots located close to the white wine control sample. Not all of the voltammetric responses were useful, since the best classifications were obtained using only the positive voltage responses. With this work, the authors demonstrated that pulse voltammetry is a simple technique to determine the quality of wines that does not require a sample pretreatment.

Adulteration of White Wines

In a study performed in Italy, the adulteration of white wines of varieties San Severo and Canterino, was evaluated by the addition of ethanol and methanol. The analysis was carried out by a multisensory array system composed by tungsten oxide semiconductors doped with Bi, Pt, Au and Pd, which produced a signal in time, while the multivariate analysis was performed using PCA and ANN. The studied adulterations included the addition of 1 and 10% of ethanol and methanol to the Canterino and San Severo varieties, respectively. Through PCA, a good classification was only obtained for the Canterino variety, but not for the San Severo variety. When the PCA classification was only performed for San Severo, the adulteration degree with methanol was more visible. Something similar occurred during the adulteration analysis with ethanol in the Canterino variety, which became more visible after the addition of 5% of ethanol. However, by ANN, it was possible to correctly classify the adulteration by adding methanol in San Severo, and ethanol in Canterino, with a predictive capability of 100% in both cases. The authors concluded that the method can be useful for the determination of adulterations during the analysis of the wine quality control (Pensa and Casano, 2004).

Others Oenological Classifications

Hron et al. (2012) performed the analysis of different varieties of white and red wines by CG-MS, determining non-flavonoid compounds as vanillic, gentisic, protocatechuic, syringic, gallic, coumaric, ferulic and caffeic acids. The multivariate analysis was performed by PCA with classic and robust biplot analysis by comparing the relative relationships of the acids in the linear and logarithmic forms. The authors found that the logarithmic robust biplot gave the best results. The gentisic and caffeic acids could be associated as chemical markers not only to varietals, but also to technological aspects linked to the winemaking; while vanillic, syringic and gallic acids were identified as possible chemical markers for red wines.

Koreňovská and Suhaj (2012) performed the classification of white wines and grapes according to the organic or conventional origin. For that, they performed the analysis of 14 elements (Ag, Ba, Ca, Cd, Cu, Fe, Hg, Mg, K, Na, Pb, Rb, Sr and Zn) by AAS, and the classification by principal component factoring (PCF) and CDA. The analysis by PCF showed a correct classification of the wines through the scores plot, where two groups were appreciated, corresponding to the organic white wines and the conventional white wines. However, by CDA, a correct classification of must and white grapes was obtained, based on the organic or conventional origin. By the correct use of the described analysis, a joint classification of grape, must and white wines was possible according to organic origin, which translates into an important tool to determine the authenticity of organic wines.

In another study, the variation in the sensory and chemical properties of Chardonnay wines from California was studied for a period of 3 months, combining five types of packaging (glass bottles with natural corks, synthetic extrusion corks, screw caps, BIB filled under normal oxygen concentrations, and BIB filled under reduced oxygen concentrations using modified atmosphere packaging (MAP) technology) and three different temperatures (10, 20, 40 °C). The aroma, flavor, mouth feel, and the color attributions were evaluated. The sensory results were correlated by the combination of PCA and PLS regression, using physical and chemical measurements, including volatile compounds, SO_2, volatile and titratable acidity, oxygen consumption, and wine color. In general, the authors found significant effects in the wine attributes with respect to the packaging-temperature (Hopfer et al., 2012).

Other authors performed a study on the effect of winemaking of Sauvignon blanc wines and grapes, using ^1H NMR spectrum and evaluating the following variables: density, alcohol, sugar, pH, total and volatile acidity, dry extract, free and total SO_2, flavonoids, flavans, phenolic compounds; and the following acids: tartaric, malic, lactic, acetic and citric acids. The studied vinifications were: traditional, cryomaceration, reductive vinification and reductive vinification with cryomaceration using PCA as chemometric tool. The PCA scores plot showed clear differences in the composition of Sauvignon blanc wines depending on the vinification method used. The authors concluded that the classification by the type of vinification can be used as a traceable characteristic in the classification of these wines (Baiano and Terracone, 2011).

Meanwhile, Ali et al. (2011) used ^1H RMN for the metabolomic analysis of white wines of Riesling and Mueller-Thurgau varietals, from the German Palatinate region. The RMN data obtained, in combination with the multivariate analysis were used to classify the quality, grape variety and vintage year of the wines. PLS was used to evaluate the quality of the wine. The compounds: proline, 2,3-butanediol, malate, quercetin, and catechin, were the indicators of good quality wines. Bidirectional orthogonal projections to latent structures (O-PLS) were used to classify the varietal and the vintage year of the wines. The Riesling variety was characterized by the high levels of catechin, caftarate, valine, proline, malate, and citrate; while the compounds quercetin, resveratrol, gallate, leucine, threonine, succinate, and lactate discriminated the Mueller-Thurgau variety. Regarding the vintage year, leucine, phenylalanine, citrate, malate, and phenolics discriminated the wines of the year 2006, while valine, proline, alanine, and succinate those from the year 2007; thus, this study generated interesting models due to the classifications obtained.

Torrea et al. (2011) compared the supplementation of inorganic nitrogen (ammonium chloride) and organic nitrogen (a combination of amino acids and ammonium chloride), to grape juices in the fermentation process of Chardonnay wines, with *Saccharomyces cerevisiae* yeasts. The effects on the volatile composition and aromatic profile were evaluated by HS-SPME with GC-MS and a panel of 12 judges. PLS models were used to evaluate the correlations between the fermentation volatile products and the sensory attributes. It was demonstrated that the supplementation with nitrogen, inorganic or organic, had a substantial impact on the formation of yeast volatile compounds, and on the perceived wine aroma. The addition of organic nitrogen increased to a greater extent the production of the acetate and medium chain fatty acid esters, and decreased to a lesser extent the formation of higher alcohols. Further, inorganic nitrogen substantially increased ethyl acetate and acetic acid content. The wines with low concentrations of nitrogen were classified as slight floral/fruity aroma by descriptors, while the wines with medium concentrations of nitrogen showed an equilibrium between the desirable and less desirable attributes. Finally, high concentrations of nitrogen produced an acetic/solvent character or higher ratings of floral/fruity attributes.

Rocha et al. (2010) carried out a study on wines and must of white grapes, Fernão-Pires variety, according to the vintage year (1998, 1999, 2000 and 2002). Their work was based on the analysis of aroma compounds by GC-MS and PCA, with the intention to find the potential varietal compounds that can be used as markers for the classification by the vintage year. In the study of wines and must, an adequate degree of grouping was observed by the PCA scores plot, because the loading described the most significant aromatic compounds of both cases. Based on the results, the authors concluded that it is possible to identify certain aromatic compounds (such as linalool and hotrienol) that can be very useful to differentiate wines and musts according to the vintage year and the winemaking technology.

Antonelli et al. (2010) performed an exploratory study to verify the applicability of reductive winemaking (RW) in two cultivars. They carried out traditional winemaking (TW) and RW with a semi-aromatic white grape (Sauvignon blanc) and a neutral white (Trebbiano romagnolo). All the phenolic parameters were greater in RW wines, while other substances such as alcohol, reducing sugars, acids, and volatile acidity were less affected by the winemaking technique. A deeper yellow color (OD 420) was a direct consequence of the higher phenolic content of RW wines, while the OD 320 was strictly related to caftaric acid integrity. The studies showed a modification of the RW wines, even in the neutral cultivar. PCA was applied to the data set, and the two first PCs explained almost 85% of the total variability, and divided TW and RW into two groups, to demonstrate that the effect of winemaking overwhelmed maturation differences. The control panelists preferred RW wines, which were characterized by a more rich and delicate aroma. The study demonstrated that a careful exclusion of air combined with an effective prevention of the oxidation, gives rise to more characteristic and pleasant wines.

Cozzolino et al. (2009) investigated the use of MIRS and multivariate data analysis techniques to classify wines from organic and non-organic production systems, analyzing four different varieties of whites (Riesling, Chardonnay, Sauvignon Blanc and Verdelho) and thirteen Australian regions (Barossa valley, Hunter Valley, Yarra Valley, Margaret River, Riverland, McLaren Valley, Clare, Canberra, Adelaide Hills, Coonawarra, Padthaway,

Langhorne Creek and Tasmania). A MIR spectrum containing information about the chemical composition and the physical state of the material under analysis produced structural information that constituted the digital fingerprint of a sample. They used discriminant models such as PCA, DPLS and LDA. The combination of MIR and chemometric techniques, in general, produced LDA models (based on PCA) that correctly classified more than 75% of the wine samples, while the DPLS models correctly classified more than 85% of the wines that belong to the organic and non-organic production systems, respectively. The authors concluded that the spectroscopic techniques can offer the possibility of performing an authenticity study of the organic production systems, without the need of the chemical or sensory analysis which are expensive and laborious. These techniques can be easily implemented in the wine industry to classify wines produced under organic systems.

Rizzon et al. (2008) determined the concentration of the principal cations in the wines from different winemaking regions of Serra Gaúcha in Brazil. Six hundred table wines were analyzed, mainly of American/hybrid varieties, and 380 wines elaborated with *Vitis vinifera* white grape. Atomic emission was used to determine K, Na, Li and Rb; and AAS for Ca, Mg, Mn, Fe, Cu and Zn. The data were subjected to ANOVA and PCA. The results showed that there was no significant difference in the concentration of Na, Cu and Zn between the different types of wines from the winemaking region of Serra Gaúcha. The table red wines had a higher concentration of Mn than the white ones whereas the latter had higher concentrations of Fe, K and Rb. Based on the analysis of these metals, PCA allowed the differentiation of wines by their variety and color.

The chemometric analysis was used to determine the effect of the varietal, geographical origin and the winemaking method, on the protein content of white wines. The analysis of protein fractions was determined by HPLC, while the data analysis was performed through PCA. From this study, the authors determined that the grape variety has a great influence on the protein profile, thus surpassing the geographical origin; while the winemaking method does not influence such profile (Sarmento et al., 2001).

A unique application using principal component similarity (PCS) was performed to classify the qualities of judges in an international wine competition (Scaman et al., 2001). The similarities analysis was used for a total of 5 judges and 21 wine-makers, who judged the quality of 42 white wines, 30 red and 25 specials. Afterwards, the results by PCS were compared with PCA. The authors suggested that PCS can be a method of interest to evaluate the abilities of the judges based on responses similarities, allowing to detect discrepant judges and to improve the quality of the wine competitions.

Huerta et al. (1998) classified samples of "Madrid" Denomination of Origin according to the color (white, rose and red) by the application of SDA. They analyzed varieties Tempranillo and Grenache of red and rose wines; and Malvar, Airén and Albillo of white wines. They studied 88 wine samples from 1993 and 1994 vintages. They performed the determination of conventional parameters such as pH, titratable acidity, volatile acidity, fixed acidity, dry extract, alcohol content, ashes content, ashes alkalinity, reducing sugars, SO_2 and volatile compounds (majority and minority ones) by different analytical techniques. They estimated that it is possible to classify the studied samples according to their composition in minority low content volatile components. They obtained a 92.5% correct classification of the samples through the following variables: 1-pentanol, benzyl alcohol, hexanoic acid, linalool, 2-phenylethanol, ethyl lactate, and trans-2-hexen-1-ol. Therefore, the authors suggested that it is possible to classify the studied samples according to their color, as a function of the content of these components

CONCLUSION

Figure 1 shows the evolution of the number of publications from 1998 to 2013, on the use of chemometric tools for the analysis of white wines. It can be observed that from 1999 to 2012 an increase was maintained, with 2012 being the most productive year, as 10 publications were released about the use of chemometric in white wines.

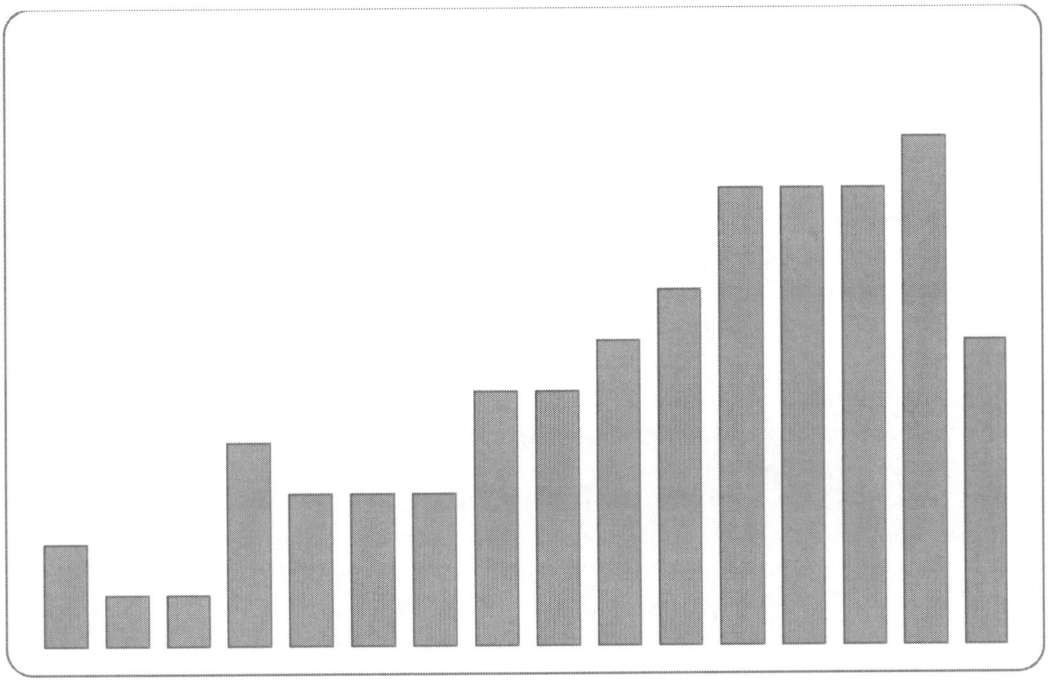

Figure 1. Historical evolution of the use of chemometric tools for the analysis of white wines.

The figure shows that the use of multivariate tools for the classification of white wines (independent of the available analysis methodologies employed: spectroscopic, spectrometric, chromatographic, electrochemical, etc.) has been growing because these tools are becoming practically indispensable for the treatment of the data. The application of electronic noses and tongues, has also expended in the last years, incorporating new detectors and sensors able to analyze flavors and aroma in wines. The use of specific chemical markers in white wines has a limited use, since even when they are found, they are only applicable to a limited and very small number of geographical regions and varieties. In addition, they require the thorough analysis of the wines and/or soils. Therefore, the chemometric analysis has been emerging as the preferred statistical tool to reach better classifications, employing accessible instrumental methods and allowing a fast and easy analysis of the results, with elevated reliability standards and immediate applicability.

REFERENCES

Aceto, M., Robotti, E., Oddone, M., Baldizzone, M., Bonifacino, G., Bezzo, G., Di Stefano, R., Gosetti, F., Mazzucco, E., Manfredi, M., Marengo, E., (2013). A traceability study on the Moscato wine chain. *Food Chem.* 138, 1914-1922.

Acevedo, F.J., Jimenez, J., Maldonado, S., Dominguez, E., Narvaez, A., (2007). Classification of wines produced in specific regions by UV-visible spectroscopy combined with support vector machines. *J. Agric. Food Chem.* 55, 6842-6849.

Aguilera, T., Lozano, J., Paredes, J.A., Alvarez, F.L., Suarez, J.I., (2012). Electronic nose based on independent component analysis combined with partial least squares and artificial neural networks for wine prediction. *Sensors* 12, 8055-8072.

Ali, K., Maltese, F., Toepfer, R., Choi, Y., Verpoorte, R., (2011). Metabolic characterization of Palatinate German white wines according to sensory attributes, varieties, and vintages using NMR spectroscopy and multivariate data analyses. *J. Biomol. NMR*. 49, 255-266.

Andreu-Navarro, A., Russo, P., Aguilar-Caballos, M.P., Fernández-Romero, J.M., Gómez-Hens, A., (2011) Usefulness of terbium-sensitised luminescence detection for the chemometric classification of wines by their content in phenolic compounds. *Food Chem.* 124, 1753-1759.

Anastasiadi, M., Zira, A., Magiatis, P., Haroutounian, S.A., Skaltsounis, A.L., Mikros, E.,(2009). 1H NMR-based metabonomics for the classification of Greek wines according to variety, region, and vintage. comparison with HPLC data. *J. Agric. Food Chem.* 57, 11067-11074.

Antonelli, A., Arfelli, G., Masino, F., Sartini, E., (2010). Comparison of traditional and reductive winemaking: influence on some fixed components and sensorial characteristics. *Eur. Food Res. Technol.* 231, 85-91.

Arvanitoyannis, I.S., Katsota, M.N., Psarra, E.P., Soufleros, E.H., Kallithraka, S., (1999). Application of quality control methods for assessing wine authenticity: Use of multivariate analysis (chemometrics). *Trends Food Sci. Tech.* 10, 321-336.

Azcarate, S.M., Cantarelli, M.A., Marchevsky, E.J., Camiña, J.M., (2013a). Evaluation of geographic origin of Torrontés wines by chemometrics. *J. Food Res.* 2, 48-56.

Azcarate, S.M., Cantarelli, M.A., Pellerano, R.G., Marchevsky, E.J., Camiña, J.M., (2013b). Classification of Sauvignon blanc white wines by chemometric methods. *J. Food Sci.* 78, C432-C436.

Baiano, A., Terracone, C., (2011). Varietal Differences among the phenolic profiles and antioxidant activities of seven table grape cultivars grown in the south of Italy based on chemometrics. *J. Agric. Food Chem.* 59, 9815-9826.

Baiano, A., Terracone, C., Longobardi, F., Ventrella, A., Agostiano, A., del Nobile, M.A., (2012) Effects of different vinification technologies on physical and chemical characteristics of sauvignon blanc wines. *Food Chem*. 135, 2694-2701.

Bellomarino, S.A., Conlan, X.A., Parker, R.M., Barnett, N.W., Adams, M.J., (2009). Geographical classification of some Australian wines by discriminant analysis using HPLC with UV and chemiluminescence detection. *Talanta* 80, 833-838.

Bevin, C.J., Dambergs, R.G., Fergusson, A.J., Cozzolino, D., (2008). Varietal discrimination of Australian wines by means of mid-infrared spectroscopy and multivariate analysis. *Anal. Chim. Acta.* 621, 19-23.

Boselli, E., Minardi, M., Giomo, A., Frega, N.G., (2006). Phenolic composition and quality of white d.o.c. wines from Marche (Italy). *Anal. Chim. Acta.* 563, 93-100.

Câmara, J.S., Alves, M.A., Marques, J.C., (2006). Multivariate analysis for the classification and differentiation of Madeira wines according to main grape varieties. *Talanta* 68, 1512-1521.

Câmara, J.S., Alves, M.A., Marques, J.C., (2007). Classification of Boal, Malvazia, Sercial and Verdelho wines based on terpenoid patterns. *Food Chem.* 101, 475-484.

Caruso, M., Galgano, F., Morelli, M.A.C., Viggiani, L., Lencioni, L., Giussani, B., Favati, F., (2012). Chemical profile of white wines produced from 'Greco bianco' grape variety in different Italian areas by nuclear magnetic resonance (NMR) and conventional physicochemical analyses. *J. Agr. Food Chem.* 60, 7−15.

Chabreyrie, D., Chauvet, S., Guyon, F., Salagoïty, M.H., Antinelli, J.-F., Medina, B., (2008). Characterization and quantification of grape variety by means of shikimic acid concentration and protein fingerprint in still white wines. *J. Agric. Food Chem.* 56, 6785-6790.

Coimbra, M.A., Goncü Alves, F., Barros, A.S., Delgadillo, I., (2002). Fourier transform infrared spectroscopy and chemometric analysis of white wine polysaccharide extracts. *J. Agric. Food Chem.* 50, 3405-3411.

Cozzolino, D., Cynkar, W.U., Shah, N., Smith, P.A., (2011) Can spectroscopy geographically classify Sauvignon blanc wines from Australia and New Zealand? *Food Chem.* 126, 673-678.

Cozzolino, D., Holdstock, M., Dambergs, R.G., Cynkar, W.U., Smith, P.A., (2009). Mid infrared spectroscopy and multivariate analysis: A tool to discriminate between organic and non-organic wines grown in Australia. *Food Chem.* 116, 761-765.

Cozzolino, D., Smyth, H.E., Cynkar, W.U., Danbers, R.G., Gishen, M., (2005). Usefulness of chemometrics and mass spectrometry-based electronic nose to classify Australian white wines by their varietal origin. *Talanta* 68, 382–387.

Cozzolino, D., Smyth, H.E., Gishen, M., (2003). Feasibility study on the use of visible and near-infrared spectroscopy together with chemometrics to discriminate between commercial white wines of different varietal origins. *J. Agr. Food Chem.* 51, 7703-7708.

de la Presa-Owens, C., Schlich, P., Davies H.D., Noble, A.C., (1998). Effect of mèthode Champenoise process on aroma of four V. vinifera varieties. *Am. J. Enol. Vitic.* 49, 289-294.

Dopico-García, M.S., Valentão, P., Jagodziñska, A., Klepczyñska, J., Guerra, L., Andrade, P.B., Seabra, R.M., (2007). Solid-phase extraction versus matrix solid-phase dispersion: Application to white grapes. *Talanta* 74, 20-31.

Du, Y.Y., Bai, G.Y., Zhang, X., Liu, M.L., (2007). Classification of wines based on combination of 1H NMR spectroscopy and principal component analysis. *Chinese J. Chem.* 25, 930-936.

Fabani, M.P., Arrúa, R.C., Vázquez, F., Diaz, M.P., Baroni, M.V., Wunderlin, D.A., (2010). Evaluation of elemental profile coupled to chemometrics to assess the geographical origin of Argentinean wines. *Food Chem.* 119, 372-379.

Fernández-Novales, J., López, M.-I., Sánchez, M.-T., Morales, J., González-Caballero, V., (2009). Shortwave-near infrared spectroscopy for determination of reducing sugar content during grape ripening, winemaking, and aging of white and red wines. *Food Res. Int.* 42, 285-291.

Francioso, L., Bjorklund, R., Krantz-Rülcker, T., Siciliano, P., (2007). Classification of multiple defect concentrations in white wine by platinum microelectrode voltammetry. *Sensors Actuat. B-Chem.* 125, 462-467.

Frías, S., Conde, J.E., Rodríguez-Bencomo, J.J., García-Montelongo, F., Pérez-Trujillo, J.P., (2003). Classification of commercial wines from the Canary Islands (Spain) by chemometric techniques using metallic contents. *Talanta* 59, 335-344.

Garrido-Delgado, R., López-Vidal, S., Arce, L., Valcárcel, M., (2009). Differentiation and identification of white wine varieties by using electropherogram fingerprints obtained with CE. *J. Sep. Sci.* 32, 3809-3816.

Garrido-Delgado, R., Arce, L., Guamán, A.B., Pardo, A., Marco, A., Valcárcel, M., (2011). Direct coupling of a gas–liquid separator to an ion mobility spectrometer for the classification of different white wines using chemometrics tools. *Talanta* 84, 471-479.

González Álvarez, M., González-Barreiro, C., Cancho-Grande, B., Simal-Gándara, J., (2011). Relationships between Godello white wine sensory properties and its aromatic fingerprinting obtained by GC–MS. *Food Chem.* 129, 890-898.

Gutiérrez, M, Llobera, A. Ipatov, A, Vila-Planas, J, Mínguez, S, Demming, S, Büttgenbach, S, Capdevila, F, Domingo, C., Jiménez-Jorquera, C., (2011) Application of an e-tongue to the analysis of monovarietal and blends of white wines. *Sensors* 11, 4840-4857.

Gutiérrez, J., Moreno-Barón, L., Pividori, M., Alegret, S., Valle, M., (2010). A voltammetric electronic tongue made of modified epoxy-graphite electrodes for the qualitative analysis of wine. *Microchim. Acta* 169, 261-268.

Héberger, K., Csomós, E., Simon-Sarkadi, L., (2003). Principal component and linear discriminant analyses of free amino acids and biogenic amines in Hungarian wines. *J. Agric. Food Chem.* 51, 8055-8060.

Hopfer, H., Ebeler, S.E., Heymann, H. (2012). The combined effects of storage temperature and packaging type on the sensory and chemical properties of Chardonnay. *J. Agric. Food Chem.* 60, 10743-10754.

Hron, K., Jelínková, M., Filzmoser, P., Kreuziger, R., Bednár, P., Bartákb, P. (2012). Statistical analysis of wines using a robust compositional biplot. *Talanta* 90, 46– 50.

Huerta, M.D., Salinas, M.R., Masoud, T., Alonso, G.L., (1998). Wine differentiation according to color using conventional parameters and volatile components. *J. Food Comp. Anal.* 11, 363-374.

Jurado, J.M., Alcázar, A., Palacios-Morillo, A., de Pablos, F., (2012) Classification of Spanish DO white wines according to their elemental profile by means of support vector machines. *Food Chem.* 135, 898-903.

Jurado, J.M., Ballesteros, O., Alcázar, A., Pablos, F., Martín, M.J., Vilchez, J.L., Navalón, A., (2008). Differentiation of certified brands of origins of Spanish white wines by HS-SPME-GC and chemometrics. *Anal Bioanal Chem.* 390, 961-970.

Kallithraka, S., Arvanitoyannis, I.S., Kefalas, P., El-Zajouli, A., Soufleros, E., Psarra, E., (2001). Instrumental and sensory analysis of Greek wines: implementation of principal components analysis (PCA) for classification according to geographical origin. *Food Chem.* 73, 501-514.

Kirsanov, D., Mednova, O., Vietoris, V., Kilmartin, P.A., Legin, A., (2012). Towards reliable estimation of an electronic tongue predictive ability from PLS regression models in wine analysis. *Talanta* 90, 109-116.

Koreňovská, M., Suhaj, M., (2012) Chemometric prediction of wines affiliation with organic and conventional production systems through their elemental profiles. *J. Food Nut. Res.* 51, 23-32.

Košir, I.J., Kidric, J., (2002). Use of modern nuclear magnetic resonance spectroscopy in wine analysis: determination of minor compounds. *Anal. Chim. Acta* 458, 77–84.

Košir, I.J., Kocjancic, M., Ogrin, N., Kidric, J., (2001). Use of SNIF-NMR and IRMS in combination with chemometric methods for the determination of chaptalisation and geographical origin of wines (the example of Slovenian wines). *Anal. Chim. Acta* 429, 195-206.

Kruzlicova, D., Mocak, J., Balla, B., Petka, J., Farkova, M., Havel, J., (2009). Classification of Slovak white wines using artificial neural networks and discriminant techniques. *Food Chem.* 112, 1046-1052.

Lampíř, L., (2013). Varietal differentiation of white wines on the basis of phenolic compounds profile. *Czech J. Food Sci.* 31, 172–179.

Liu, L., Cozzolino, D., Cynkar, W.U., Dambergs, R.G., Janik, L., O'Neill, B.K., Colby, C.B., Gishen, M., (2008). Preliminary study on the application of visible–near infrared spectroscopy and chemometrics to classify Riesling wines from different countries. *Food Chem.* 106, 781-786.

Louw, L., Roux, K., Tredoux, A., Tomic, O., Naes, T., Nieuwoudt, H., van Rensburg, P., (2009). Characterization of selected South African young cultivar wines using FTMIR spectroscopy, gas chromatography, and multivariate data analysis. *J. Agric. Food Chem.* 57, 2623-2632.

Lozano, J., Santos, J.P., Gutiérrez, J., Horrillo, M.C., (2007). Comparative study of sampling systems combined with gas sensors for wine discrimination. *Sensors Actuat B-Chem.* 126, 616-623.

Lozano, J., Fernández, M.J., Fontecha, J.L., Aleixandre, M., Santos, J.P., Sayago, I., Arroyo, T., Cabellos, J.M., Gutiérrez, F.J., Horrillo, M.C., (2006). Wine classification with a zinc oxide SAW sensor array. *Sensors Actuat B-Chem.* 120, 166-171.

Lozano, J., Santos, J.P., Sayago, I., Gutiérrez, J., Horrillo, M.C., (2004). Identification of typical wine aromas by means of an electronic nose. Proceedings of IEEE, 345-348.

Lozano, J., Santos, J.P., Gutiérrez, J., Horrillo, M.C., (2005a). Comparison among several fiber coating for a solid phase microextraction (SPME) based electronic nose applied to wine discrimination. Proceedings of IEEE, 1339-1342.

Lozano, J., Santos, J.P., Horrillo, M.C., (2005b). Classification of white wine aromas with an electronic nose. *Talanta* 67, 610–616.

Martelo-Vidal, M.J., Domínguez-Agis, F., Vázquez, M., (2013). Ultraviolet/visible/near-infrared spectral analysis and chemometric tools for the discrimination of wines between subzones inside a controlled designation of origin: a case study of Rías Baixas. *Austral. J. Grape Wine Res.* 19, 62–67.

Martí, M.P., Boqué, R., Busto, O., Guasch, J., (2005). Electronic noses in the quality control of alcoholic beverages. *TrAC-Trends Anal. Chem.* 24, 57-66.

Paneque, P., Álvarez-Sotomayor, M.T., Clavijo, M., Gómez, I.A., (2010). Metal content in southern Spain wines and their classification according to origin and ageing. *Microchem. J.* 94, 175-179.

Pensa, M., Casano, G., (2004). Recognition of adulteration of Italian wines by thin-film multisensor array and artificial neural networks. *Anal. Chim. Acta* 509, 159–177.

Pereira, A.C., Reis, M.S., Saraiva, P.M., Marques, J.C., (2011). Madeira wine ageing prediction based on different analytical techniques: UV–vis, GC-MS, HPLC-DAD. *Chem. Intel. Lab. Sys.* 105, 43-55.

Pereira, A.C., Reis, M.S., Saraiva, P.M., Marques, J.C., (2010). Analysis and assessment of Madeira wine ageing over an extended time period through GC–MS and chemometric analysis. *Anal. Chim. Acta* 660, 8-21.

Pigani, L., Foca, G., Ionescu, K., Martina, V., Ulrici, A., Terzi, F., Vignali, M., Zanardi, C., Seeber, R., (2008). Amperometric sensors based on poly(3,4-ethylenedioxythiophene)-modified electrodes: Discrimination of white wines. *Anal. Chim. Acta.* 614, 213-222.

Ražić, S., Čokeša, Đ., Sremac S., (2007). Multivariate data visualization methods based on elemental analysis of wines by atomic absorption spectrometry. *J. Serb Chem Soc.* 72, 1487-1492.

Riu-Aumatell, M., Loä Pez-Barajas, M., Loä Pez-Tamames, E., Buxaderas, S., (2002). Influence of yield and maturation index on polysaccharides and other compounds of grape juice. *J. Agric. Food Chem.* 50, 4604-4607.

Rizzon, L.A.., Salvador, M.B.G., Miele, A., (2008). Teores de cátions dos vinhos da Serra Gaúcha. *Ciênc. Tecnol. Aliment.* 28, 635-641.

Rocha, S.M., Coutinho, P., Coelho, E., Barros, A.S., Delgadillo, I., Coimbra, M.A., (2010) Relationships between the varietal volatile composition of the musts and white wine aroma quality. A four year feasibility study. *LWT - Food Sci Tech.* 43, 1508-1516.

Rocha, S.M., Coutinho, P., Barros, A., Delgadillo, I., Coimbra, M.A., (2006). Rapid tool for distinction of wines based on the global volatile signature. *J. Chromatogr. A.* 1114, 188-197.

Rocha, S., Coutinho, P., Barros, A., Coimbra, M.A., Delgadillo I., Dias Cardoso, A., (2000). Aroma potential of two Bairrada white grape varieties: Maria Gomes and Bical. *J. Agric. Food Chem.* 48, 4802-4807.

Sánchez Arribas, A., Martínez-Fernández, M., Chicharro, M., (2012). The role of electroanalytical techniques in analysis of polyphenols in wine. *TrAC-Trends in Anal Chem.* 34, 78-96.

Santos, J.P., Arroyo, T., Aleixandre, M., Lozano, J., Sayago, I., García, M., Fernández, M.J., Arés, L., Gutiérrez, J., Cabellos, J.M., Gilb, M., Horrillo, M.C. (2004). A comparative study of sensor array and GC–MS: application to Madrid wines characterization. *Sensors Actuat B-Chem.* 102, 299–307.

Santos, J.P., Lozano, J., Aleixandre, M., Arroyo, T., Cabellos, J.M., Gil, M., Horrillo, M.C., (2010). Threshold detection of aromatic compounds in wine with an electronic nose and a human sensory panel. *Talanta* 80, 1899-1906.

Sarmento, M.R., Oliveira, J.C., Slatner, M., Boulton, R.B., (2001). Comparative quantitative analysis of the effect of cultivar, wine growing region and vinification method on the protein profiles of some white wines. *Int. J. Food Sci. Tech.* 36, 759-766

Saurina, J., (2010). Characterization of wines using compositional profiles and chemometrics. *TrAC-Trend Anal. Chem.* 29, 234-245.

Scaman, C., Dou, J., Cliff, M., Yuksel, D., King, M., (2001). Evaluation of wine competition judge performance using principal component similarity analysis. *J. Sens. Stud.* 16, 287-300.

Smyth, H., Cozzolino, D., (2013). Instrumental methods (spectroscopy, electronic nose, and tongue) as tools to predict taste and aroma in beverages: advantages and limitations. *Chem. Rev.* 113, 1429–1440.

Šnuderl, K., Darinka Brodnjak-Vončina, J.M., Sedláčková, B., (2009). Classification of white varietal wines using chemical analysis and sensorial evaluations. *Acta Chim. Slov.* 56, 765-772.

Son, H.S., Hwang, G.S., Ahn, H.J., Park, W.M., Lee, C.H., Hong, Y.S., (2009). Characterization of wines from grape varieties through multivariate statistical analysis of 1H NMR spectroscopic data. *Food Res. Int.* 42, 1483-1491.

Šperková, J., Suchánek, M., (2005). Multivariate classification of wines from different Bohemian regions (Czech Republic). Food Chem. 93, 659-663.

Torrea, D., Varela, C., Ugliano, M., Ancin-Azpilicueta, C., Leigh Francis, I., Henschke, P.A., (2011). Comparison of inorganic and organic nitrogen supplementation of grape juice – Effect on volatile composition and aroma profile of a Chardonnay wine fermented with Saccharomyces cerevisiae yeast. *Food Chem.* 127, 1072-1083.

Tredoux, A., de Villiers, A., Májek, P., Lynen, F.d.r., Crouch, A., Sandra, P., (2008). Stir bar sorptive extraction combined with gc-ms analysis and chemometric methods for the classification of South African wines according to the volatile composition. *J. Agric. Food Chem.* 56, 4286-4296.

Urbano, M., Luque de Castro, M.D., Pérez, P.M., García-Olmo, J., Gómez-Nieto, M.A., (2006). Ultraviolet–visible spectroscopy and pattern recognition methods for differentiation and classification of wines. *Food Chem.* 97, 166-175.

Vilanova, M., Genisheva, Z., Masa A., Oliveira, J.M., (2010). Correlation between volatile composition and sensory properties in Spanish Albariño wines. *Microchem. J.* 95, 240-246.

Villagra, E., Santos, L.S., Vaz, B.G., Eberlin, M.N., Felipe Laurie, V., (2012). Varietal discrimination of Chilean wines by direct injection mass spectrometry analysis combined with multivariate statistics. *Food Chem.* 131, 692-697.

Weldegergis, B.T., de Villiers, A., Crouch, A.M., (2011). Chemometric investigation of the volatile content of young South African wines. *Food Chem.* 128, 1100-1109.

Welke, J.E., Zanus, M., Lazarotto, M., Schmitt, K.G., Zini, C.A., (2012). Volatile characterization by multivariate optimization of headspace-solid phase microextraction and sensorial Evaluation of Chardonnay base wines. J. *Braz. Chem. Soc.* 23, 678-687.

Zamúz, S., Vilanova, M., (2006). Volatile compounds after spontaneous fermentation of musts from Vitis vinifera cv. Albariño grapes cultivated in different origins from Rías Baixas AOC, Spain. *Flavour Frag. J.* 21, 743-748.

In: Wine
Editor: Youssef El Rayess

ISBN: 978-1-63321-048-6

Chapter 11

RECOVERY OF PHENOLIC COMPOUNDS FROM BY-PRODUCTS OF THE WINEMAKING CHAIN

Maurizio Aceto*, Marco Clericuzio and Bruno Burlando
Dipartimento di Scienze e Innovazione Tecnologica,
Università degli Studi del Piemonte Orientale, Alessandria, Italy

ABSTRACT

The beneficial effects of wine drinking are at present acknowleged by the scientific community, for the reason of the polyphenolic content of wine which causes a lower incidence of cardiovascular diseases. The health treasure kept in wine derives from its raw materials; grapes, in fact, contain high amounts of polyphenols. However, only a fraction of the polyphenolic content of grapes is actually transferred to wine. After the winemaking process, indeed, only a fraction of total grape polyphenols is found in wine, while the rest remains in grape skin, stem and seeds.

The by-products of winemaking are then extremely rich in compounds that, due to their relevant biological properties, could be exploited for industrial purposes, with applications in the nutritional, medical and cosmetic fields, as well as development of advanced solar cells and related devices. Hence, the exploitation of such a polyphenol treasure, which is commonly disregarded, could provide an economic boost to the whole industry of winemaking countries or regions. An additional achievement obtained by the treatment of winemaking residuals is the possibility of alleviating problems of residual disposal. Most of these materials, in fact, should be treated as special wastes, according to current regulations.

Keywords: By-products, pomace, stem, seeds, health, nutritional, cosmetic

---

* Corresponding author: Maurizio Acato. Dipartimento di Scienze e Innovazione Tecnologica, Università degli Studi del Piemonte Orientale, viale T. Michel, 11-15121 Alessandria. E-mail: maurizio.aceto@mfn.unipmn.it.

INTRODUCTION

Healthful Properties of Wine Polyphenols

As widely known, the beneficial effects of moderate wine drinking, exemplified by the concept of the French paradox, derive from the raw matter with which it is produced. Grapes contain high amounts of polyphenols, which, according to several scientific studies (Fresco et al., 2006; Soleas et al., 2006; Xia et al., 2010) display multiple biological properties such as antioxidant, antiinflammatory, anti-atherosclerosis, cardioprotective and cancer protective effects. Since along the chain from grapes to bottled wine a fraction of the original polyphenols is maintained unaltered and some new polyphenols originate during fermentation and aging, these beneficial effects, though to a lesser extent, arise also in wine and have suggested a scientific basis to explain the French paradox. The relationship between grapes, wine and polyphenols has been recently reviewed by Garrido and Borges (2013).

The medicinal and nutritional value of grapes and wine has been known to mankind for thousands of years. Ancient Egyptians used to eat grapes at least 6.000 years ago, and several ancient Greek philosophers praised the healing power of grapes and wine. European folk healers made an ointment from the sap of grapevines to treat skin and eye diseases, while grape leaves were used to stop bleeding, inflammation, and pain, such as the kind brought on by hemorrhoids (Badet, 2011). Unripe grapes were used to treat sore throats, and dried grapes (raisins) were used for constipation and thirst. Round, ripe, sweet grapes were used to treat a range of health problems including cancer, cholera, smallpox, nausea, eye infections, and skin, kidney, and liver diseases. Most of these beneficial properties are unquestionably linked to polyphenols. Due to the great number of relevant biological properties of these compounds, the therapeutic properties of grapes have been recognized and widely studied.

Distribution of Polyphenols in Grapes

Grapes are rich in polyphenols, which constitute an important legacy. But how is this legacy exploited by man? We must consider that the different parts of grapes (i.e., pulp, skin and seeds) contain different polyphenols and in different amounts. In particular, the pulp, the poorest part, is only rich in non-flavonoid polyphenols; the skin is rich in proanthocyanidins with high polymerization degrees, stilbenes, flavonols and, of course, anthocyanins; seeds, the richest part, contain high amounts of flavan-3-ols and proanthocyanidins with low polymerization degrees. In numbers, seeds account for a polyphenolic content ranging from 46 to 69% in terms of GAE/Kg, skin contains from 25 to 50 % and pulp accounts for 6% (Zoecklein et al., 1995). Indicative concentration values may be found in Pinelo et al. (2006).

In addition, it must be considered that stems, usually discarded in winemaking, are rich in proanthocyanidins with low polymerization degrees and stilbenes. The most representative polyphenolic compounds in the various parts of grapes are shown in figure 1.

The percentage of total polyphenols passing from grapes to wine depends strongly on the processes involved along the winemaking chain.

Polyphenols contained in pulp are quantitatively transferred to must and therefore to wine.

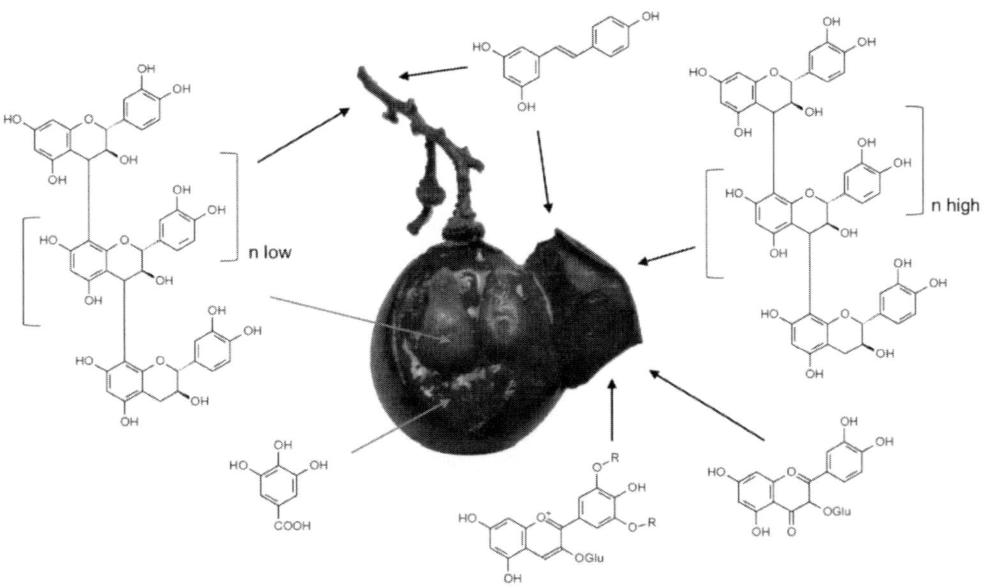

Figure 1. The most representative polyphenols in grapes.

Skins and seeds (in particular when grapes are processed for red winemaking), during must fermentation stay in contact with the fermenting wine for several days and therefore are subjected to a mild, prolonged ethanolic extraction so that their polyphenols are partially transferred to wine, providing red wine with a high polyphenolic content.

Polyphenols contained in stems, on the contrary, are nearly all retained in the raw matter, because stems are usually discarded.

The polyphenolic composition in grapes is subjected to large variations upon processing in the winery. Microbial activity and oenological practices vary the original concentration of several polyphenols. New polyphenols may be generated by fermentation and storage in oak barrels. Polyphenols originated of these latters are mostly constituted of non-flavonoid type. Wine aging will also modify the polyphenolic composition because these compounds, apart from extraction from wood, undergo different transformations, such as oxidation processes, condensation and polymerisation reactions, all phenomena related to color and other sensorial characteristics such as flavor, bitterness and astringency of the final wine. As a consequence, the polyphenolic content in wine is much more complex and variegated than in grapes. A thorough description of the factors influencing the polyphenolic content in the passage from grapes to wine can be found in Garrido and Borges (2013).

The By-Products of the Winemaking Chain

Where Have All the Polyphenols Gone?

We know that grapes and wine can have healthful properties, and that a large number of scientific publications have indicated polyphenols as responsible of these properties.

What is relatively less known is the fact that the precious legacy contained in grapes, i.e., the high content of polyphenols, is transferred only in a moderate percentage to wine. In fact, after the winemaking process, a variable percentage, ranging from 10 to 70% of total grape polyphenols is found in wine, while the rest remains in grape skins, stems and seeds, collectively identified as by-products of the wine chain (WCBP thereafter).

The usual destinations of WCBP can be various, as it will be detailed further on, but in rare cases this implies the possibility to exploit on large scale the above mentioned therapeutic potential of the left-over polyphenols.

The reasons for this decrease from grapes to wine may be very logical from the winemaker point of view: the largest part of polyphenols is kept in seeds, but in the form of proanthocyanidins characterized by high astringency, so that oenological practices are applied in order to avoid their passage from grapes to wine. Another significant percentage is lost from skins: the extraction of anthocyanins would be optimal at very low pH, i.e., less than 2, but these conditions would not be suitable for the successive fermentation reactions; therefore, uncomplete extraction of anthocyanins from skins takes place.

Finally, a large part of potentially available polyphenols is kept in stems, which are almost completely discarded before fermentation.

To understand how much of the potential polyphenols content is lost, we must compare the polyphenolic content of grapes vs. wine. Amerine and Joslyn (1987) determined the total amount of polyphenols in different parts of grapes and in red and white wine (Table 1).

Using the total polyphenolic content is not completely correct, as the distributions of polyphenols are quite different when passing from grapes to wine.

One way is to put the focus on specific compounds, e.g., to compare the content of anthocyanins, proanthocyanidins or stilbenes, but ultimately the most reasonable way is to express the total content of polyphenols in terms of Gallic Acid Equivalents or GAE. Such a comparison, of course, suffers from some inaccuracy. First of all, it is just an estimate, since it reduces different classes of compounds to the activity of a single molecule. Secondly, as said before the original concentration of some polyphenols may be varied and new polyphenols may arise, very different in structure from gallic acid.

Nevertheless, comparison of GAE values is a good way to reason about different matrices. Another way is to consider the antioxidant activity.

Table 1. Comparison of polyphenolic content in grapes and wines

Item	Weight (%)	Water (%)[c]	Total polyphenols (%)[c]
Grape stems	2-6[a]	60-80	1-4
Grape berry	95-97[a]	-	-
Grape pulp	85-87[b]	60-85	< 1
Grape skins	5-12[b]	70-80	1-2
Grape seeds	0-5[b]	30-40	5-8
Red wine		85-90	0.1-0.3
White wine		85-90	0.01-0.03

Adapted from Amerine Joslyn, 1987.

[a]% on grape bunch.

[b]% on grape berry.

[c]% on total item weight.

GAE values for grapes have large variations, but typical values can be 5,500 and 4,000 mg GAE/Kg for red and white grapes respectively (Singleton and Esau, 1969). In view of this, the gross polyphenolic composition estimated in mg GAE/L for a typical red table wine may reach 2,500 for a young wine and 1,300 for an aged wine (Zoecklein et al., 1995). Corresponding levels are 215 and 190-290 for, respectively, a young white wine and an aged one.

Several studies have been carried out on the comparison of grapes and wines. Some of them are concerned by analysis of whole berries, but others used separate analysis of skins and/or seeds extracts, since these are the parts in the berry that contribute mostly to polyphenols in wine. The second type of study, indeed, actually underestimate the decrease of polyphenols from grapes to wine since stems and/or skins and/or seeds are not considered.

Kennedy (2008), in an experiment carried out on cv. Pinot Noir, found that only 9% of total tannins contained in seeds and skin tissues were extracted in wine during production; maceration, as is usually desired, favoured the extraction from skin at the expense of seeds. Gambacorta et al. (2011) compared the polyphenolic content in red grapes of Aglianico, Montepulciano, Nero di Troia and Sangiovese cultivars and in the corresponding wines produced in Apulia (Southern Italy) using different winemaking technologies. It resulted that extraction of polyphenols in wine from these cultivars was very high and ranging 65-70%, with the notable exception of Nero di Troia which wine was richer in polyphenols possibly due to a major content of proanthocyanidins.

Fragoso et al. (2011) studied the passage of polyphenols from grapes to wines in five red grape cultivars grown in Spain: Tempranillo, Merlot, Syrah, Cariñena and Cabernet Sauvignon. Samples were collected at three different ripening stages covering the grapes development from veraison (i.e., first stage) until harvest (complete ripeness and last stage) of 2009 vintage. Authors found that the total polyphenolic contents in grapes ranged among 1,000 and 1,970 mg GAE/Kg fresh matter, total anthocyanins ranged among 480 and 1,000 mg malvidin-3-glucoside/Kg and total condensed tannins ranged among 1,350 and 3,200 mg catechin/Kg. The corresponding values for wines were 480-1,060 mg GAE/L for total polyphenols, 140-470 mg malvidin-3-glucoside/L for total anthocyanins and 280-760 mg catechin/L for total condensed tannins. By calculating ratios of wine figures/grapes figures, it results that from 30 to 55% of grapes polyphenols and anthocyanins are extracted into wine, while extraction of condensed tannins ranges from 10 to 45%. These results are not surprising, since it is apparent that the extraction conditions carried out in the analytical characterization are much stronger and effective than what it is possible to apply in the oenological processes. This is particularly true for tannins, which are contained mainly in seeds: during winemaking, seeds are not crushed, unlike what occurs in the accelerated process of grape extraction previously to analytical measurement.

De Nisco et al. (2013) compared the polyphenolic content in grapes of *Vitis vinifera* cv. Aglianico, a variety typical of Suthern Italy, and in the wine obtained from it. Authors showed that total flavonols, anthocyanins and resveratrol isomers (trans- + cis-) in skin grapes were respectively 15-, 14- and 100-fold higher than in the corresponding wine.

On the other hand, the total antioxidant capacity was found to be about 2 to 4 times higher in wine than in berry skin exctracts, possibly due to the increased presence of anthocyanins derivatives in wine.

The studies cited before demonstrate clearly that a large percentage of the original polyphenolic content of grapes (and grapevine) is not transferred to wine.

It is difficult to express this percentage; according to the studies, it can range between 90% in cases in which stems are entirely discarded, and 30% in cases in which extraction from skins has been optimized. In the most favourable cases, though, the percentage of loss is underestimated (i.e., the transfer of polyphenols to wine is overestimated) since it does not take into account the amount contained in stems, which is obvious from the oenologist point of view, but must be considered in view of the potential recovery of the precious polyphenolic compounds from the remains after grapes pressing: skins and seeds above all, but also stems and vine leaves and shoots. These materials, here defined WCBP, may constitute a real treasure if adequately exploited.

To have an idea of the amount of potential recovery of valuable resources from WCBP, let us do some calculations. According to data by Organisation Internationale de la Vigne et du Vin (OIV), the world production of grapes in 2009 has been 68.17 million tons (Castellucci et al., 2009). The production of unprocessed grapes intended for table consumption accounted for 21.18 million tons, to which 1.21 million tons of raisins, i.e., dried grapes, must be added, that means other 4.84 million tons of grapes (it takes 4 kg of fresh grapes to obtain 1 kg of raisins). It can be calculated that the amount of pressed grapes has been 42.15 million tons, 61.8% of world grape production (Mazza et al., 1993, estimated this value reaching 80%). This percentage has lowered in the last years, passing from 69.9 in 2000 to 61.8 in 2009 as a consequence of the increase in the production of grapes intended for table consumption; in a parallel way, though, the world production of grapes increased from 60.03 million tons in 2000 to 68.17 in 2009, so that the amount of pressed grapes, indeed, has kept fairly constant with a slight increase (from 41.96 million tons in 2000 to 42.15 in 2009). It has been estimated (Jackson, 1994; Pinelo et al., 2005; Arvanitoyannis et al., 2006b) that from 13 to 20% of the amount of pressed grapes finally results in waste or WCBP, which from these figures gives 5-8 million tons of potentially exploitable matter, though other estimates (Torres and Bobet, 2001; Peralbo-Molina and Luque de Castro, 2013b) report higher values such as up to 14.5 million of tons solely in Europe; it must be considered that no statistic data actually exist on WCBP, so that data reported in the literature may be rather inconsistent. At any rate, this represents unquestionably an enormous amount of matter from which, apart from other substances, precious polyphenolic compounds could be extracted; it must be noted that these data are actually underestimated, because they do not take into account the potential contribution of stems, leaves, shoots and minor waste materials such as diatomites.

Lorem ipsum dolor sit amet, consectetuer adipiscing elit. Maecenas porttitor congue massa. Fusce posuere, magna sed pulvinar ultricies, purus lectus malesuada libero, sit amet commodo magna eros quis urna.

Definition

WCBP are defined as the natural residues generated along the winemaking chain. According to EU Regulation no. 1234/07 (European Union, 2007), WCBP are:

- pomaces or grape marc: the residue from the pressing of fresh grapes, whether or not fermented, which contain mostly skins, seeds and some stems;

- piquette: the liquid product obtained by (a) the fermentation of untreated grape marc macerated in water; (b) leaching fermented grape marc with water;
- lees: the residue (a) accumulating in vessels containing wine after fermentation, during storage or after authorized treatment; (b) obtained from filtering or centrifuging the product referred to in (a); (c) accumulating in vessels containing grape must during storage or after authorized treatment; (d) obtained from filtering or centrifuging the product referred to in (c);

Other materials to be considered as WCBP, though not explicitly cited by CE Regulation n. 1234/07, are the grapevine leaves and shoots, which do not enter at all in the winemaking chain, of course, since they are not collected during harvesting, but nevertheless show interesting features for the same reasons for which WCBP are interesting, i.e., they are rich in polyphenols as it will be described further on.

Waste streams are generated when water is used to clean and sterilize winery equipment; these effluents are rich in organic matter. Finally, further WCBP can be considered diatomites or kieselgur which are siliceous sedimentary rocks reduced to fine powder and used for filtration of wine before bottling; these materials can retain some polyphenols.

The whole of WCBP and the point of the winemaking chain in which they are generated are shown in the flow diagram of Figure 2.

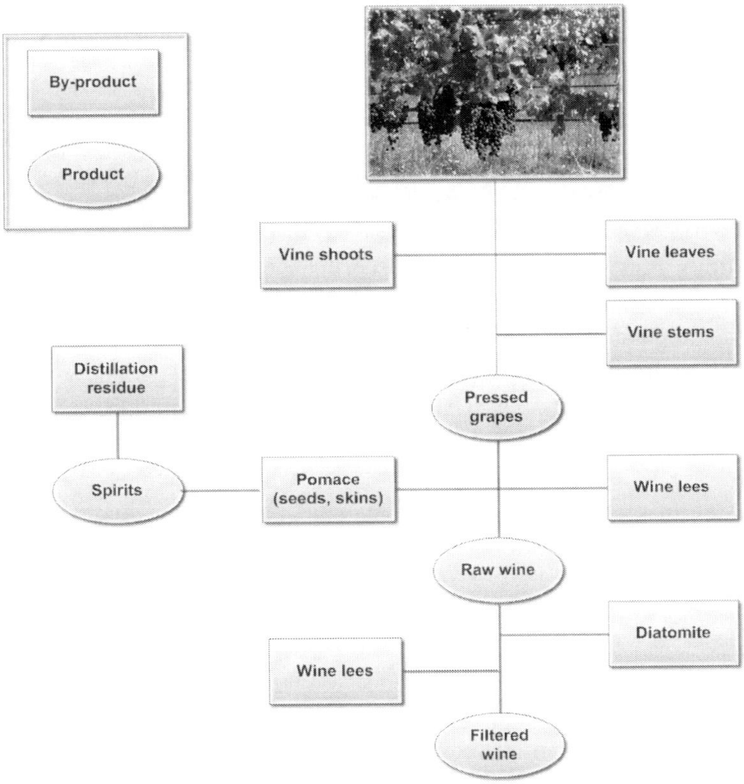

Figure 2. The by-products of winemaking chain.

The disposal of WCBP must be performed according to local regulations. In European Union, Regulation no. 479/2008 (2008) on the common organization of the wine market determined that pomaces and lees must be sent to alcohol distilleries, to produce exhausted grape pomace and a liquid waste (vinasse). In alternative, waste disposal in landfills can be choosen, but wine companies must bear the cost of the disposal (or pay a fine imposed by authorities in case of fraudulent disposal).

In Italy, until 2010, pomaces had to be used only for distillation of grappa, a typical Italian grape-based pomace brandy. After experimentation upon 2009 harvest, a recent Decree 7407 of 04/08/2010 by *Ministero delle politiche agricole alimentari e forestali* (2010) has been stated which introduces an important change. It determines that different new uses can be envisaged for disposal of WCBP, apart from the traditional delivery to distilleries:

- delivery to distilleries: this is specific of pomaces;
- direct agronomical use by spreading materials on agricultural crops as soil ameliorants in the limit of 3,000 Kg/Ha;
- undirect agronomical use by preparation of fertilizers;
- energetic use as biomass for biogas production or as fuel for energy production;
- pharmaceutical use by preparation of natural drugs;
- cosmetic use by preparation of natural cosmetics;
- food use as enocyanin, an extract from berry skins (see later) or as additive to food products (to be authorized, upon request, byDirezione generale delle politiche internazionali e dell'Unione europea).

This regulation (and similar ones all over the world) opens new frontiers for the exploitation of materials that until now were considered as a burden and not as a resource.

As it will be detailed further on, several applications of WCBP have been developed and patented in the last few years.

There are other issues to find new ways of disposal for WCBP. From the oenological point of view, the control on elimination of WCBP according to regulations has two reasons: to avoid the fraudulent use of WCBP in re-fermentation processes for production of bad quality wines and to preserve the quality standards of viticultural productions which could be compromised by eccessive pressing. There is, though, another important issue which is the environmental one. Cultivation of wine grapes and production of wine are associated with a large number of environmental concerns, among which the generation and management of waste streams; waste effluents are potential oxygen sources due to their organic content. Pomaces can initiate a number of environmental hazards ranging from surface and ground water pollution to foul odors; they attract flies and pests and can easily spread diseases. Leachates from pomaces can cause oxygen depletion in the soil and infiltrate surface, soil and ground waters.

For the reasons explained above, a correct disposal of WCBP is mandatory. The different treatment methods of WCBP have been reviewed by Arvanitoyannis et al. (2006b). A thorough discussion on the costs of recycling vs. valorization of WCBP has recently been published by Devesa-Rey et al. (2011).

Utilization of WCBP

WCBP are a potential reservoir of polyphenols, as will be clearer from the following paragraph. These materials, indeed, have been utilized since long time in order to exploit their organic nature or to extract some valuable components.

The use as soil ameliorant and as fuel must have been usual in the past by farmers and nowadays some studies suggest the potential use of WCBP as fertilizers (Bustamante et al., 2009; Rubio et al., 2013), though others suggested that high levels of polyphenols could cause germination problems in crops (Negro et al., 2003) and that prior conditioning treatments are necessary (Bustamante et al., 2008).

The most common use of pomaces is the production of spirits such as Italian *grappa* or French *marc*, which have more than secular tradition. The practice of distillation of alcohol-containing substrates in order to yield spirits is as ancient as 4,000 B.C., but we have no indication that pomaces were used. The product named *grappa* in Italy is specifically characterized by its derivation from grape pomace.

It was possibly conceived as part of the studies of the Scuola Medica Salernitana that in 11th century sat down rules on the concentration of alcohol by means of distillation. Pomaces were soon taken into consideration for grappa production, even if a spirit made from pomaces is explicitely cited since 15th century only, and written records on pomaces distillation are datable to 16th in studies of Jesuit priests.

On a more scientific basis, a large amount of research has been devoted to this subject in the last 30 years and some specific publications have been issued (Arvanitoyannis et al., 2006a; Oreopoulou and Russ, 2007; Zhao, 2007; Peralbo-Molina and Luque de Castro, 2013b). Particularly promising seems to be the extraction of chemicals. Extraction of residual tartaric acid from wastewaters has been proposed for its use in food, cosmetic and pharmaceutical preparations (Smagge et al., 1992; Andres et al., 1997).

Other substrates useful for food applications that could be extracted from pomaces are ethanol, dietary fibers and citric acid (Schieber et al., 2001), while oils could be extracted from seeds (Mazza, 1995). Grape seed oil has recently become very popular because of its recognized health properties, in particular itsvery high level of linoleic acid, an essential fatty acid for humans. Oil content in grape seeds ranges from 13% to 18% for all varieties. It can be recovered by pressing seeds or by chemical extraction with solvents. The product is rich in compounds with antioxidant potential, such as polyunsaturated lipids, tocopherols and phytosterols (Crews et al., 2006). Apart from health properties, its burnpoint (190-230 °C) is higher than that of olive oil, which would make it suitable for cooking.

Fumiyiwa and Ough (1982) evaluated the possibility that grape pomace could be used as animal feed. Deiana et al. (2009) indicated that grape stems could be used, after removal of ashes, to produce activated carbon. Nowadays one among the most important uses of WCBP is as fuel to feed biomass stations (Corona and Nicoletti, 2010).

Grape pomaces have recently found many applications in the field of biotechnology (Zhao, 2007). They are suitable substrates for Solid-State Fermentation (SSF), a modern approach used for several processes such as bioremediation or biodegradation of toxic compounds, detoxification of agricultural wastes, biotransformation of crops, biopulping and preparation of new high value products, such as secondary metabolites, organic acids, pesticides, aromatic compounds, fuels, and enzymes.

SSF involves the growth of microorganisms on wet solid supports in the absence (or near absence) of free water; substrates such as grape pomaces provide conditions close to the natural habitats of microorganisms, allowing them to produce certain enzymes and metabolites in a more efficient way.

Characterization of the Polyphenolic Content of WCBP

WCBP have attracted considerable attention since at least 30 years, even at the industrial level, due to the potential high recovery of polyphenols. The extraction and the utilization of polyphenols from fresh grapes was already known but it has been shown in several studies that polyphenols can be extracted, and therefore profitably used, from the residues of the processes used to transform grapes into wine, obtaining substrates with significant antioxidant power that can be exploited in different fields. Xia et al. (2010) summarized the antioxidant potential of extracts from pomace and from different parts of grapes. As clearly appears from Table 2, the antioxidant potential of WCBP and of grape components is generally higher than the one of wine.

The composition of WCBP has been defined in several studies. In order to evaluate the potential exploitation of these materials, the content of valuable compounds, i.e., polyphenols and other antioxidants, must be extracted and determined.

As for grapes, though, the polyphenolic content of WCBP can vary widely.

Table 2. Comparison of polyphenolic parameters in grapes and wines

Resource	TEAC[a]	FRAP[b]	DPPH[c]	ORAC[d]
Grape pomace	0.91 g/L (EC_{50})		0.20 g/L (EC_{50})	
grape seed			>663 µmol TE/g	
defatted grape seed	36.36 mol TE/100 g	21.6 mol TE/100 g	-	-
whole seed	76.3 mol TE/100 g	58.04 mol TE/100 g	-	-
grape seed	281.3 µmol TE/g	-	16.8-92 mmol TE/g	42.18 mmol TE/g
grape skin	236.1 µmol TE/g	-	15.7-113.3 mmol TE/g	36.40 mmol TE/g
grape seed	281.3 µmol TE/g			
grape leaf	236.1 µmol TE/g			
grape skin	12.8 µmol TE/g			
grape flesh	2.4 µmol TE/g			
grape juice	25 mmol TE/L	32 mmol Fe^{2+}/L	15 mmol TE/L	
wine		8.8 µmol TE/g	22.9-26.7 µmol TE/g	
wine		3.098 mg TE/L	70.7% inhibition	10.72 µmol/L

[a] TEAC: Trolox equivalent antioxidant capacity assay.
[b] FRAP: ferric reducing antioxidant power assay.
[c] DPPH: 2,2-diphenyl-1-picrylhydrazyl method.
[d] ORAC: oxygen radical absorbance capacity assay.

In addition to the same factors influencing the amount of polyphenols and their distribution in grapes, other factors play important roles such as processing and storage conditions, and extraction and analytical methods used in the characterisation (Lafka et al., 2007).

To demonstrate this variability, Thimothe et al. (2007) showed that the polyphenolic content in Pinot Noir pomace was slightly higher than in the grapes.

Grape Pomace

Grape pomace is composed by pressed skins, seeds and (partially) stems in percentages of respectively about the 72%, 17% and 11% of total dry matter. Most of its polyphenolic content derives from the fresh materials, which were described in paragraph 1.3; therefore pomace is expected to have mostly contribution of anthocyanins (in the case of red grapes), hydroxycinnamic acids, flavanols and flavonol glycosides from skins, and flavan-3-ols and their polymers, the precious proanthocyanidins, from seeds. Resveratrol and other stilbenes are also likely to occur, but it is not easy to determine their exact amount due to the high reactivity.

The characterization studies on pomaces started at least 40 years ago. The work by Rice (1976) is among the first scientific studies on the potential recovery of chemicals from WCBP. This matter has been recently reviewed by Yu and Ahmedna (2012) and by Fontana et al. (2013) with an eye to biological properties and potential applications. Deng et al. (2011) reported values of total polyphenolic compounds from grape pomaces together with from fresh grape skins, grape berries and other fruit WCBP, as determined in own and different studies.

A comprehensive chemical composition analysis of pomaces from red and white grapes was performed by Bravo and Saura-Calixto (1998). All samples were found to be rich in protein (12% to 14% dry matter), fat (7% to 12%), and minerals (6% to 9%), with lower amounts of soluble sugars (about 3%) and polyphenols (4% to 5%). Condensed tannins ranged from 27% in red grape skins to 20% in white grape skins to 16% in white grape seeds.

Llobera and Cañellas (2007) evidenced that WCBP from Manto Negro variety, native of Mallorca, had a high content in dietary fibre and associated polyphenols, particularly for grape stems which has a polyphenolic content of 116 g GAE/Kg dry matter, four times higher than that for the grape pomace (26.3 g GAE/Kg dry matter); according to these values both WCBP can be considered excellent antioxidant dietary fibre products.

Tseng and Zhao (2013) found the following composition in grape pomace from *Vitis vinifera L.* cv. Pinot Noir (Table 3).

An interesting study was that of Ratnasooriya et al. (2010). Authors compared the content of some classes of polyphenols in fresh grapes and corresponding pomaces of some table and wine grape cultivars grown in Nova Scotia (Canada). The total concentrations of anthocyanins, flavan-3-ols, flavonols and stilbenes were found to be systematically higher (up to 2-fold) in pomaces than in fresh fruits; this can be partly explained with the loss of juice, but it strongly suggests that pomaces are very promising reservoirs of polyphenols.

Many characterization studies have been devoted, apart from the whole of grape pomaces, to grape seeds contained in them. The most precious compounds kept in seeds are proanthocyanidins, which hold tremendous interest because of their strong antioxidant activity.

Table 3. Composition of grape pomace from Pinot Noir

Parameter	% composition (DM)[a]
Moisture content	5.63 ± 0.10
Ash	5.07 ± 0.05
Protein	10.32 ± 0.22
Lipid	11.09 ± 0.33
Soluble sugar	3.89 ± 0.30
Pectin	3.68 ± 0.05
Condensed tannin	12.11 ± 1.17
Dietary fibre	61.32 ± 1.69
Total phenolic compound (mg GAE/g)	67.74 ± 6.91
Radical scavenge activity (mg AAE/g)	37.46 ± 1.86
Radical scavenge activity (mg TE/g)	91.78 ± 4.58

[a]DM: dry matter.

Proanthocyanidins containing two or more monomers chemically linked together are called oligomeric proanthocyanidins (OPCs); it has been reported (Bagchi et al., 1997) that the antioxidant power of OPCs in grape seeds is approximately twice that of vitamin E and four times that of vitamin C, a feature which opened the way to many applications in nutraceutical, cosmetic and pharmaceutical fields, as it will be detailed later. Waterhouse et al. (2000) quantified different fractions of proanthocyanidins from grape seed extracts with HPLC. Monomeric, oligomeric (2-7 subunits) and polymeric (8-24 subunits, and similar to 24+ subunits) fractions were discriminated; percentages ranged from 5% to 30% for monomers, 17% to 63% oligomers, 11% to 39% polymers and 2% to 50% for the large (24+) polymers. Mattivi et al. (2002; 2008) evaluated that the average extractable flavanol and proanthocyanidin content of red grapes, measured by means of the vanillin assay, varies in the range 0.87–3.65 g/kg according to variety. González-Paramás et al. (2004) determined the total amounts of flavanols in grape pomaces and in seeds obtained from different red and white grape varieties grown in Spain. In all the cases, seeds presented greater concentration (from 2- to 7-fold) in total flavanols and individual proanthocyanidins than complete pomaces. An important finding in this study was that dried seeds obtained as a final winery waste were only 10% lower in total flavanols with respect to original seed, despite the high temperatures used in the drying process. This suggests that seeds can represent a very cheap source for extraction of antioxidant flavanols. Maier et al. (2009) characterized the polyphenolic profile in seeds of seven grape cultivars grown in Southern Germany. Authors determined polyphenols both in seeds and in the press residues of grape seed oil production, finding that residues also are a polyphenol-rich by-product with high antioxidant activity; this result is not surprising, considering that polyphenols are hardly soluble in fatty oil, but it is nevertheless interesting, considering that press residues are usually considered a waste and treated consequently. Another interesting product that can be obtained from press residues of grape seed (or from fresh grape seeds) is grape seed flour: it is a crushed and ground product which maintains almost unaltered the beneficial features of seeds, i.e., rich content in proanthocyanidins and great antioxidant and antiradical activity. Several studies correlated the total polyphenolic contents with the antioxidant activity of pomace extracts (Kähkönen et al., 2001; Alonso et al., 2002; Chidambara Murthy et al., 2002; Negro et al., 2003).

In particular Larrauri et al. (1996; 1997), while determining the amount of extractable polyphenols and the antioxidant activity of white and red grape pomaces, found that red grape skin antioxidant activity was 1.5-2-fold lower than that of red wine.

Other studies have focused attention on the characterization of representative polyphenolic compounds. Such detailed data are of high significance since individual compounds may differ considerably in their bioavailability and bioactivity. In a comprehensive study by Kammerer et al. (2004), a screening by HPLC-MS was carried out on pomaces originated from both red and white wines; several anthocyanins, hydroxybenzoic and hydroxycinnamic acids, catechins and flavonols, and 2 stilbenes were identified and quantified, with no main differences between red and white grape varieties apart from lack of anthocyanins in white grape pomace.

For what concerns resveratrol and other stilbenes, it is known that a part of the original content in grapes is extracted in wine, so that grape pomaces are generally poorer that fresh skins. Moreover, exposure of such matter to light and oxygen in the winery during the winemaking process may result in oxidation of the highly reactive molecules. As a consequence, the content of stilbenes in pomaces highly depends on the storage conditions and on the extraction procedure used in the determination. Kammerer et al. (2004) found a resveratrol content ranging from 11.1 to 123.0 mg/kg dry matter in pomaces from red and white grapes cultivated in Germany.

Careri et al. (2003; 2004) in two studies on the content of resveratrol isomers in red wine, grapes and relative WCBP obtained from Nero d'Avola variety (Southern Italy), found that grape pomace contained up to half of the original resveratrol content in fresh skins.

A further by-product derived from grape pomace is the residue following ethanol distillation. When pomaces are treated in distilleries in order to obtain spirits such as Italian *grappa*, only volatile compounds are effectively extracted from skins and seeds, among which very few polyphenols. The drastic conditions of the distillation process cause degradation of many molecules, nevertheless valuable compounds can still be found. Pinelo et al. (2005) showed that the polyphenolic content in grape pomaces and in distilled pomaces were nearly identical. The total amount was expressed in GAE, i.e., equivalents of gallic acid, as usual, so that it is not possible to verify what and how much of the original compounds were still present; nevertheless it is interesting to note that the antiradical activity, expressed as DPPH radical-scavenging assay, was found to be unaltered before and after distillation. Similar conclusions could be drawn from the study by Cruz et al., (2004) in which authors applied different extraction procedures and subsequent chromatographic fractionation to waste samples, registering remarkable antioxidant activity in the fractions collected. Díaz et al. (2012) applied processes included in a patented application (Moure et al., 2010) in order to recover polyphenols from residues of distillation of white grapes pomace; from 1 L of residue they were able to extract 42 g dry weight of product with a polyphenolic content of 44.7% expressed as GAE and good antioxidant and radical scavenging capacity. Peralbo-Molina et al. (2012; 2013a) demonstrated that a significant amount of polyphenols can still be recovered from pomace residues with application of proper extraction systems. The total amount was found to be 4.2 mg GAE/mL of extract from white grape skins and 1.5 mg GAE/mL from red grape skins; 3.4 mg GAE/mL from white grape seeds and 2.7 mg GAE/mL from red grape seeds. Therefore, the residue of distillation contains more than appreciable polyphenolic content, but its usual destiny is delivery to landfills or use as a heating source.

Finally, it must be cited a recent, promising and largely unexplored field in the study of polyphenols in natural products: Non-Extractable Polyphenols (NEPP). The concept of NEPP was deepened by Saura-Calixto in the late '80s (1988) and it refers to the fraction of polyphenols that is usually unreleased from plant tissues both during treatment processes (e.g., from grapes upon pressing and maceration) and when substrates are extracted for analytical characterization. This is due to the ability of aromatic rings and hydroxyl groups of polyphenols to form macromolecules (i.e., proanthocyanidins or tannins) as well as to bind to polysaccharides, proteins and cell walls by different mechanisms, including covalent bonds, hydrogen bonding, and hydrophobic and hydrophilic interactions. It is obvious that the extraction conditions during pressing and maceration of fresh grapes are unsuitable for release of NEPP into must (and indeed they are not intended for, since it is not a matter of oenological interest); from the analytical point of view, the common procedures developed to characterize the content and composition of polyphenols in foods involve mostly use of aqueous organic solvents for HPLC analysis, isolation of specific compounds and biological assays; these conditions are able to evidence low molecular weight compounds only, such as olygomeric proanthocyanidins. As a consequence, grape pomaces do contain highly significant amounts of NEPP such as polymeric polyphenols or low molecular weight polyphenols bound to protein, polysaccharides, or cell walls, which can be released upon strong acidic or enzymatic treatments. To evaluate how underestimated is the potential of NEPP, Saura-Calixto (2012) reported that in the past decade over 30,000 papers were published focused exclusively on Extractable Polyphenols (EPP), whereas not much than 30 papers were dealing with NEPP. However, it is important to evaluate the relative ratio NEPP vs. EPP in natural products. Saura-Calixto et al. (2007) estimated that the amount of NEPP in the mean daily intake of Spanish people was almost double that of EPP. Pérez-Jiménez et al. (2009) determined NEPP content in grape pomaces from red grapes (cv. Cencibel) and white grape (cv. *Thompson seedles*) after chemical or enzymatic hydrolysis. Due to the great structural complexity of the macromolecules involved in NEPP, authors evaluated some alternatives to be used as calibration standards, eventually determining that Carob pod tannins extract beared more resemblance to the profile of NEPP from the food samples analyzed. In this way, they determined that NEPP content was 240 g/Kg dry matter in red grape skins, 100 g/Kg in red grape seeds and in overall pomace, while 5 g/Kg were found in white grapes.

According to these studies, NEPP could be more abundant than extractable polyphenols in much food, including grapes and therefore grape WCBP. These results show apparently that once more, WCBP contain a potential treasure of biologically active molecules, though largely underestimated. The development of more powerful analytical methods, involving mass spectrometry (Monagas et al., 2010), could help in evaluating the possibilities yielded by these complex natural products.

Stems

Unlike skins and seeds, stems retain almost totally their polyphenols since they are almost totally discarded in the destemming process. For this reason, stems have attracted scarce attention for what concern their composition and possible exploitation in both oenological and post-oenological practices.

Extracts from stems, nevertheless, show significant antioxidant power.

Nearly 80% of stem polyphenols are tannins (Souquet et al., 2000), more precisely proanthocyanidins with low polymerization degrees (much lower than skin tannins). Some studies (Yan et al., 2001; Yan et al., 2002; Makris et al., 2007) have also indicated stems as a major source of stilbenic metabolites, among which resveratrol, with percentages largely higher than those found in grape berries. Püssa et al. (2006) reported concentrations of resveratrol in the range 1-3 mg/g dry weight in stems of grapevine varieties cultivated in Estonia, to be compared with 0.05-0.1 mg/g of resveratrol in fresh grape skins and 0.2-7.7 mg/L of *trans*-resveratrol in red wines. Up to 0.3 g/Kg dry weight of *trans*-resveratrol, i.e., 100-fold the amount present in berries, were found in stems by Balik et al. (2008) who determined the content of stilbenes in the different parts of grape bunch of grapes varieties grown in Southern Moravia (Czech Republic). Other relevant polyphenolic compounds in stems are flavan-3-ols, mainly catechin and epicatechin.

From these numbers, it is apparent that stems may be a significant, low-cost source of *trans*-resveratrol, proanthocyanidins and other polyphenols.

Wine Lees

While most of the interest in WCBP is focused on grape pomace, in the last few years a moderate number of studies suggested that wine lees also can be exploited for the recovery of valuable compounds. Lees are mainly composed by microorganisms (yeasts), tartaric acid, inorganic matter and polyphenolic compounds, among which anthocyanins. Alonso et al. (2002) evaluated the antioxidant activities of grape pomaces, stems and lees from grapes of Spanish red and white varieties, and compared them with the amount of total polyphenols. Authors found that, as expected, both antioxidant activity and total polyphenols in lees were lower than in pomaces and stems. Values of total polyphenols ranged from 10 to 200 mg GAE/g dry matter in lees, against 200-500 in stems and in pomaces; antioxidant activity was accordingly lower. The results are nevertheless interesting and somewhat unexpected. Higher percentage recoveries were found by Pérez-Serradilla and Luque de Castro (2011) by application of a more efficient extraction system. These results suggest that lees must be seen as promising substrates for recovery of polyphenols. A thorough review on the role of wine lees in oenology, their composition and the possibilities given by their reutilization, was presented by Pérez-Serradilla and Luque de Castro (2008).

Vine Lees

The polyphenolic content in grapevine leaves has been rarely studied but it was found to be rich in phenolic acids, flavonols, flavanols, anthocyanins (Schneider et al., 2008) and, interestingly, in stilbenes and the corresponding glucosides (Pezet et al., 2003). The concentration of trans-resveratrol was found to be in the range of 2.5-20.9 mg/kg though other authors suggested that the 3-O-glucoside was indeed present (Schoedl et al., 2011; Schoedl et al., 2012) instead of the aglycone form. Balik et al. (2008) found up to 21 mg/Kg dry matter of trans-resveratrol and up to 61 mg/Kg dry matter of trans-piceid in leaves of grapes varieties grown in Southern Moravia (Czech Republic); these values were systematically higher than the content of these stilbenes in berries.

Pastrana-Bonilla et al. (2003) found that in Muscadine grapes variety, the total amount of polyphenols in leaves equals that in skins and is 6-fold lower than in seeds, but interestingly the antioxidant capacity of leaves is only 20% lower than the antioxidant capacity of seeds, so that the ratio of antioxidant capacity/total phenolics was 6-fold higher for leaves than for the seeds, possibly due to a higher antioxidant capacity of the major phenolics in leaves or to the presence of other unknown antioxidant compounds different from phenolics. The great antioxidant capacity of muscadine leaves was found to be at least twice the value for the leaves of other berry plants.

As in berry skin, also in leaves some polyphenolic compounds, mainly hydroxycinnamic acids and flavonoids, display UV-screening action (Kolb et al., 2001).

Vine Shoots

Vine shoots are of course totally excluded from harvest and do not enter in fermentation vats. For this reason, data concerning their composition is quite rare (Jiménez et al., 2006) and even scarcer is the information concerning their polyphenolic content. The composition of vine shoots is characterized by three main fractions: cellulose, hemicellulose, and lignin; this last is a high molecular mass crosslinked polymer whose units are characterized by a polyphenolic structure. Lignin can be hydrolyzed to release aromatic polyphenolic compounds. Delgado-Torre et al. (2012a) determined the polyphenolic content of vine shoots of different *Vitis vinifera* cultivars cultivated in Spain; total amount ranged among 3,300 and 5.800 μg GAE/g of vine-shoots. Polyphenols identified were mostly simple phenols and phenolic acids such as pyrogallol, gallic acid and protocatechuic acid, and flavan-3-ols such as catechin. Balik et al. (2008) found comparable amounts of stilbenes (trans-resveratrol and trans-piceid) and hydroxycinnamic acids (caftaric acid) in vine shoots and vine leaves of grapes varieties grown in Southern Moravia (Czech Republic).

According to the chemical features, vine shoots could constitute a further source of polyphenolic compounds to be used in the nutraceuticals, cosmetics, pharmacological, oenological, and food additive industries, even considering the high availability of inexpensive matter: an estimated total of 11.2-16 million tons of vine shoots is produced each year (Peralbo-Molina and Luque de Castro, 2013b).

Diatomites

Diatomite (or diatomaceous earth or kieselgur/kieselguhr), is a naturally occurring siliceous sedimentary rock that can be crumbled into a fine white powder. The composition of diatomite is 80-90% silica, 2-4% alumina and 0.5-2% iron oxides. Its use in filtration systems is nowadays fully acknowledged in oenology for removal of particles in musts and wines. The disposal of diatomite waste may results in a heavy environmental issue if performed outside local regulations, since its water dispersion, which has a pappy consistency, can clog up pipings. Therefore, alternative disposal ways are under study. In a research carried out in Piedmont by Università degli Studi del Piemonte Orientale in collaboration with Regione Piemonte and other local authorities, named FAR FOS - Farine Fossili (Regione Piemonte, 2011), it has been verified whether diatomite waste could be spread in crops as ameliorant, in alternative to disposal in landfills as special waste (which is a cost for winemakers). Besides analytical tests carried out to exclude the dangerousness of diatomite wastes, it was determined their residual polyphenolic content by means of Folin-Ciocalteau colorimetric method. Values between 0.3 and 1.8 g GAE/Kg dry matter were found, with higher values for diatomites used in the filtration of red grapes wines, as expected. These values are far lower than those found in pomaces and lees, but are nonetheless interesting.

Utilization of WCBP with an Eye to Polyphenols

Long time before many scientific studies had demonstrated the chemical potential of WCBP, mankind already knew how to use them in order to exploit the high content of polyphenols. An ancient industrial practice exploiting residual polyphenols, much previously than suggested by science, has been the recovery of anthocyanins from grapes skins. In 1879 Antonio Carpenè, an Italian chemist and oenologist, and his coworker Enrico Comboni, developed an industrial process to extract anthocyanins from grapes skins. This product is known as Enocyanin and it is used, since then, as a natural food dye.

Nowadays, enocyanin production has been patented by many authors (Marakakis, 1982) and it is recognized in the industrial world as a precise chemical product with CAS No. 11029-12-2. European Union allows its use in drinks, marmalades, candies, ice creams and pharmaceutical products (European Union, 1994).

More recently, it has become very popular the production of nutraceutical, cosmeceutical and pharmaceutical preparations based on polyphenols extraction from WCBP. Particular concern has been given to the possibility of substituting synthetic antioxidants, the utilization of which in the several fields is common and restricted, by natural polyphenolic compounds which frequently show a higher antiradical and antioxidant activity. This possibility increases also the ability to respond to consumers' needs of safer products.

Applications in the Food Industry

The most traditional use of wine pomaces is distillation and production of spirits. In this case the role of polyphenols in WCBP is quite marginal since, upon distillation, they tend to remain in pomace residues; in spirits, in fact, only tannins coming from wood and only in aged spirits can be observed.

From a less traditional point of view, a great number of applications have been proposed for use of grape WCBP in food industry. On the wave of the beneficial effects of wine drinking, described in several thousands of scientific publications, ADA (American Dietetic Association) has suggested a consumption of wine between 250 to 500 mL/day (ADA, 2004). This value, indeed, is not feasible in most populations, apart from the potential for addiction and other serious diseases (risk of hypertension, liver disease, breast cancer and weight gain). To overcome this difficulty, new products have entered the food market, which come from grapes, wine or WCBP. In fact, it has been possible to incorporate grape-derived extracts to foods and beverages in order to obtain health appeal. It has been demonstrated (Tseng and Zhao, 2013) that grape pomaces, thanks to the high content of polyphenolic compounds and dietary fibre, meet the requirements for being defined antioxidant dietary fibre (ADF), a concept first proposed by Saura-Calixto (1998). ADF defines a matter of natural origin in which 1 g has both free radical-scavenging capacity (as determined with 2,2-diphenyl-1-picrylhydrazyl or DPPH method) equivalent to at least 50 mg vitamin E and dietary fibre content higher than 50% dry matter from the natural constituents.

Due to the commercial significance of these developments, preparations based on grape extracts have rapidly been patented. Examples are the following: a concentrated food ingredient with extracts from green tea, grape skin and grape seeds was patented in US by Wild and Sass (2006); Shrikhande et al. (2007) patented a process in which extraction of monomeric and oligomeric proanthocyanidins has been maximized to the detriment of polymeric proanthocyanidins; the extract was used to prepare formulated dry mix products to be used in beverages, nutraceutical capsules and supplements containing vitamins and minerals.

Another application in the field of alcoholic beverages is the addition of pomace extracts to juice or wine. The use of additives is a common practice in the wine industry for improving the sensory profile of wines and reducing defects. Examples of internationally approved oenological additives are wood chips, enzymes, oenological tannins and the already cited Enocyanin or additive E-163, obtained from red grape skins. It is obvious that the use of natural- and vegetal-based additives should be better perceived by consumers with respect of synthetic ones. In a patent application by Perlman et al. (2008), inventors proposed the fortification of grape juice with a pomace extract. The beverage was tested at different extract concentrations for sensorial acceptance and antioxidant activity. Dehydrated waste grape skins were proposed for addition to white wines in order to produce rosé wines (Pedroza et al., 2011) or to young red wines in order to ameliorate their quality parameters of color and polyphenolic compounds before bottling (Pedroza et al., 2013). A similar application was proposed by Delgado-Torre et al. (2012b) who compared the polyphenolic profiles of oak chips, frequently employed in the aging of wines or spirits, and vine shoots of different vine varieties cultivated in Spain. Authors considered a representative panel of polyphenolic compounds of interest from an oenological point of view, finding that hydroalcoholic extracts from some vine shoots were similar to those from oak chips. Therefore, the selected vine shoots varieties could be considered suitable as an alternative to oak barrels or oak chips in the aging process of wines and spirits.

Apart from the production of spirits or the fortification of wines, pomace or its extract can be added to food products in order to extend their shelf-life, due to the antioxidant and antimicrobial action yielded by polyphenols, or simply to improve food color, aroma and taste. Ozkan et al. (2004) demonstrated that grape pomace extracts at 1% and 2.5% concentrations displayed antibacterial effects against 14 pathogenic bacteria, suggesting the potential use for food additive against the deterioration of stored foods. Sánchez-Alonso et al. (2007) showed that the addition of a preparate produced from red grape pomace considerably delayed lipid oxidation in minced horse mackerel muscle during the first 3 months of frozen storage. Goñiet al. (2009) demonstrated that grape pomace concentrates, administered to chickens at various levels, were effective in inhibiting lipid oxidation of chilled and long-term frozen stored chicken patties.

Even more promising is the exploitation of extracts from grape seeds, which contain a high level of oligomeric proanthocyanidins. Grape seed extracts (GSE) have been sold in the market as dietary supplements in Japan since 1993, added to different food and beverages. In the United States, the positive effects of addition of GSE to foodstuffs are guaranteed by the recognition of generally recognized as safe (GRAS) status which was acknowledged to preparations containing oligomeric proanthocyanidins in 2003 (Smith et al., 2003). GRAS assessment of a substance or a preparation is a procedure in which a panel of experts of the US Flavor and Extract Manufacturers Association evaluates a combination of relevant data on that substance/preparation in view of its use in the food industry.

To appreciate the importance of this feature, it must be noted that items granted with GRAS status are not considered to be food additives and are therefore excluded from mandatory premarket approval by the US Food and Drug Administration.

The action of seeds proanthocyanidins in the inhibition of oxidative reactions is reputed higher than that of tocopherol and ascorbic acid and can limit oxidation of lipids and vitamins in foodstuffs, as well as generation of off-flavors and colors. Another interesting property of grape seed extracts is that they can act as emulsifiers in oil-and-water systems.

With strong scientific support, commercial products based on GSE are nowadays sold in the food industry market. Two renowned examples are ActiVin®, commercialized by US Dry Creek Nutrition Incorporated (Modesto, California), which was granted GRAS status in 2001, and Gravinol®, commercialized since 1993 by Japan's Kikkoman Biochemifa Company and granted GRAS status in 2003.

Also from grape seeds another product, grape seeds flour (i.e., the powdery residue of seeds after milling), can be obtained as a suitable additive to be used in the food industry. Several products containing this derivative are being sold nowadays; it has become part of gluten-free diets. Rosales Soto et al. (2012) evaluated the antioxidant activity and the consumer acceptance of food products containing grape seed flour produced from pomace of Merlot and Cabernet Sauvignon grape varieties. Flour was included in the preparation of cereal bars, pancakes and noodles. The best balance among antioxidant activity and consumer acceptability was found in cereal bars containing 5% Merlot grape seed flour.

A further, highly technological development in the application of pomace extracts to food industry is the encapsulation of polyphenols into nanoemulsions, in order to favor their delivery through the biological membranes and to prevent degradation caused by oxidation (Sessa et al., 2013). Nanoemulsions are formulated with natural ingredients such as sunflower oil or palm oil. Particularly interesting is the application of this process to resveratrol because its physicochemical properties, with concern to the high reactivity and low solubility in aqueous and lipid phases, must be seen as limiting factors for its bioavailability and the efficacies of the desired health-beneficial effects (Sessa et al., 2011).

Applications in the Medical Field

The interest for WCBP in the medical field has grown considerably after the inception of the French paradox linkage with red wine and grapes consumption, and immediately after, several scientific studies have been carried out in order to demonstrate that not only wine and grapes, but also pomaces show therapeutical effects. Soon after, pharmaceutical preparations have been patented containing pomace or grape seeds extracts (Masquelier, 1987) in which the strong antioxidant and antiradical properties of proanthocyanidins were exploited. Proanthocyanidins, in fact (but also anthocyanins), exert capillaroprotective, antinflammatory and anticoagulant properties, which are useful for the treatment of circulatory insufficiency, varicose veins and hemorrhoids (Costantini et al., 1999; Flesch et al., 1998).

Today, WCBP are being sold to the rapidly growing dietary supplement industry. Several applications of WCBP in the medical field are described in a review by Yu and Ahmedna (2012).

As it is easy to foresee, the lion's share in nutraceutical applications is again performed by grape seeds extracts, or GSEs, and their oligomeric proanthocyanidins. The great comercial interest to grape seeds is demonstrated by the fact that, on a total of 42 grape health products commercialized in the United States, 22 of them were obtained from grape seeds, as reported by Shrikhande (2000). Health properties displayed by oligomeric proanthocyanidins-based products such as ActiVin® and Gravinol® are strongly supported by scientific research (Banerjee and Bagchi, 2001; Kalfin et al., 2002; Vinson et al., 2002; Kim et al., 2010; Su and D'Souza, 2011). GSEs antioxidant power is stronger than vitamin C, E and β-carotene.

GSEs are cytotoxic to breast, lung and stomach adenocarcinoma cells, and protect mouth keratinocytes from *in vitro* apoptosis induced by tobacco (Bagchi et al., 2002). They also protect hepatic cells from cytotoxicity induced by chemotherapeutic drugs, through a modulation of genes which regulate cell cycle and apoptosis, such as Bcl2, p53 and c-myc, and inhibit the cytochrome P450 2E1, thus limiting the hepatotoxic effects of various drugs.

Also, GSEs protect the human organism from age-related diseases, like degradation of joint collagen and tissue damage caused by sun exposure (Yamakoshi et al., 2002). A grape seed extract has promoted the expression of VEGF (vascular endothelial growth factor) in keratinocytes, thus favouring skin angiogenesis and wound healing. A similar extract, administered orally to guinea pigs has reduced the pigmentation caused by direct exposure to ultraviolet irradiation, probably acting via the inhibition of melanin synthesis and melanocyte proliferation (Yamakoshi et al. 2003). Recently, Carrieri et al. (2013) demonstrated the ability of GSEs obtained from different table grape varieties, to act as antithrombotics agents by inhibiting the synthesis of tissue factor by monocytes/macrophages activated by inflammatory agents.

Grape seed oil also has found innumerous applications in the medical field. It is easily absorbed by the skin, on which it exerts emollient, mildly astringent and hydrating effects. It is particularly suitable for stress-induced skin damage, particularly in the periocular region, because of its high regenerative properties.

The oil forms a thin film on the skin and is therefore used as a vehicle of essential oils in aromatherapy. It is also useful in the treatment of acne (Burlando et al. 2010). Duncan and Bernhard (2010) developed a mixture composition useful in wound healing, containing extracts from red wine grape seeds and American elderberry fruit, ascorbid acid source, tannin and gallic acid microbe source and protein source.

In the medical field, not less interesting than GSEs, and certainly supported by a higher amount of scientific research, are the stilbenes. A major part of scientific studies on the beneficial effects of grapes, wine and WCBP have been focused on resveratrol and its isomers and glycosides (Fremont, 2000; Aleixandre et al., 2013). The large number of health benefits displayed by resveratrol has recently been reviewed by Guilford and Pezzuto (2011). Several pharmaceutical preparations containing resveratrol are now available on the market.

Another interesting feature of polyphenols is their antioxidant power, which contrasts aging processes induced by UV radiations and free radicals. Proanthocyanidins have also induced a rise in collagen synthesis in fibroblast cell cultures. Resveratrol has a tightening and firming action on the skin and favours tissue renewal, thus acting similarly to retinol (Fructus 2004).

The topic application of resveratrol to nude mice has protected the skin from phototoxicity and tumorigenesis induced by UV-B radiations (Aziz et al., 2005). In melanoma cells, the compound has inhibited cell growth and induced apoptosis, but in vivo tests have not confirmed its efficacy as a therapeutic agent for melanoma. Proanthocyanidins and resveratrol seem able to improve wound healing, while topic application of resveratrol to mice has been found useful against herpes (Docherty et al. 2004, Khanna et al. 2002). Resveratrol has also shown antimicrobial effects against bacteria and dermatophytes that infect the skin (Chan 2002).

Recently, major focus has been given to the potential of Non-Extractable Polyphenols or NEPP, with particular concern to Non-extractable Proanthocyanidins (NEPA). Some studies have showed the beneficial effects of ingestion of foods containing NEPA, evidencing that they are not bioaccesible in the small intestine but travel through the gastrointestinal tract as insoluble substrates reaching the colon, where they release single polyphenols and different bioavailable metabolites by the action of bacterial microbiota (Saura-Calixto, 2012). Mateos-Martín et al. (2011) showed that NEPA behave in vivo as a source of progressively released polyphenolics that can come into contact with the intestinal walls and are bioavailable for at least 24 h after ingestion.

Applications with a high degree of technology are those carried out at nanoscale, as in the food field. In the medical field, a new research domain is open by production of nutraceuticals incorporating polyphenols in various inorganic or organic nanocapsules, with the result of enhancing their antioxidant properties. Sessa et al. (2011; 2013) have suggested the application of nanoencapsulation of polyphenols from pomaces in lipids in order to increase cell permeability and bioavailability. Iosub et al. (2012) studied the optical properties of anthocyanins in order to evaluate the possible enhancement of their properties as chromophores or fluorescent active biomolecules after encapsulation in sol-gel systems (Lacatusu et al., 2009). In this view, anthocyanins could be used as high eco-friendly constituents for electro-chromic applications in bio-imaging for diagnostic and therapy.

Applications in the Cosmetic Field

Red grape extracts and active principles are widely used in cosmetic products, as also shown by the number of related patent publications (Baumann, 2007). The importance of grapes in skin care is testified by winetherapy, which originally developed in the region of Bordeaux, France, and then diffused to other countries. Winetherapy basically consists of baths or applications of wine and grape extracts rich in polyphenols in order to stimulate microcirculation and strengthen blood vessels.

In this field also, a great interest is devoted to GSEs and to their proanthocyanidin content. Proanthocyanidins exert protective effect on the UVB-promoted photo-oxygenation of polyunsaturated fatty acids, with potency comparable to that of vitamin E; this suggests the use of proanthocyanidins as adjuvants of physical/chemical sunscreens in skin protection from photodamage (Carini et al., 1998).

The stabilizing action of proanthocyanidins on capillary vessels gives a contribution to the maintenance of skin elasticity and improves disorders linked to skin circulation, such as couperose, or to an insufficient draining of subdermal tissue, like cellulite (Lis-Balchin, 1999).

Hughes-Formella et al. (2007) investigated the anti-inflammatory and skin-hydrating properties of dietary supplements and topical formulations containing oligomeric proanthocyanidins.

Also of interest is grape seed oil. The use of this oil in cosmetics is widely supported on the large number of commercial products which include it in their composition. Grape seeds flour is another product of interest in this field, nowadays used in many cosmetics, creams and masks in Europe.

Applications As Energy Renewable Sources

Besides food, medical and cosmetic uses, a latest promising idea about the exploitation of polyphenols obtained from WCBP concerns their use in electronic devices such as third-generation photovoltaic cells. The most suitable polyphenols in this concern are anthocyanins; some studies have therefore deeply investigated their electronic properties, correlating spectral characteristics with trichromatic parameters and applying with quantum chemistry computation (Iosub et al., 2012). These studies were inspired by the introduction of Dye Sensitized Solar Cells (DSSC) that were firstly developed at the Ecole Polytechnique Federale de Lausanne, Switzerland, by Grätzel and O'Regan in 1991. DSSC are semiconductor photovoltaic devices that directly convert solar radiation into electric current by the combination of nanostructured electrodes and efficient charge injection dyes. Natural dyes such as betalains, carotenoids, chlorophyll and flavonoids - anthocyanins among them - have been tested as suitable sensitizers (Narayan, 2012). Calogero et al. (2012) tested the efficiency of different natural photosensitizers, among which Giacchè grapes, an autochthonous variety growing in Lazio (Italy); authors found that the sensitization activity of anthocyanins extracted from Giacchè grapes was lower than those extracted from other vegetables and fruits such as Sicilian prickly pear, blackberry, red Sicilian orange, red mulberry, radicchio and eggplant. Similar results were found by Alhamed et al. (2012) and by Ahmadian (2011) which found blueberry and black raspberry extracts far more suitable than grapes, possibly due to their higher amount of anthocyanins. According to these authors, the variegated composition of anthocyanins in grapes, with more than 20 compounds, is a confounding factor that limits understanding the low suitability of grapes for application to DSSC. Jasim (2011) attributed the fact that grape anthocyaninsare not good sensitizers to the lack of carbonyl (-C=O) or hydroxyl (-OH) groups capable of complexing to the Ti(IV) sites on the TiO_2 surfacedeposited on the anode to activate electronic conduction. Nevertheless, it is possible that some grape variety, with a particular anthocyanins distribution rich in cyanidin, could result suitable in the future.

Other Applications

There are a few other applications that are being developed lastly, which do not belong to the previous fields. Ping et al. (2011) verified the suitability of aqueous extracts of grape pomace, in the presence of alkali, as wood adhesives. Bettini et al. (2012) extracted anthocyanins from grape skins and transferred them onto solid substrates in combination with an oligophenylenevinylene derivative; these substrates were then used to develop multilayers colorimetric chemical sensors for the detection of a commercial herbicide. Oligomer was added in the film to promote the stability of the anthocyanin moieties when exposed to UV-Vis radiations and to enhance, therefore, their durability.

Conclusion

Winemaking is generally thought as a high level exploitation of agricultural products. Yet, a large amount of the energy needed for growth and ripening of grapes, in terms of both solar energy and working activities carried out by humans and machineries, is underused or converted into disposable excess, with further costs for recycling. This material includes winemaking waste like stems and grape seeds, but also portions of the wine that do not enter the winemaking process, like stems and vine leaves and shoots.

In summary, winemakers currently discard huge amounts of potentially valuable resources as winemaking residuals. The main traditional use, the exploitation of grape pomaces to produce spirits, is certainly an interesting and noble way to use wastes, but it must be taken into account the fact that in many wine-growing countries the production and commercialization of such beverages has decreased in recent years, due to a combination of changing tastes and drinking habits, and strict environmental legislation for distilleries, that has made all but the highest value products non-viable. For these reasons, alternative uses for WCBP must be foreseen, with an eye to the production of commercially valuable products, and another to the proper management of waste.

The possibilities are several and promising, as the several works cited have demonstrated. WCBP are extremely rich in compounds that could be potentially used on large scale for industrial purposes. The extraction of the polyphenolic fraction would yield high amounts of raw materials that could be marketed in this form or further refined in order to obtain more valuable substances, like pure anthocyanins or proanthocyanidins. Possible users for these products are already well represented in the market, including producers and distributors of nutraceutical, cosmeceutical and pharmaceutical preparations, as well as developers and manufacturers of advanced solar cells and related devices. Besides, such an achievement could also alleviate problems of residual disposal to landfills.

Hence, the exploitation of such a chemical treasure, which is commonly disregarded, could provide an economic boost to the whole industry of winemaking. A market for these products is already available and active in some parts of the world, hence, the planning and development of polyphenols extraction and refining processes stands as a missing link for a quite innovative production chain. It is possible foreseeing wineries selling their WCBP rather than paying others to dispose of it. One industry's trash can become another industry's treasure.

References

ADA, American Dietetic Association (2004). Position of the American Dietetic Association: Functional Foods. *J. Am. Diet. Assoc.* 104, 814-826.

Ahmadian, R., (2011). Estimating the impact of dye concentration on the photoelectron-chemical performance of anthocyanin-sensitized solar cells: a power law model. *J. Photonics Energy* 1, 011123.

Aleixandre, J. L., Aleixandre-Tudó, J. L., Bolaños-Pizzaro, M., Aleixandre-Benavent, R., (2011). Mapping the Scientific Research on Wine and Health (2001–2011). *J. Agr. Food Chem.* DOI: 10.1021/jf404394e.

Alhamed, M., Issa, A. S., Wael Doubal, A., (2012). Studying of natural dyes properties as photo-sensitizer for Dye Sensitized Solar Cells (DSSC). *J. Electron Devices* 16, 1370-1383.

Alonso, A. M., Guillen, D. A., Barroso, C. G., Puertas, B., Garcia, A., (2002). Determination of antioxidant activity of wine byproducts and its correlation with polyphenolic content. *J. Agric. Food Chem.* 50, 5832-5836.

Amerine, A., Joslyn, M. A., (1987). *Composition of grapes and distribution of phenolics from table wines, the technology of their production*. Berkeley, CA: University of California Press.

Andres, L., Riera, F., Alvarez, R., (1997). Recovery and concentration by electrodialysis of tartaric acid from fruit juice industries waste waters. *J. Chem. Technol. Biot.* 70, 247-252.

Arvanitoyannis, I. S., Ladas, D., Mavromatis, A., (2006). Potential uses and applications of treated wine waste: a review. *Int. J. Food Sci. Tech.* 41, 475-487.

Arvanitoyannis, I. S., Ladas, D., Mavromatis, A., (2006). Wine waste treatment methodology. *Int. J. Food Sci. Tech.* 41, 1117-1151.

Aziz, M. H., Reagan-Shaw, S., Wu, J., Longley, B. J., Ahmad, N., (2005). Chemoprevention of skin cancer by grape constituent resveratrol: relevance to human disease? *Faseb J.* 19, 1193-1195.

Badet, C., (2011). Antibacterial activitiy of grape (*Vitis vinifera*, *Vitis rotundifolia*) seeds. In: V. R. Preedy, R. R. Watson and V. B. Patel (Eds.), *Nuts and seeds in health and disease prevention* (pp. 545-552), Amsterdam: The Netherlands, Elsevier.

Bagchi, D., Garg, A., Krohn, R. L., Bagchi, M., Tran, M. X., Stohs, S. J., (1997). Oxygen free radical scavenging abilities of vitamins C and E, and A grape seed proanthocyanidin extract in vitro. *Res. Commun. Mol. Path.* 95, 179-189.

Bagchi, D., Bagchi, M., Stohs, S., Ray, S. D., Sen, C. K., Preuss, H. G., (2002). Cellular protection with proanthocyanidins derived from grape seeds. *Ann. N.Y. Acad. Sci.* 957, 260-270.

Balik, J., Kyselakova, M., Vrchotova, N., Triska, J., Kumsta, M., Veverka, J., Hic, P., Totusek, J., Lefnerova, D., (2008). Relations between polyphenols content and antioxidant activity in vine grapes and leaves. *Czech J. Food Sci.* 26, S25-S32.

Banerjee, B., Bagchi, D., (2001). Beneficial effects of a novel IH636 grape seed proanthocyanidin extract in the treatment of chronic pancreatitis. *Digestion* 63, 203-206.

Baumann, L. S., (2007). Less-known botanical cosmeceuticals. *Dermatol. Ther.* 20, 330-342.

Bettini, S., Valli, L., Santino, A, Martinelli, C, Farinola, G. M., Cardone, A, Sgobba, V., Giancane, G., (2012). Spectroscopic investigations, characterization and chemical sensor application of composite Langmuir-Schäfer films of anthocyanins and oligophenylene-vinylene derivatives. *Dyes Pigm.* 94, 156-162.

Bravo, L., Saura-Calixto, F., (1998). Characterization of dietary fiber and the in vitro indigestible fraction of grape pomace. *Am. J. Enol. Viticult.* 49, 135-141.

Burlando, B., Verotta, L., Cornara, L., Bottini-Massa, E., (2010). *Herbal principles in cosmetics: properties and mechanisms of action*. Boca Raton, FL: CRC Press.

Bustamante, M. A., Moral, R., Paredes, C., Perez-Espinosa, A., Moreno-Caselles, J., Perez-Murcia, M. D., (2008). Agrochemical characterisation of the solid by-products and residues from the winery and distillery industry. *Waste Manage.* 28, 372-380.

Bustamante, M. A., Paredes, C., Morales, J., Mayoral, A. M., Moral, R., (2009). Study of the composting process of winery and distillery wastes using multivariate techniques. *Bioresource Technol.* 100, 4766-4772.

Calogero, G., Yum, J. H., Sinopoli, A., Di Marco, G., Grätzel, M., Nazeeruddin, M. K., (2012). Anthocyanins and betalains as light-harvesting pigments for dye-sensitized solar cells. *Sol. Energy* 86, 1563-1575.

Careri, M., Corradini, C., Elviri, L., Nicoletti, I., Zagnoni, I., (2003). Direct HPLC analysis of quercetin and trans-resveratrol in red wine, grape, and winemaking byproducts. *J. Agr. Food Chem.* 2003, 51, 5226-5231.

Careri, M., Corradini, C., Elviri, L., Nicoletti, I., Zagnoni, I., (2004). Liquid Chromatography-Electrospray Tandem Mass Spectrometry of cis-resveratrol and trans-resveratrol: development, validation, and application of the method to red wine, grape, and winemaking byproducts. *J. Agr. Food Chem.* 2004, 52, 6868-6874.

Carini, M., Maffei Facino, R., Aldini, G., Callonii, M. T., Bombardelli, E., Morazzoni, P., (1998). The protection of polyunsaturated fatty acids in micellar systems against UVB-induced photo-oxidation by procyanidins from *Vitis vinifera L.*, and the protective synergy with vitamin E. *Int. J. Cosmetic Sci.* 20, 203-15.

Carrieri, C., Milella, R. A., Incampo, F., Crupi, P., Antonacci, D., Semeraro, N., Colucci, M., (2013). Antithrombotic activity of 12 table grape varieties. Relationship with polyphenolic profile. *Food Chem.* 140, 647-653.

Castellucci, F., Iasiello, B., Magalhaes, V., Aigrain, P., (2013). *OIV Vine and Wine Outlook 2008-2009.* Available from: URL: http://www.oiv.int/oiv/files/OIVStatistiques2008-2009.pdf.

Chan, M. M. Y., (2002). Antimicrobial effect of resveratrol on dermatophytes and bacterial pathogens of the skin. *Biochem. Pharmacol.* 63, 99-104.

Chidambara Murthy, K. N., Singh, R. P., Jayaprakasha, G. K., (2002). Antioxidant activities of grape (*Vitis vinifera*) pomace extracts. *J. Agr. Food Chem.* 50, 5909-5914.

Corona, G., Nicoletti, G., (2010). Renewable energy from the production residues of vineyards and wine: evaluation of a business case. *New Medit.* 9, 41-47.

Costantini, A., De Bernardi, T., Gotti, A., (1999). Clinical and capillaroscopic evaluation of chronic uncomplicated venous insufficiency with procyanidins extracted from *Vitis vinifera*. *Minerva Cardioangiol.* 47, 39-46.

Crews, C., Hough, P., Godward, J., Brereton, P., Lees, M., Guiet, S., Winkelmann, W., (2006). Quantitation of the main constituents of some authentic grape-seed oils of different origin. *J. Agr. Food Chem.* 54, 6261-6265.

Cruz, J. M., Domínguez, H., Parajó, J. C., (2004). Assessment of the production of antioxidants from winemaking waste solids. *J. Agr. Food Chem.* 52, 5612-5620.

Deiana, A. C., Sardella, M. F., Silva, H., Amaya, A., Tancredi, N., (2009). Use of grape stalk, a waste of the viticulture industry, to obtain activated carbon. *J. Hazard. Mater.* 172, 13-19.

Delgado-Torre, M. P., Ferreiro-Vera, C., Priego-Capote, F., Pérez-Juan, P. M., Luque de Castro, M. D., (2012). Comparison of accelerated methods for the extraction of phenolic compounds from different vine-shoot cultivars. *J. Agr. Food Chem.* 60, 3051-3060.

Delgado-Torre, M. P., Priego-Capote, F., Luque de Castro, M. D., (2012). Oenological Purposes by Superheated Liquid Extraction and High-Resolution Liquid Chromatography–Time-of-Flight/Mass Spectrometry Analysis. *J. Agr. Food Chem.* 60, 3409-3417.

Deng, Q., Penner, M. H., Zhao, Y., (2011). Chemical composition of dietary fiber and polyphenols of five different varieties of wine grape pomace skins. *Food Res. Int.* 44, 2712-2720.

De Nisco, M., Manfra, M., Bolognese, A., Sofo, A., Scopa, A., Tenore, G. C., Pagano, F., Milite, C., Russo, M. T., (2013). Nutraceutical properties and polyphenolic profile of berry skin and wine of *Vitis vinifera* L. (cv. Aglianico). *Food Chem.* 140, 623-629.

Devesa-Rey, R., Vecino, X., Varela-Alende, J. L., Barral, M. T., Cruz, J. M., Moldes, A. B., (2011).Valorization of winery waste vs. the costs of not recycling. *Waste Manage.* 31, 2327-2335.

Díaz, B., Gomes, A., Freitas, M., Fernandes, E., Nogueira, D. R., González, J., Moure, A., Levoso, A., Vinardell, M. P., Mitjans, M., Domínguez, H., Parajó, J. C., (2012). Valuable Polyphenolic Antioxidants from Wine Vinasses. *Food Bioprocess Tech.* 5, 2708-2716.

Docherty, J. J., Fu, M. M., Hah, J. M., Sweet, T. J., Faith, S. A., Booth, T., (2004). Effect of resveratrol on herpes simplex virus vaginal infection in the mouse. *Antiviral Res.* 67, 155-162.

Duncan, B. R., Bernhard, B. B., (2010). Mixture composition and method useful for topical and internal application. *US Patent No. 7678393*. Washington, DC: US Patent and Trademark Office.

European Union, (1994). European Parliament and Council Directive No. 94/36/EC on colors for use in foodstuffs. *Official Journal of the European Union* 10/09/1994, L237.

European Union, (2007). Council Regulation No. 1234/2007 establishing a common organisation of agricultural markets and on specific provisions for certain agricultural products (Single CMO Regulation). *Official Journal of the European Union* 16/11/2007, L299, Annex III, Part IIIa.

European Union, (2008). Council Regulation No. 479/2008 on the common organization of the market in wine. *Official Journal of the European Union* 6/6/2008, L148, Annex VI.

Flesch, M., Schwarz, A., Böhm, M., (1998). Effects of red and white wine on endothelium-dependent vasorelaxation of rat aorta and human coronary arteries. *Am. J. Physiol.* 275, H1183-H1190.

Fontana, A. R., Antoniolli, A., Bottini, R., (2013). Grape pomace as a sustainable source of bioactive compounds: extraction, characterization, and biotechnological applications of phenolics. *J. Agr. Food Chem.* 61, 8987-9003.

Fragoso, S., Guasch, J., Aceña, L., Mestres, M., Busto, O., (2011). Prediction of red wine color and phenolic parameters from the analysis of its grape extract. *Int. J. Food Sci. Tech.* 46, 2569-2575.

Fremont, L., (2000). Biological effects of resveratrol. *Life Sci.* 66, 663-673.

Fresco, P., Borges, F., Diniz, C., Marques, M. P., (2006). New insights on the anticancer properties of dietary polyphenols. *Med. Res. Rev.* 26, 747-766.

Fructus, A., (2004). Composition pour des soins de la peau, et contenant un ou des oligomeres de resveratrol, en particulier de l'epsilone-viniferine, et ou certains de leurs derives.

France Patent No. FR 2844715. Paris, France: Institut National de la Propriété Industrielle.

Fumiyiwa, O., Ough, C. S., (1982). Grape pomace: possibilities as animal feed. *Am. J. Enol. Viticult.* 33, 44-46.

Gambacorta, G., Antonacci, D., Pati, S., La Gatta, M., Faccia, M., Coletta, A., La Notte, E., (2011). Influence of winemaking technologies on phenolic composition of Italian red wines. *Eur. Food Res. Technol.* 233, 1057-1066.

Garrido, J., Borges, F., (2013). Wine and grape polyphenols - A chemical perspective. *Food Res. Int.* DOI: 10.1016/j.foodres.2013.08.002.

Goñi, I., Sayago-Ayerdi, S. G., Brenes, A., Viveros, A., (2009). Antioxidative effect of dietary grape pomace concentrate on lipid oxidation of chilled and long-term frozen stored chicken patties. *Meat Sci.* 83, 528-533.

González-Paramás, A. M., Esteban-Ruano, S., Santos-Buelga, C., de Pascual-Teresa, S., Rivas-Gonzalo, J. C., (2004). Flavanol content and antioxidant activity in winery by-products. *J. Agr. Food Chem.* 52, 234-238.

Guilford, J. M., Pezzuto, J. M., (2011). Wine and health: a review. *Am. J. Enol. Viticult.* 62, 471-486.

Hughes-Formella B., Wunderlich, O., Williams, R., (2007). Anti-inflammatory and skin-hydrating properties of a dietary supplement and topical formulations containing oligomeric proanthocyanidins. *Skin Pharmacol. Physiol.* 20, 43-49.

Iosub, I., Kajzar, F., Makowska-Janusik, M., Meghea, A., Tane, A., Rau, I., (2012). Electronic structure and optical properties of some anthocyanins extracted from grapes. *Opt. Mater.* 34, 1644-1650.

Jackson, R. S., (1994). *Wine science principles and applications*. San Diego, CA: Academic Press.

Jasim, K. E., (2011). Dye Sensitized Solar Cells - working principles, challenges and opportunities. In: L. A. Kosyachenko (Ed.), *Solar Cells - Dye-Sensitized Devices*. Rijeka, Croatia: InTech, pp. 171-204.

Jiménez, L., Angulo, V., Ramos, E., De la Torre, M. J., Ferrer, J. L., (2006). Comparison of various pulping processes for producing pulp from vine shoots. *Ind. Crop. Prod.* 23, 122-130.

Kalfin, R., Righi, A., Del Rosso, A., Bagchi, D., Generini, S., Guiducci, S., Matucci Cerinic, M., Das, D. K., (2002). Activin, a grape seed-derived proanthocyanidin extract, reduces plasma levels of oxidative stress and adhesion molecules (ICAM-1, VCAM-1 and E-selectin) in systemic sclerosis. *Free Radical Res.* 36, 819-825.

Kammerer, D., Claus, A., Carle, R., Schieber, A., (2004). Polyphenol screening of pomace from red and white grape varieties (*Vitis vinifera* L.) by HPLC-DAD-MS/MS. *J. Agr. Food Chem.* 52, 4360-4367.

Kennedy, J. A., (2008). Grape and wine phenolics: Observations and recent findings. *Cienc. Investig. Agrar.* 35, 107-120.

Khanna, S., Venojarvi, M., Roy, S., Sharma, N., Trikha, P., Bagchi, D., Bagchi, D., Sen, C. K., (2002). Dermal wound healing properties of redox-active grape seed proanthocyanidins. *Free Radical Bio. Med.* 33, 1089-1096.

Kim, Y. J., Kim, Y. A., Yokozawa, T., (2010). Attenuation of oxidative stress and inflammation by gravinol in high glucose-exposed renal tubular epithelial cells. *Toxicology* 270, 106-111.

Kolb, C. A., Kaser, M. A., Kopecky, J., Zotz, G., Riederer, M., Pfundel, E. E., (2001). Effects of natural intensities of visible and ultraviolet radiation on epidermal ultraviolet screening and photosynthesis in grape leaves. *Plant Physiol.* 127, 863-875.

Kähkönen, M. P., Hopia, A. I., Heinonen, M., (2001). Berry phenolics and their antioxidant activity. *J. Agric. Food Chem.* 49, 4076-4082.

Lacatusu, I., Badea, N., Bojin, D., Iosub, S., Meghea, A., (2009). Novel fluorescence nanostructured materials obtained by entrapment of an ornamental bush extract in hybrid silica glass. *J. Sol-Gel Sci. Technol.* 51, 84-91.

Lafka, T. I., Sinanoglou, V., Lazos, E. S., (2007). On the extraction and antioxidant activity of phenolic compounds from winery wastes. *Food Chem.* 104, 1206-1214.

Larrauri, J. A., Ruperez, P., Saura-Calixto, F., (1996). Antioxidant activity of wine pomace. *Am. J. Enol. Viticult.* 47, 369-372.

Larrauri, J. A., Ruperez, P., Saura-Calixto, F., (1997). Effect of drying temperature on the stability of polyphenols and antioxidant activity of red grape pomace peels. *J. Agr. Food Chem.* 45, 1390-1393.

Li, Y., Han, L., Ma, R., Xu, X., Zhao, C., Wang, Z., Chen, F., Hu, X., (2012). Effect of energy density and citric acid concentration on anthocyanins yield and solution temperature of grape peel in microwave-assisted extraction process. *J. Food Eng.* 109, 274-280.

Lis-Balchin, M., (1999). Parallel placebo-controlled clinical study of a mixture of herbs sold as a remedy for cellulite. *Phytother. Res.* 13, 627-629.

Llobera, A., Cañellas, J., (2007). Dietary fibre content and antioxidant activity of Manto Negro red grape (*Vitis vinifera*): pomace and stem. *Food Chem.* 101(2), 659-666.

Maier, T., Schieber, A., Kammerer, D. R., Carle, R., (2009). Residues of grape (*Vitis vinifera L.*) seed oil production as a valuable source of phenolic antioxidants. *Food Chem.* 112, 551-559.

Makris, D. P., Boskou, G., Andrikopoulos, N. K., Kefalas, P., (2007). Characterisation of certain major polyphenolic antioxidants in grape (*Vitis vinifera* cv. Roditis) stems by liquid chromatography-mass spectrometry. *Eur. Food Res. Technol.* 226, 1075-1079.

Marakakis, P., (1982). *Anthocyanins as food colors*. New York, NY: Academic Press.

Masquelier, J., (1987). Plant extract with a procyanidins content as a therapeutic agent having radical scavenger effect and use thereof. *US Patent No. 4698360*. Washington, DC: US Patent and Trademark Office.

Mateos-Martína, M. L., Pérez-Jiménez, J., Fuguet, E., Torres, J. L., (2011). Non-extractable proanthocyanidins from grapes are a source of bioavailable (epi)catechin and derived metabolites in rats. *Brit. J. Nutr.* 108, 290-297.

Mattivi, F., Zulian, C., Nicolini, G., Valenti, L., (2002). Wine, biodiversity, technology and antioxidants. In: D. K. Das and F. Ursini (Eds.), *Ann. N.Y. Acad. Sci.* 957, 37-56.

Mattivi, F., Vrhovsek, U., Masuero, D., Trainotti, D., (2008). Differences in the amount and structure of extractable skin and seed tannins amongst red grape varieties. *Aust. J. Grape Wine Res.*15, 27-35.

Mazza, G., Miniati, E., (1993). *Anthocyanins in fruits, vegetables, and grains*. Boca Raton, FL: CRC Press.

Mazza, G., (1995). Anthocyanins in grapes and grape products. *CRC Crit. Rev. Food Sci.* 35, 341-371.

Monagas, M., Quintanilla-López, J. E., Gómez-Cordovés, C., Bartolomé, B., Lebrón-Aguilar, R., (2010). MALDI-TOF MS analysis of plant proanthocyanidins. *J. Pharm. Biomed. Anal.* 51, 358-372.

Moure Varela, A., Domínguez González, H., Parajó Liñares, J. C., González Muñoz, M. J., Díaz Reinoso, B., Conde Piñeiro, E., Conde Piñeiro, M. J., González López, N., Levoso Touceda, A., Castro Gonzalez, M., Vidal Canto, E., González Mendez, J., (2010). Procedimiento de recuperación de compuestos antioxidantes presentes en efluentes de destileria de vino. *Spain Patent No. ES2345748*. Madrid: Spain.

Narayan, M. R., (2012). Review: Dye sensitized solar cells based on natural photosensitizers. *Renew. Sust. Energ. Rev.* 16, 208-215.

Negro, C., Tommasi, L., Miceli, A., (2003). Phenolic compounds and antioxidant activity from red grape marc extracts. *Bioresour. Technol.* 87, 41-44.

Oreopoulou, V., Russ, W., (2007). Utilization of by-products and treatment of waste in the food industry. *ISEKI-Food series*. New York, NY: Springer.

Ozkan, G., Sagdic, O., Baydar, N. G., Kurumahmutoglu, Z., (2004). Antibacterial activities and total phenolic contents of grape pomace extracts. *J. Sci. Food Agric.* 84, 1807-1811.

Pastrana-Bonilla, E., Akoh, C. C., Sellappan, S., Krewer, G., (2003). Phenolic content and antioxidant capacity of Muscadine grapes. *J. Agric. Food Chem.* 51, 5497-5508.

Pedroza, M. A., Carmona, M., Salinas, M. R., Zalacain, A., (2011). Use of dehydrated waste grape skins as a natural additive for producing rosé wines: study of extraction conditions and evolution. *J. Agric. Food Chem.* 59, 10976-10986.

Pedroza, M. A., Carmona, M., Alonso, G. L., Salinas, M. R., Zalacain, A., (2013). Pre-bottling use of dehydrated waste grape skins to improve color, phenolic and aroma composition of red wines. *Food Chem.* 136, 224-236.

Peralbo-Molina, A., Priego-Capote, F., Luque de Castro, M. D., (2012). Comparison of extraction methods for exploitation of grape skins residues from ethanol distillation. *Talanta* 101, 292-298.

Peralbo-Molina, A., Priego-Capote, F., Luque de Castro, M. D., (2013). Characterization of grape seed residues from the ethanol-distillation industry. *Anal. Methods* 5, 1922-1929.

Peralbo-Molina, A., Luque de Castro, M. D., (2013). Potential of residues from the Mediterranean agriculture and agrifood industry. *Trends Food Sci. Tech.* 32, 16-24.

Pérez-Jiménez, J., Arranza, S., Saura-Calixto, F., (2009). Proanthocyanidin content in foods is largely underestimated in the literature data: An approach to quantification of the missing proanthocyanidins. *Food Res. Int.* 42, 1381-1388.

Pérez-Serradilla, J. A., Luque de Castro, M. D., (2008). Role of lees in wine production: a review. *Food Chem.* 111, 447-456.

Pérez-Serradilla, J. A., Luque de Castro, M. D., (2011). Microwave-assisted extraction of phenolic compounds from wine lees and spray-drying of the extract. *Food Chem.* 124, 1652-1659.

Perlman, D., Ramonas, L., (2008). Astringency-compensated polyphenolic antioxidant-containing comestible composition. *US Patent Application No. 20080044539*. Washington, DC: US Patent and Trademark Office.

Pezet, R., Perret, C., Jean-Denis, J. B., Tabacchi, R., Gindro, K., Viret, O., (2003). δ-Viniferin, a resveratrol dehydrodimer: one of the major stilbenes synthesized by stressed grapevine leaves. *J. Agric. Food Chem.* 51, 5488-5492.

Pinelo, M., Rubilar, M., Jerez, M., Sineiro, J., Núñez, M. J., (2005). Effect of solvent, temperature, and solvent-to-solid ratio on the total phenolic content and antiradical activity of extracts from different compounds of grape pomace. *J. Agric. Food Chem.* 53, 2111-2117.

Pinelo, M., Arnous, A., Meyer, A. S., (2006). Upgrading of grape skins: significance of plant cell-wall structural components and extraction techniques for phenol release. *Trends Food Sci. Tech.* 17, 579-590.

Ping, L., Pizzi, A., Guo, Z. D., Brosse, N., (2011). Condensed tannins extraction from grape pomace: Characterization and utilization as wood adhesives for wood particleboard. *Ind. Crop. Prod.* 34, 907-914.

Püssa, T., Floren, J., Kuldkepp, P., Raal, A., (2006). Survey of grapevine *Vitis vinifera* stem polyphenols by Liquid Chromatography-Diode Array Detection-Tandem Mass Spectrometry. *J. Agric. Food Chem.* 54, 7488-7494.

Ratnasooriya, C. C., Rupasinghe, H. P. V., Jamieson, A., (2010). Juice quality and polyphenol concentration of fresh fruits and pomace of selected Nova Scotia-grown grape cultivars. *Can. J. Plant Sci.* 90, 193-205.

Regione Piemonte (2011). Progetto "FAR FOS" Farine Fossili.

Repubblica Italiana, Ministro delle politiche agricole alimentari e forestali. (2010). Modifiche all'articolo 5 del decreto ministeriale 27/11/2008, n. 5396, concernente "Disposizioni di attuazione dei regolamenti (CE) n. 479/2008 e n. 555/2008 per quanto riguarda l'applicazione della misura della distillazione dei sottoprodotti della vinificazione". *Gazzetta Ufficiale*, Serie generale 212 del 10 Settembre 2010. Roma: Istituto Poligrafico e Zecca dello Stato.

Rice, A. C., (1976). Solid waste generation and by-product recovery potential from winery residues. *Am. J. Enol. Vitic.* 27, 21-26.

Robinson, J., (2006). *The Oxford companion to wine*. New York, NY: Oxford University Press.

Rosales Soto, M. U., Brown, K., Ross, C. F., (2012). Antioxidant activity and consumer acceptance of grape seed flour-containing food products. *Int. J. Food Sci. Tech.* 47, 592-602.

Rubio, R., Perez-Murcia, M. D., Agullo, E., Bustamante, M. A., Sanchez, C., Paredes, C., Moral, R., (2013). Recycling of agro-food wastes into vineyards by composting: agronomic validation in field conditions. *Commun. Soil Sci. Plant Anal.* 44, 502-516.

Sánchez-Alonso, I., Jiménez-Escrig, A., Saura-Calixto, F., Borderías, A. J., (2007). Effect of grape antioxidant dietary fibre on the prevention of lipid oxidation in minced fish: Evaluation by different methodologies. *Food Chem.* 101, 372-378.

Saura-Calixto, F., (1988) Effect of condensed tannins in the analysis of dietary fiber in carob pods. *J. Food Sci.* 53, 1769-1771.

Saura-Calixto, F., (1998). Antioxidant dietary fiber product: a new concept and a potential food ingredient. *J. Agric. Food Chem.* 46, 4303-4306.

Saura-Calixto, F., Serrano, J., Goñi, I., (2007). Intake and bioaccessibility of total polyphenols in a whole diet. *Food Chem.* 101, 492-501.

Saura-Calixto, F., (2012). Concept and health-related properties of nonextractable polyphenols: the missing dietary polyphenols. *J. Agric. Food Chem.* 60, 11195-11200.

Schieber, A., Stintzing, F. C., Carle, R., (2001). By-products of plant food processing as a source of functional compounds - recent developments. *Trends Food Sci. Tech.* 12, 401-413.

Schneider, E., von der Heydt, H., Esperester, A., (2008). Evaluation of Polyphenol Composition in Red Leaves from Different Varieties of *Vitis vinifera*. *Planta Med.* 74, 565-572.

Schoedl, K., Forneck, A., Sulyok, M., Schuhmacher, R., (2011). Optimization, in-house validation and application of a liquid chromatography-tandem mass spectrometry (LC-MS/MS)-based method for the quantification of selected polyphenolic compounds in leaves of grapevine (*Vitis vinifera* L.). *J. Agric. Food Chem.* 59, 10787-10794.

Schoedl, K., Schuhmacher, R., Forneck, A., (2012). Studying the polyphenols of grapevine leaves according to age and insertion level under controlled conditions. *Sci. Hortic.* 141, 37-41.

Sessa, M., Tsao, R., Liu, R. H., Ferrari, G., Donsi, F., (2011). Evaluation of the stability and antioxidant activity of nanoencapsulated resveratrol during in vitro digestion. *J. Agric. Food Chem.* 59, 12352-12360.

Sessa, M., Casazza, A., Perego, P., Tsao, R., Ferrari, G., Donsi, F., (2013). Exploitation of Polyphenolic Extracts from Grape Marc as Natural Antioxidants by Encapsulation in Lipid-Based Nanodelivery Systems. *Food Bioprocess Tech.* 6, 2609-2620.

Shrikhande, A. J., (2000). Wine by-products with health benefits. *Food Res. Int.* 33, 469-474.

Shrikhande, A. J., Wang, H., Kupina, S. A., (2007). Grape extract, dietary supplement thereof, and processes therefor. *US Patent No. 20070071871*. Washington, DC: US Patent and Trademark Office.

Singleton, V. L., Esau, P., (1969). *Phenolic substances in grapes and wines and their significance*. New York, NY: Academic Press.

Smagge, F., Mourgues, J., Escudier, J., Conte, T., Molinier, J., Malmary, C., (1992). Recovery of calcium tartrate and calcium malate in effluents from grape sugar production by electrodialysis. *Bioresour. Technol.* 39, 85-189.

Smith, R. L., Cohen, S. M., Doull, J., Feron, V. J., Goodman, J. I., Marnett, L. J., Portoghese, P. S., Waddell, W. J., Wagner, B. M., Adams, T. B., (2003). GRAS Flavoring Substances. *Food Technol.* 57, 46-59.

Soleas, G. J., Grass, L., Josephy, P. D., Goldberg, D. M., Diamandis, E. P., (2006). A comparison of the anticarcinogenic properties of four red wine polyphenols. *Clin. Biochem.* 39, 492-497.

Souquet, J. M., Labarbe, B., Le Guernevé, C., Cheynier, V., Moutounet, M., (2000). Phenolic composition of grape stems. *J. Agric. Food Chem.* 48, 1076-1080.

Su, X. W., D'Souza, D. H., (2011). Grape seed extract for control of human enteric viruses. *Appl. Environ. Microbiol.* 77, 3982-3987.

Thimothe, J., Bonsi, I. A., Padilla-Zakour, O. I., Koo, H., (2007). Chemical characterization of red wine grape (*Vitis vinifera* and *Vitis* interspecific hybrids) and pomace phenolic extracts and their biological activity against *Streptococcus mutans*. *J. Agric. Food Chem.* 55, 10200-10207.

Torres, J. L., Bobet, R., (2001). New Flavanol Derivatives from Grape (*Vitis vinifera*) Byproducts. Antioxidant Aminoethylthio-Flavan-3-ol Conjugates from a Polymeric Waste Fraction Used as a Source of Flavanols. *J. Agric. Food Chem.* 49, 4627-4634.

Tseng, A., Zhao, Y., (2013). Wine grape pomace as antioxidant dietary fibre for enhancing nutritional value and improving storability of yogurt and salad dressing. *Food Chem.* 138, 356-365.

Vinson, J. A., Mandarano, M. A., Shuta, D. L., Bagchi, M., Bagchi, D., (2002). Beneficial effects of a novel IH636 grape seed proanthocyanidin extract and a niacin-bound chromium in a hamster atherosclerosis model. *Mol. Cell. Biochem.* 240, 99-103.

Waterhouse, A. L., Ignelzi, S., Shirley, J. R., (2000). A comparison of methods for quantifying oligomeric proanthocyanidins from grape seed extracts. *Am. J. Enol. Vitic.* 51, 383-389.

Wild, H. P., Sass, M., (2006). Concentrate comprising green tea, grape skin extract and grape extract, the production thereof and use of the same. *US Patent No. 7087259*. Washington, DC: US Patent and Trademark Office.

Xia, E., Deng, G., Guo, Y., Li, H., (2010). Biological activities of polyphenols from grapes, *Int. J. Mol. Sci.* 11, 622-646.

Yamakoshi, J., Saito, M., Kataoka, S., Kikuchi, M., (2002). Safety evaluation of proanthocyanidin-rich extract from grape seeds. *Food Chem. Toxicol.* 40, 599-607.

Yamakoshi, J., Otsuka, F., Sano, A., Tokutake, S., Saito, M., Kikuchi, M., Kubota, Y., (2003). Lightening effect on ultraviolet-induced pigmentation of guinea pig skin by oral administration of a proanthocyanidin-rich extract from grape seeds. *Pigm. Cell Res.* 16, 629-638.

Yan, K. X., Terashima, K., Takaya, Y., Niwa, M., (2001). A novel oligostilbene named (+)-viniferol A from the stem of *Vitis vinifera* 'Kyohou'. *Tetrahedron* 57, 2711-2715.

Yan, K. X., Terashima, K., Takaya, Y., Niwa, M., (2002). Two new stilbenetetramers from the stem of *Vitis vinifera* 'Kyohou'. *Tetrahedron* 58, 6931-6935.

Yu, J., Ahmedna, M., (2012). Functional components of grape pomace: their composition, biological properties and potential applications. *Int. J. Food Sci. Tech.* 48, 221-237.

Zhao, Y., (2007). Utilization of berry processing by-products. In: Y. Zhao (Ed.), *Berry fruit value-added products for health promotion* (pp. 387-410). Boca Raton: FL, CRC Press.

Zoecklein, B. W., Fugelsang, K. C., Gump, B. H., Nury, F. S., (1995). *Wine analysis and production*. New York, NY: ChapmanHall.

In: Wine
Editor: Youssef El Rayess

ISBN: 978-1-63321-048-6

Chapter 12

EXTRACTION AND PURIFICATION OF PHENOLIC COMPOUNDS FROM BY-PRODUCTS OF THE WINEMAKING PROCESS

Sami Yammine[1,*], Rémy Ghidossi[1] and Martine Mietton-Peuchot[1]
[1]Université de Bordeaux, USC 1219 Oenologie/INRA/IPB/UB,
EA 4577, Unité de recherche Oenologie, ISVV,
Villenave d'Ornon, France

ABSTRACT

Several million tons of agri-food solid waste is produced annually and disposed through various methods leading to serious environmental problems. One of the most sought-after ingredients in by-products is the family of polyphenols. Mainly, it is the process of extraction that influences the yield of polyphenols compounds derived from grape by-products. As different compounds have varying levels of solubility in different solvents, there is no uniform or complete procedure that is suitable for the extraction of all polyphenols. Solubility of the compounds is highly dependent on the polarity of the solvent, with much higher extraction yields possible in less polar solvents, such as methanol and ethanol.

Up to now, numerous methods of extraction have been developed with the objective of obtaining bioactive compounds from natural sources with higher yields and lower costs. Such is the case of extraction with organic solvents, like methanol, ethanol and acetone.

Also these traditional extraction methods were fully studied, having several drawbacks, mainly being toxic, flammable and/or expensive. New approaches such as pressurized liquid extraction, ultrasound-assisted extraction, microwaves assisted solvent extraction, supercritical or subcritical fluid extraction, pulse-electric fields (PEF) and high voltage electrical discharges (HVED) have been proposed as alternatives to the conventional procedure. On the other hand, further separation of individual components from grape crude extracts is a field that has not received much attention, in particular concerning its application to large-scale, industrial purposes. This book chapter provides

[*] E-mail: sami.yammine@u-bordeaux.fr.

an updated and comprehensive overview on phenolic extraction and purification methods, varying from extraction and purification parameters, recent work, and future trends.

Keywords: Polyphenol, extraction, purification, wine by-products

INTRODUCTION

Worldwide, several million tons of agri-food solid waste is produced annually and disposed through various methods, including incineration, land applications, and landfills. Although the wastes are mainly composed of biodegradable constituents, their disposal causes serious environmental problems. Many investigations on the re-use of this waste have been aimed at converting the materials into food ingredients, bio-fuels, and other value-added applications.

One of the most sought-after ingredients in by-products is the family of polyphenols. From residual sources they could be used to increase the stability of foods, since they prevent lipid oxidation. Polyphenols protect against oxidative damage in living systems using their capability to scavenge reactive oxygen species. In addition, significant attention has been paid to applications of polyphenols in the food, cosmetic and pharmaceutical industries because of their similar functionality requirements for the products.

It is mainly the process of extraction that influences the yield of polyphenols compounds derived from plant materials. Up to now, numerous methods of extraction have been developed with the objective of obtaining bioactive compounds from natural sources with higher yields and lower costs. Such is the case of extraction with organic solvents, like methanol, ethanol and acetone.

These traditional extraction methods have several drawbacks; they are time consuming, laborious, have low selectivity and/or low extraction yields. Moreover, these traditional techniques employ large amounts of toxic solvents. Although ethanol is classified as a GRAS (Generally-Recognized-As-Safe) solvent, its utilization must be strictly accounted for under legal statues.

In addition, the whole process generally requires further purification of the extracts by solid-phase (i.e., Sep-Pak C18), liquid/liquid extraction (i.e., ethyl acetate), or size exclusion chromatography (i.e., Sephadex LH-20) for their final separation.

As different compounds have varying levels of solubility in different solvents, there is no uniform or complete procedure that is suitable for the extraction of all polyphenols. Solubility of the compounds is highly dependent on the polarity of the solvent, with much higher extraction yields possible in less polar solvents, such as methanol and ethanol. Overall, alcohols have been found to be better solvents at room temperature than water. Since the solvents utilized in the extraction process are usually toxic, flammable and/or expensive, extraction techniques have focused on reducing the use of these solvents.

Approaches such as pressurized liquid extraction, ultrasound-assisted extraction, microwaves assisted solvent extraction, supercritical or subcritical fluid extraction and electrical treatments have been proposed as alternatives to the conventional procedure and will be reviewed in this chapter book.

CONVENTIONAL EXTRACTION TECHNIQUES (SOLID-LIQUID EXTRACTION)

Solvent extractions are the most commonly used procedures to prepare extracts from wine by-products due to their ease of use, efficiency, and wide applicability. The solubility of phenolic compounds is governed by its chemical nature, as well as the polarity of the solvents used. Recovery of the various classes of polyphenols, or of individual phenols varying from simple (e.g., phenolic acids, anthocyanins) to highly polymerized substances (e.g., Tanins), has been studied thoroughly. Grape pomaces, or seeds alone, have been extracted using a wide variety of solvents, such as methanol, ethanol, acetone, water and ethyl acetate (Table 1). Diverse results were obtained depending on the parameter of interest, i.e., total polyphenols (TP), total yield of extracted material, total antioxidant activity, etc. It is generally known that the yield of chemical extraction depends on the type of solvents and their pH, extraction time and temperature, solid to-solvent ratio as well as on the chemical composition and physical characteristics of the by-product. The type of solvent has been the most investigated factor. Ethyl acetate was reported as one of the best solvents for the extraction of polyphenols from grape seed (Pekic et al., 1998). On the other hand, alcoholic solvents have been commonly employed to extract phenolic substances from natural sources. Alcoholic solvents give quite a high yield of total extract even if they are not highly selective for different types of phenols. As seen in the table 1 acetone, when used alone, gives the highest yields among studied solvents in polyphenols extraction (Yalimaz et al., 2006).

However, mixtures of alcohols and water have revealed to be more efficient in extracting phenolic constituents than the corresponding mono-component solvent system. By increasing the amount of solvent, more specifically ethanol in water, this has lead to an increased total polyphenol extraction (Spigno et al., 2007). The optimum extraction was reached through the use of an ethanol/water solution having a percentage of ethanol between 50 and 70% (Yalimaz et al., 2006; Bozan et al., 2008). Later studies confirmed the results and specified a 66% ethanol solution (Rajha et al. 2014).

The acidity of the extraction solvent plays an important role. Increasing the acidity of the extraction solvent, lead to the denaturation of the cell membrane, which simultaneously dissolves anthocyanin, and stabilizes them (Ju and Howard 2003). To obtain the best yields of anthocyanin extraction, weak organic acid, or low concentration strong acid have been used. An acidic pH, between 1.0 and 2.5, increases both total yields and phenolic contents. For example 0.1% HCl, 1% acetic acid, and tartaric acid have been often used in the solid-liquid extraction process (Makris et al. 2007, Hashim et al. 2013). In addition, sulfured water has also been proposed as an extraction solvent seeking a reduction of the use of organic solvents (Ju and Howard 2003). However, excessive acidity of the medium leads to hydroxylation of labile, acyl, and sugar residues during the concentration step (Revilla et al. 1998).

The recovery of phenolic compounds is also influenced by time and temperature. An increase in temperature can cause a higher solubility yet this is accompanied by degradation of polyphenols due to their oxidation. For example several studies have shown an increase of total polyphenol yield at 50°C in comparison to room temperature extraction (Guerrero et al. 2008, Bozan et al. 2008,). However, an increase in the temperature should be coupled by shorter extraction times, lower than 8h, to avoid polyphenols degradation and/or polymerization.

Table 1. Comparison of several Solid-liquid extraction works, modified version from Spigno et al. 2007

Raw material	*Extraction procedure*	*Polyphenols yield (%) Weight % of the initial material*	*Extract purity (%)*	*References*
Fresh red grape skins	Skins of 100 grapes/60ml solvent: Methanol 16h/–25 C+80% Methanol 4h/Troom+ 50% Methanol 4h/Troom+ deionised water 16h/–25 C + 75% acetone 1h/Troom	(GAE) 0.25		Revilla et al. (1998)
	Methanol/12 HCl (99:1) 4h/25 C + MetOH/12N HCl (99:1) 20h/25 C + Methanol/12 N HCl (99:1) 4 h/25 C	0.24		
Fresh grapes	100 grapes/60ml solvent (as above)	0.43–0.51		
White grape seeds	10g, 8h with 40ml Ethanol:water (9:1), T NR	(Catechins) 0.5	47	Pekić et al. (1998)
Red grape marc	250g (50g d.m.)/1L Ethanol:water (1:1) 5–30′,T (NA)	0.11 (HPLC)	0.011	Bonilla et al. (1999)
Fresh grape seeds	240 mg crushed seeds/10ml Methanol:water (4:1), 16 T room 1h T<30 C, with ultrasonication 30mg and near critical CO2	(Catechins) 0.078		Palma and Taylor (1999)
Red grape seeds	50g degreased powder, 5h at 60/70 C, twice 150ml	(Catechins)		Jayaprakasha et al. (2001)
	Ethanol:water (9:1)	0.86	43	
	Ethanol:water (17:3)	1.35	54	
	Ethanol:water (4:1)	1.56	50.5	
Grape seeds	100g powder with 100ml 100% EtOH, 24h	(tannins) 0.53		Ahn et al. (2002)
Red grape seeds	100g degreased powder, 8h with 150ml hot (T NR) Acetone:water:acetic acid (90:9.5:0.5)	(Catechins) 2.5	46	Jayaprakasha et al. (2003)

Raw material	*Extraction procedure*	*Polyphenols yield (%)*	*Extract purity (%)*	*References*
Red grape marc	1g/3mL 80% EtOH with 0.5% 0.1N HCl, Temprature and time (NA)	(GAE) 0.59	0.14	Negro et al. (2003)
Seeds from marc		2.4	0.28	
Skins from marc		0.37	0.11	
Fresh grape seeds	100g degreased powder/200ml 8h:	(GAE)		Baydar et al. (2004)
	Acetone:water:acetic acid (90:9.5:0.5),	9.6	62.8	
	Ethanol:Methanol:water (60:30:10)	10.8	66.8	
Grape marc	5g freeze dried+NRml MetOH 75%, 15′ ultrasonic bath (TNR), twice. Liquid concentrated, water dissolved and	0.08 – 0.14	flavan-3-ols	Gonzáles-Paramás et al. (2004)
Distilled marc	extracted 3 times with EA 2g freeze-dried for seeds	0.06		
Grape seeds		0.15		
Distilled seeds		0.15–0.18		
Distilled grape marc	1 kg washed and dried/8 kg solvent in autoclave:	(GAE)		Cruz et al. (2004)
	Water, 4% H2SO4, 130 C, 90 min	12.5		
	Water, 100 C 5h	13.3		
	Water, 130 C, 90 min	18.3		
Grape seeds	90 min, 50°C, Methanol	0.08 (GAE)		Pinelo et al. (2005)
Grape stems	Solvent to Solid ratio 1:1	0.04		
Grape peels		0.98		
Grape marc	50g/100ml water or 70% Ethanol or 70% Methanol, 1–12–24h, Troom	(GAE) 0.035–1.36		Lapornik et al. (2005)
Grape seeds	Milled grape seed, (0.16 - 0.125) mm, 50% Ethanol, 50 C, 40 min	(GAE)		Bucic´-Kojic´et al. (2006)
		6.6		
Grape seed	Seed powder, solvent at a ratio of 1:10 (w/v), Sonicated for 15 min, Temp. (NA)	(GAE)		Yalimaz et al. (2006)
	Ethanol 60%	3		
	Methanol 60%	2.82		
	Acetone 50%	4		

Table 1. (Continued)

Raw material	*Extraction procedure*	*Polyphenols yield (%)*	*Extract purity (%)*	*References*
Grape seeds Grape stems Grape peels	Ground by-product (solvent-to-solid ratio 5.5%), 700 rpm 15 min, 0.1% HCl in MeOH/acetone/water (6/3/1, v/v/v)	13.76 7.47 0.97		Makaris et al. 2007
Red grape marc Red grape stems	50g powder 60 C, 24h, with 200ml: EtOH EA:water (9:1)	(GAE) 0.27 0.33	 6 22.34	Spigno et al. (2007)
Grape seeds	Freeze-dried 2 g of defatted 50 mL of 0.5% acetic acid in acetone:water (70:30) at 50 C for 2 h	(GAE) 15.4		Bozan et al. (2008)
Distilled grape pomace	Air-dried, 27 g, extracted in a laboratory extractor with continous flow, Water, 4ml/min	(GAE) 1.53		Guerrero et al. (2008)
Grape marc	93 minutes at 94°C and in 66% Ethanol/water solvent	(GAE) 0.5		Rajha et al. (2014)

Since the temperatures are higher than 70°C, they lead to fast anthocyanins degradation (Patras et al., 2010).

Moreover, these traditional techniques consume large solvent, are labor intensive, may have contamination issues and may require an evaporation step prior to analysis to remove excess solvent. Several alternative approaches will be presented in this chapter.

Non-Conventional Extraction Techniques

As the solvents utilized in the extraction process are usually toxic, flammable and/or expensive, alternative extraction techniques have focused on reducing the use of these solvents. Approaches such as pressurized liquid extraction, ultrasound-assisted extraction, microwaves assisted solvent extraction, supercritical or subcritical fluid extraction, electrical treatment, pulse-electric fields (PEF) and high voltage electrical discharges (HVED) have been proposed as alternatives to the conventional procedure. Nowadays, among the extraction methods able to overcome the above-mentioned drawbacks, PEF, HVED and Subcritical Water Extraction (SWE) are the most promising processes.

Supercritical Fluid Extraction (SFE)

By forcing a fluid in a chamber above critical temperature and pressure, it becomes a supercritical fluid. Under these conditions, the fluid gains a density similar to that of a liquid but a viscosity similar to that of a gas and diffusivity between the two states. Therefore, this state is defined as an intermediary between the liquid and the gaseous state, or as a state in which the fluid is compressible (i.e., similar behavior to a gas) even though possessing a density similar of a liquid and, therefore, similar solvating power. These properties provide several advantages over traditional extraction methods. SFE can diffuse easily through solid materials and can therefore give faster extraction yields (Anklam et al., 1998). One of the main characteristics of a supercritical fluid is the possibility of modifying the density of the fluid by changing its pressure and/or its temperature. Since density is directly related to solubility, then by altering the extraction pressure, the solvent strength of the fluid can be modified (Valle and Aguilera, 1999; Raventós et al., 2002). Other advantages, as compared to other extraction techniques, are the higher efficiency of the extraction process (in terms of increasing yields and lower extraction times), and the possibility of direct coupling with analytical chromatography techniques such as gas chromatography (GC) or supercritical fluid chromatography (SFC).

As for the solvents, there is a wide range of compounds that can be used as supercritical fluids; carbon dioxide at pressures higher than 7.4 MPa and temperatures higher than 31 °C is a convenient supercritical (SC) fluid often employed in the extraction of molecules from natural sources. Its use in the extraction of polyphenols from the winemaking industry by-products has been widespread: it has been used both alone and in conjunction with organic solvents (Table 2).

The use of supercritical fluid alone or coupled to an organic solvent was studied using SC extraction applied on grape seeds. The two modalities compared were: one by SC CO_2 alone, and the second by SC CO_2 added with MeOH (5:1). The former extract contained lipophilic molecules as fats and sterols, while the latter was rich in polyphenols (Palma et al. 1999).

Arce et al. (2001) considered the influence of extraction variables: pressure, temperature, and flow-rate on grape marc. The solvent power of SC-CO2 below 27 MPa was too low to dissolve polar compounds, however, increasing the pressure, temperature, and flow rate showed a more significant extraction of polyphenols. The optimal polyphenol extraction was achieved with a value of 35 MPa at 50°C and a flow rate of 4ml/min.

Farìas-Campomanes et al. (2013) have reported a study on the economic feasibility of an industrial plant for polyphenols extraction from grape Bagasse, employing SC CO_2 coupled with 10% EtOH. The main compounds identified in the bagasse extracts were syringic, vanillic, gallic, p-hydroxybenzoic, protocatechuic and p-coumaric acids, as well as quercetin. Rutin and resveratrol were not identified in the extracts.

Table 2. Supercritical liquid extraction parameters, polyphenol yields and extracts purity adapted from DIÄAZ-REINOSO et al., 2006

Raw material	CO_2 SCFE conditions *Pressure (MPa) /Temprature (K) /* Modifier (E, ethanol; M, methanol, % v/v) / Time (h) / flow rate (l/h).	Extraction yield Weight % of the initial material	Reference
Grape marc	35/323/M (5)/0.25/0.24	flavonols (Res, C, EC) phenolic acids (SIA, GeA, SA, SnpA)	Arce et al., 2001
Grape seed	45.4/308.2/M (5:1)/0.25/–	GA/C/EC	Palma et al.,1999
Grape seed	20/313/M (5)/–/30 20/313/M (15)/–/30 30/313/M (5)/–/30 30/313/M (15)/–/30	GA (0.05); PCA (0.12); PCAD (0.20) GA (2.43); PCA (0.75); MGG (0.04); PCAD (0.22); C (0.21); EC (0.05) GA (0.21); PCA (0.12); PCAD (0.21) GA (2.71); PCA (0.77); MGG (0.08); PCAD (0.24); C (0.33); EC (0.05)	Murga et al., 2000
Grape seed	65.4/353/M (30)/1/0.12 65.4/353/M (35)/1/0.12 65.4/353/M (40)/1/0.12	C (60)/EC (59) C (72)/EC (73) C (77)/EC (79)	Hadolin et al., 2004
Grape skin	15/313/E (7.5)/0.75/0.12	Res (35.6–170.6 mg/kg)	Pascual-Martí et al., 2001
Grape bagasse	20/313/E (10)/3/(9.8 ± 0.4 × 10^{-5} kg/s) 35/313/E (10)/3/(9.8 ± 0.4 × 10^{-5} kg/s)	GA (0.448), PCAD (0.091), V (0.43), p-HBA (0.175), SA (0.95), PC (0.07), Q (0.015) GA (0.3), PCAD (0.112), V (0.42), p-HBA (0.143), SA (1.078), PC (0.053), Q (0.027)	Farías-Campomanes et al., 2013

Res, resveratrol; C, catechin; EC, epicatechin; GA, gallic acid; GeA, gentisic acid; SA, syringic acid; SnpA, sinapic acid; PCAD, protocatechualdehyde; PCA, protocatechuic acid; RA, rosmarinic acid; MGG, monogalloyl glucose; V, vanillic acid; p-HBA, p-hydroxybenzoic acid; Q, quercetin; PC, p-Coumaric acid.

From the kinetic parameters and the economic analysis, the best balance between the cost of manufacturing and the extraction yield of the bagasse extract was obtained at conditions of 40°C, 20 MPa and 180 min extraction time. The estimated cost of manufacturing using a 0.5 m^3 SFE plant was US$ 133.16 per kilogram of extract, with an expected phenolics concentration of at least 23 g per kilogram of extract.

However, due to the application of high pressure in these techniques, the requirements of instrumentation are high which in turn raises the cost of these methods on the industrial scale, often outweighing the technical benefits.

SUBCRITICAL WATER EXTRACTION (SWE)

Subcritical water is obtained at temperatures between 100°C and 374°C (the critical temperature) under high pressure (usually from 1 to 6 MPa) to maintain the water in the liquid state (Figure 1).

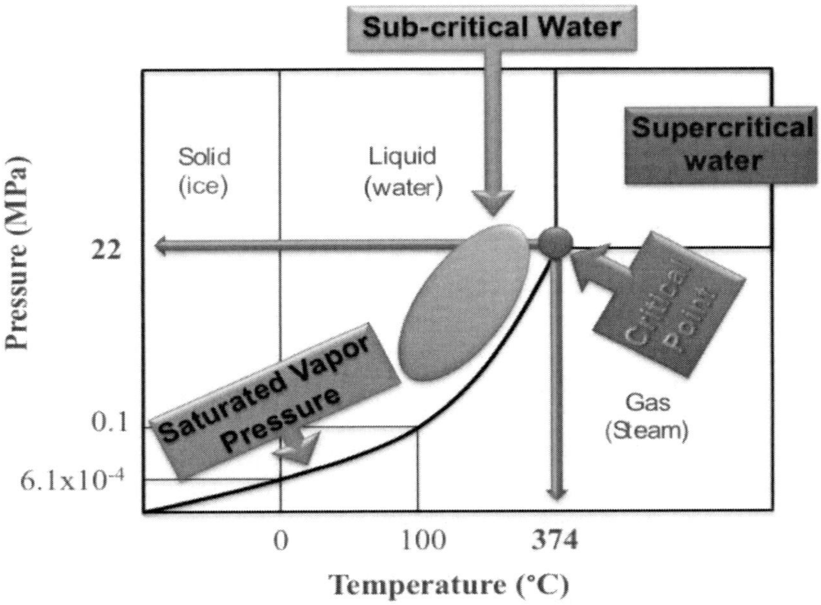

Figure 1. The Phase Diagram of Water.

Superior mass transfer properties of subcritical water lead to high diffusivity and hence higher extraction efficiency. The solubility and diffusivity of water at moderate pressure is, therefore, comparable with organic solvents. The dielectric constant of water at ambient conditions is 80; however, it can be decreased to 56 and 27 (similar to organic solvents) by increasing the pressure to 5 MPa and temperature to 100°C and 250°C, respectively (Y. Yang et al. 1997).

The ability to use water as a solvent, together with the simple experimental equipment required and its potential for on-line coupling with different techniques have boosted the number of applications of the technique. Subcritical water extraction (SWE) has also been widely used to extract different compounds from wine by-product matrices. Studies have been conducted to compare SWE to traditional extraction methods.

García-Marino et al. (2006) have used subcritical water for the extraction of catechins and proanthocyanidins from grape seeds. In their study, sequential extractions were performed and subcritical water was used at 10.3 MPa, within the temperature range of 50–150°C. The recovery of catechin and epicatechin was enhanced two-fold, compared to conventional methanol:water (75:25) extraction using ultrasonic bath.

Luque-Rodríguez et al. (2007) employed superheated ethanol–water mixture for the extraction of anthocyanins and other phenolic compounds from grape pomace. In this study, neutral ethanol:water and acidified ethanol:water with 0.8% (v/v) HCl were used for extraction. The yield of extraction was increased 7 and 12 fold by increasing pressure and temperature from atmospheric conditions to 8 MPa and 120 ◦C when using neutral and acidified solvents, respectively.

Monrad et al. (2010) studied the effect of accelerated solvent extraction on recovery of anthocyanins from grape pomace at different temperatures (40–140 °C) and ethanol to water ratios (10–70%, v/v).

The results of this study showed that the optimum ranges for temperature and ethanol concentration are 80–120°C and 50–70% (v/v) respectively. It is concluded that despite lower extraction yield, using lower concentration of ethanol in hydroethanolic solvent can substantially decrease running cost of the extraction process.

Aliakbarian et al. (2012) studied different extraction temperatures (100-140°C) and pressures (8-15 MPa).

Extractions with significantly higher polyphenols, flavonoids and antioxidant activity were achieved when using subcritical water extraction compared to conventional methods. The optimum extraction conditions and the desirability of model were at 140°C and 11.6 MPa. The pressure has negligible effect on polyphenol extraction in the range of the tested pressures.

Casazza et al. (2010) have made an interesting survey of polyphenols extraction from grape wastes employing MeOH or EtOH, both at ambient pressure and at high temperature coupled with high pressure extraction conditions, with and without the aid of ultrasounds or microwaves. The highest TP content was achieved working in the subcritical extraction, and using MeOH as solvent, for a contact time of 90 minutes. In 2012, the same authors have used MeOH at high pressures and temperature and the optimal total polyphenol content was obtained at 150°C using an extraction time of 270 min.

Ko et al. (2014) studied the optimal conditions for subcritical water extraction conditions in relation with various flavonoids, and the influence of extraction parameters on the chemical structure. At high temperature (250°C) during 30 minutes, the majority of flavonoids were degraded.

Similar to supercritical extraction, a major disadvantage of subcritical water extraction is the high operating pressure, which requires expensive equipment. Thus, the cost of the process is relatively high, but the high purity of extracts and the efficiency of the process may lead to the future development on the industrial scale. Also, the possibility of fine-tuning the selectivity of polyphenol extraction through a small change in water temperature thus varying the dielectric constant is another advantage of subcrtical water extraction (Smith, 2002; Ramos et al., 2002).

PULSED ELECTRICAL FIELD / HIGH VOLTAGE ELECTRICAL DISCHARGE

Pulsed Electric Fields (PEF) treatment is an innovative and promising method for non-thermal process and a good alternative to conventional cell membrane permeabilization techniques. During PEF treatment, vegetal tissue is placed between two electrodes and an electric field of moderate intensity (0.1-10KV/cm) is applied in form of repetitive very short voltage pulses (from 10µs up to 1ms). The duration or/and number of pulses is limited in order to exclude any important rise in the temperature (3-5°C).

PEF induces a transmembrane potential difference across the cell membrane. When the potential difference reaches a critical value, called the breakdown potential, localized electrical breakdown of the membrane occurs and cell permeability increases. The effectiveness of the PEF treatment depends not just on the value of electric field strength (E) and total PEF duration (Tpef), but also on other parameters like impulse length, form, and frequency. Hence, the permeabilization of plant cell membrane may significantly improve the mass transfer in subsequent processes such as expression, dehydration and rehydration, osmotic dehydration or extraction of plant metabolites. The potential of PEF technology has been demonstrated to improve mass transfer while extracting different compounds from grape pomace. In high voltage electric discharge (HVED) technology, the energy (several tens of kilojoules) is used to induce the electrical breakdown of water. The discharge phenomenon is accompanied by the formation of gas bubbles and high-amplitude pressure shock waves (up to 10 kbar) that are responsible for fragmentation of particles. Thus, this technology allows the recovery of intracellular compounds (Boussetta et al., 2009a).

Recent experiments have confirmed the positive effect of PEF on polyphenols extraction improvement from grape seeds (Liu et al., 2011). Boussetta et al. (2009b) have investigated the effects of PEF on the efficiency of aqueous extraction of total soluble matter and polyphenols from grape skins of Chardonnay grapes. The evolution of the total polyphenols was increased by 5% in PEF-treated samples (E = 1300 V/cm, $Tpef$= 0.1s). After the aqueous extraction at ambient temperature the quantity of extracted polyphenols from PEF treated samples was increased by about 11%. However, the most effective in terms of extraction enhancement was the application of HVED at 40 kV/cm which increased the polyphenol extraction by 33%.

ULTRASOUND ASSISTED EXTRACTION (UAE)

Ultrasound is acoustic (sound) energy in the form of waves having a frequency above the human hearing range. The highest frequency that the human ear can detect is approximately 20 thousand cycles per second (20 kHz). This is where the sonic range ends, and where the ultrasonic range begins. The enhancement of extraction efficiency of organic compounds by ultrasound is attributed to the phenomenon of cavitation produced in the solvent by the passage of an ultrasonic wave. Cavitation bubbles are produced and compressed during the application of ultrasound. The increase in the pressure and temperature caused by the compression leads to the collapse of the bubble. With the collapse of bubble, a resultant ''shock wave'' passes through the solvent enhancing the mixing. Ultrasound also exerts a mechanical effect, allowing greater penetration of solvent into the sample, by increasing the contact surface area between solid and liquid phase. This effect coupled with the enhanced mass transfer and significant cell disruption of cells via cavitation bubble collapse, increases the release of intracellular components into the medium (Wang and Weller, 2006). Two general designs of ultrasound-assisted extractors can be found: the ultrasonic baths and the closed extractors fitted with an ultrasonic horn transducer. The mechanical effects of ultrasound induce a greater penetration of solvent into cellular materials and improve mass transfer. Ultrasound in extraction can also disrupt biological cell walls, facilitating the release of contents. Therefore, efficient cell disruption and effective mass transfer are cited as two major factors leading to the enhancement of extraction with ultrasonic power (Mason et al., 1996). It is necessary to take into account plant characteristics such as moisture content and particle size, and the used solvent for the extraction in order to obtain an effecient and effective ultrasound assisted extraction. Furthermore, many factors govern the action of ultrasound including frequency, temperature, and sonication time.

Ultrasound assisted extraction (UAE) has been employed for the extraction of polyphenols from grape seeds. The optimization of aqueous EtOH concentration, extraction time and temperature of the process, operating with fixed values of ultrasound frequency and power where studied (Ghafoor et al., 2009).

Food Science Australia has studied the use of high-powered ultrasound assisted systems for extraction of bioactive compounds, from food by-products. Principle targets have been polyphenols and carotenoids. The ultrasound extraction trials have demonstrated improvements in extraction yield ranging between 10 and 35% from red grape marc (Vilkhu et al., 2007). Carrera et al. (2011) showed that ultrasound-assisted extraction allows the quantitative and reproducible extraction of the phenolic compounds (total phenolic, total anthocyanins and condensed tannins) present in grape, in a short time (6 min), employing ethanol–water as extracting solvent. Compared with usual method, ultrasound-assisted extraction produced similar or even higher recoveries.

Ultrasound-assisted extraction is an inexpensive, simple and efficient alternative to conventional extraction techniques. The main benefits of the use of ultrasound in solid–liquid extraction include the increase of extraction yield (up to 35%) and faster kinetics (half the time) (Wu et al., 2001). Ultrasound can also reduce the operating temperature allowing the extraction of thermolabile compounds. Compared with other novel extraction techniques such as microwave-assisted extraction, the ultrasound apparatus is cheaper and easier to operate (Wang and Weller, 2006).

Microwave Assisted Extraction (MAE)

Microwave assisted extraction is more advanced than the traditional solvent extraction method as it heats the matrix internally and externally without a thermal gradient and as such, functional compounds can be extracted efficiently and protectively. MAE works on the principle that moisture inside the cells is heated due to the microwave effect and evaporates producing tremendous pressure on the cell wall (Wang and Weller, 2006). The wall eventually ruptures, facilitating the leaching of the constituents into the surrounding solvent and improving the yield of the phytoconstituents (Mandal et al., 2007). The benefits of MAE are many, including shorter extraction times and dramatically reduced solvent volumes.

According to Ghafoor et al. (2009), MAE extracts possess a substantially higher level of polyphenols (almost double) in comparing to UAE extracts. Also, the times required by the two methods in achieving the same extraction quantities, showed that MAE (4.6 min) has definite advantages over UAE (29 min). On the other hand, Bucić-Kojić et al. (2009) presented divergent results, with traditional methods (solvent extraction) producing a slight increase in total polyphenol content relative to MAE. However, the time required by the solvent extraction technique (200 min) was much higher comparing to that needed by UAE (30 min). This increase in reaction time will lead to concomitant increases in power usage and other ancillary factors which will serve to increase the carbon footprint of this alternative method.

Extract Purification

Further separation of individual components from grape crude extracts is a field that has not received much attention, in particular concerning its application to large-scale, industrial purposes.

Solid-Phase Extraction (SPE)

Solid phase extraction (SPE) is a widely used technique in which the sample matrix passes through a column containing sorbent material (solid phase) on which the analytes are retained. Then, analytes can be selectively removed from the column through the correct choice of wash solvents. Finally, the analytes may be selectively recovered from the column by an elution solvent, resulting in a highly purified extract.

Solid-phase extraction (SPE) has been employed in the concentration of polyphenols, low molecular weight phenols and organic acids from wine and grape extracts (Hashim et al., 2013, Dopico-Garcìa et al., 2007). The separation of nonpolymeric from polymeric phenols in red wine was also possible through the use of SPE (Pinelo et al., 2006). The uses of these purification methods have so far been restricted to analytical purposes only, due to the fact that preparative preparations would be too expensive (given the cost of the stationary phases) at industrial scale.

Counter Current Chromatography / Size Exclusion Chromatography

Countercurrent chromatography is a form of liquid - liquid chromatography that uses two immiscible liquid phases without any solid supporting matrix. The chromatographic process in countercurrent chromatography is comparable to the repeated partitioning of analyte between two immiscible phases by vigorous mixing in a separating funnel. The analytes separate due to their differing affinities for the stationary phase (Berthod, 2002).

Modern commercial counter-current chromatographs have been applied to polyphenols separation. Due to the absence of any solid stationary phase, adsorption losses are minimized and 100% sample recovery is guaranteed.

The liquid nature of the stationary phase means that the compounds retained in this phase can be easily recovered by simply pumping this phase out (Valls et al., 2009). CCC was proposed in the preparative isolation of polyphenols from grape seeds extracts (Delaunay et al., 2002, Köhler et al., 2008).

Resin Adsorption

Similar to solid phase separation, polyphenols in resin adsorption are retained by micro-beads and then specifically eluted using a wash solvent. Shrikhande et al. 2003 in their patent have proposed copolymer of trimethylolpropane trimethacrylate as an adsorbent resin to maximize the concentration and purification of the beneficial polyphenolic substances.

MEMBRANE PROCESSES

A completely different approach consists in using membrane techniques, such as nanofiltration, ultrafiltration, reverse osmosis, and others (Massot et al., 2008). Employing membranes having different structure and pore sizes, it is possible to achieve good recovery of polyphenols from grape wastes, with at the same time, a useful fractionation in chemical classes, according to their molecular masses (Crespo and Brazinha, 2010). This process has the clear advantage of being of low environmental impact, as it utilizes little amounts of solvents, mild temperatures and low pressures. Accordingly, ultrafiltration of the aqueous EtOH extract from grape seeds significantly increased the polyphenols final recovery (Nawaz et al., 2006): this simple method turned out to be low-cost, efficient and environmental friendly.

Few experiments present the use of membrane processing as a method that reduces the use of toxic organic solvents and concentrates the final product. Membrane technologies have been presented as alternative techniques to purify and fractionate extracts from winery wastes with the aim of obtaining proanthocyanic fractions with different degrees of polymerization.

Using different successive cut-off of the membranes, molecules could be separate (Figure 2).

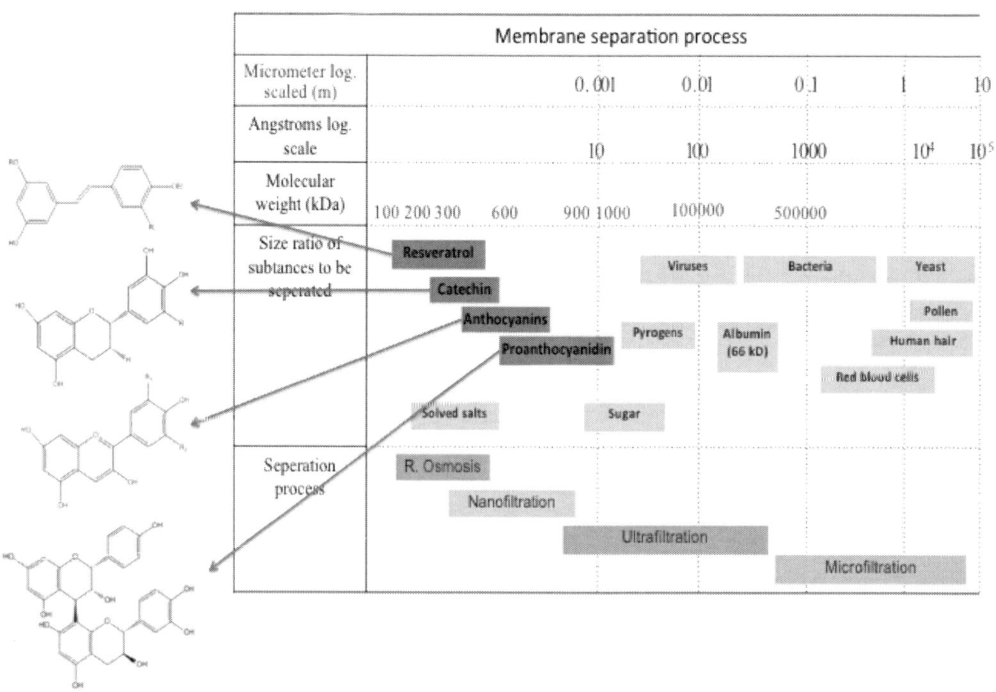

Figure 2. Scheme relating the membrane separation processes with the size of different solutes.

Some results exist in the literature concerning extraction from seeds. With the concentration using of microfiltration (0.22μm), the procedure provided high extraction rates, high extraction selectivity, short extraction time and significant labor savings (Liu, 2011).

CONCLUSION

Phenolic compounds are a much diversified group of phytochemicals that are widely distributed by-product in the wine making chain. Nowadays, there is a growing interest in the high added value substances, which could be used as food components or as specific preventive pharmaceuticals. These properties of polyphenols and their health benefits have intensified research efforts to discover and utilize methods for their extraction and separation.

Subsequently, the extraction and purification of these compounds is difficult task due to the large number of compounds present in grape by-product and the difficulties of extractability.

This chapter presented several conventional solid-liquid extraction conditions of polyphenols from grape seeds and skins. The use of the alternative technologies has a series of advantages: clean, quick, cheap, efficient, they have the possibility of automation. Extract purification is a field that has not received much attention but will in the coming years, as technology and innovation in the field slowly unravels how best to make use of wine by-products.

REFERENCES

Ahn, H. S., Jeon, T. I., Lee, J. Y., Hwang, S. G., Lim, Y., Park, D. K., (2002). Antioxidative activity of persimmon and grape seed extract: in vitro and in vivo. *Nutrition Res.*, 22, 1265-1273.

Aliakbarian, B., Fathi, A., Perego, P., Dehghani, F., (2012). Extraction of antioxidants from winery wastes using subcritical water. *J. Supercrit. Fluid,* 65, 18-24.

Anklam, E., Berg, H., Mathiasson, L., Sharman, M., Ulberth, F. (1998). Supercritical fluid extraction (SFE) in food analysis: a review. *Food Addit. Contam.,*15, 729-750.

Arce, L., Lista, A. G., Ríos, A., Valcárcel, M., (2001). Screening of polyphenols in grape marc by on-line supercritical fluid extraction–flow through sensor. *Anal. lett.*, 34, 1461-1476.

Baydar, N. G., Özkan, G., Sağdiç, O., (2004). Total phenolic contents and antibacterial activities of grape (< i> Vitis vinifera</i> L.) extracts. *Food Control,* 15, 335-339.

Berthod, A. (2002). Countercurrent chromatography, (Vol. 38). Elsevier.

Bonilla, F., Mayen, M., Merida, J., Medina, M., (1999). Extraction of phenolic compounds from red grape marc for use as food lipid antioxidants. *Food Chem.*, 66, 209-215.

Boussetta, N., Lanoisellé, J. L., Bedel-Cloutour, C., Vorobiev, E., (2009). Extraction of soluble matter from grape pomace by high voltage electrical discharges for polyphenol recovery: effect of sulphur dioxide and thermal treatments. *J. Food Eng.,* 95, 192-198.

Boussetta, N., Lebovka, N., Vorobiev, E., Adenier, H., Bedel-Cloutour, C., Lanoiselle, J. L., (2009). Electrically assisted extraction of soluble matter from Chardonnay grape skins for polyphenol recovery. *J. Agric. Food Chem.*, 57, 1491-1497.

Bozan, B., Tosun, G., Özcan, D., (2008). Study of polyphenol content in the seeds of red grape (< i> Vitis vinifera</i> L.) varieties cultivated in Turkey and their antiradical activity. *Food Chem.*, 109, 426-430.

Bucić-Kojić, A., Planinić, M., Tomas, S., Bilić, M., Velić, D., (2007). Study of solid–liquid extraction kinetics of total polyphenols from grape seeds. *J. Food Eng.*, 81, 236-242.

Bucić-Kojić, A., Planinić, M., Tomas, S., Jakobek, L., Šeruga, M., (2009). Influence of solvent and temperature on extraction of phenolic compounds from grape seed, antioxidant activity and colour of extract. *Int. J. Foof Sci. Tech.*, 44, 2394-2401.

Carrera, C., Ruiz-Rodríguez, A., Palma, M., Barroso, C. G., (2012). Ultrasound assisted extraction of phenolic compounds from grapes. *Anal. Chim. Acta*, 732, 100-104.

Casazza, A. A., Aliakbarian, B., Mantegna, S., Cravotto, G., Perego, P., (2010). Extraction of phenolics from< i> Vitis vinifera</i> wastes using non-conventional techniques. *J. Food Eng.*, 100, 50-55.

Crespo, J. G., Brazinha, C., (2010). Membrane processing: Natural antioxidants from winemaking by-products. *Filtr. Separat.*, 47, 32-35.

Cruz, J. M., Domínguez, H., Parajó, J. C., (2004). Assessment of the production of antioxidants from winemaking waste solids. *J. Agric. Food Chem.*, 52, 5612-5620.

Delaunay, J. C., Castagnino, C., Chèze, C., and Vercauteren, J., (2002). Preparative isolation of polyphenolic compounds from< i> Vitis vinifera</i> by centrifugal partition chromatography. *J. Chromatogr. A*, 964, 123-128.

Dopico-García, M. S., Valentao, P., Guerra, L., Andrade, P. B., Seabra, R. M., (2007). Experimental design for extraction and quantification of phenolic compounds and organic acids in white "Vinho Verde" grapes. *Anal. Chim. Acta*, 583(1), 15-22.

Farias-Campomanes, A. M., Rostagno, M. A., Meireles, M. A. A., (2013). Production of polyphenol extracts from grape bagasse using supercritical fluids: Yield, extract composition and economic evaluation. *J. Supercrit. Fluid*, 77, 70-78.

García-Marino, M., Rivas-Gonzalo, J. C., Ibáñez, E., García-Moreno, C., (2006). Recovery of catechins and proanthocyanidins from winery by-products using subcritical water extraction. *Anal. Chim. Acta*, 563(1), 44-50.

Ghafoor, K., Choi, Y. H., Jeon, J. Y., Jo, I. H., (2009). Optimization of ultrasound-assisted extraction of phenolic compounds, antioxidants, and anthocyanins from grape (Vitis vinifera) seeds. *J. Agric. Food Chem.*, 57, 4988-4994.

González-Paramás, A. M., Esteban-Ruano, S., Santos-Buelga, C., de Pascual-Teresa, S., and Rivas-Gonzalo, J. C. (2004). Flavanol content and antioxidant activity in winery byproducts. *J. Agric. Food Chem.*, 52, 234-238.

Guerrero, M. S., Torres, J. S., Nuñez, M. J., (2008). Extraction of polyphenols from white distilled grape pomace: optimization and modelling. *Bioresource Technol.*, 99, 1311-1318.

Hadolin, M., Rižner Hraš, A., Bauman, D., Knez, Ž. (2004). Isolation and concentration of natural antioxidants with high-pressure extraction. *Inn. Food Sci. Emerg.*, 5, 245-248.

Hashim, S. N., Schwarz, L. J., Boysen, R. I., Yang, Y., Danylec, B., Hearn, M. T., (2013). Rapid solid-phase extraction and analysis of resveratrol and other polyphenols in red wine. *J. Chromatogr. A*, 1313, 284-290.

Jayaprakasha, G. K., Selvi, T., Sakariah, K. K., (2003). Antibacterial and antioxidant activities of grape (< i> Vitis vinifera</i>) seed extracts. *Food Res. Int.*, 36, 117-122.

Jayaprakasha, G. K., Singh, R. P., Sakariah, K. K., (2001). Antioxidant activity of grape seed (< i> Vitis vinifera</i>) extracts on peroxidation models in vitro. *Food Chem.*, 73, 285-290.

Ju, Z. Y., Howard, L. R., (2003). Effects of solvent and temperature on pressurized liquid extraction of anthocyanins and total phenolics from dried red grape skin. *J. Agric. Food Chem.*, 51, 5207-5213.

Ko, M. J., Cheigh, C. I., Chung, M. S. (2014). Relationship analysis between flavonoids structure and subcritical water extraction (SWE). *Food chem.*, 143, 147-155.

Köhler, N., Wray, V., Winterhalter, P., (2008). Preparative isolation of procyanidins from grape seed extracts by high-speed counter-current chromatography *J. Chromatogr. A*, 1177, 114-125.

Lapornik, B., Prošek, M., Golc Wondra, A., (2005). Comparison of extracts prepared from plant by-products using different solvents and extraction time. *J. Food Eng.*, 71, 214-222.

Liu, D., Vorobiev, E., Savoire, R., Lanoisellé, J. L., (2011). Intensification of polyphenols extraction from grape seeds by high voltage electrical discharges and extract concentration by dead-end ultrafiltration. *Sep. Purif. Technol.*, 81, 134-140.

Luque-Rodríguez, J. M., Luque de Castro, M. D., Pérez-Juan, P., (2007). Dynamic superheated liquid extraction of anthocyanins and other phenolics from red grape skins of winemaking residues. *Bioresource technol.*, 98, 2705-2713.

Makris, D. P., Boskou, G., Andrikopoulos, N. K., Kefalas, P., (2008). Characterisation of certain major polyphenolic antioxidants in grape (Vitis vinifera cv. Roditis) stems by liquid chromatography-mass spectrometry. *Eur. Food Res. Technol.*, 226, 1075-1079.

Mandal, V., Mohan, Y., Hemalatha, S., (2007). Microwave assisted extraction-an innovative and promising extraction tool for medicinal plant research. *Pharmacogn. Rev.*,1, 7.

Mason, T. J., Paniwnyk, L., Lorimer, J. P., (1996). The uses of ultrasound in food technology. *Ultrasonics sonochemistry*, 3, S253-S260.

Massot, A., Mietton-Peuchot, M., Peuchot, C., Milisic, V., (2008). Nanofiltration and reverse osmosis in winemaking. *Desalination*, 231, 283-289.

Monrad, J. K., Howard, L. R., King, J. W., Srinivas, K., Mauromoustakos, A., (2010). Subcritical solvent extraction of anthocyanins from dried red grape pomace.*J. Agric. Food Chem.*, 58, 2862-2868.

Murga, R., Ruiz, R., Beltrán, S., Cabezas, J. L., (2000). Extraction of natural complex phenols and tannins from grape seeds by using supercritical mixtures of carbon dioxide and alcohol. *J. Agric. Food Chem.*, 48, 3408-3412.

Nawaz, H., Shi, J., Mittal, G. S., Kakuda, Y., (2006). Extraction of polyphenols from grape seeds and concentration by ultrafiltration. *Sep. Purif. Technol.*, 48, 176-181.

Negro, C., Tommasi, L., Miceli, A., (2003). Phenolic compounds and antioxidant activity from red grape marc extracts. *Bioresource Technol.*, 87, 41-44.

Palma, M., Taylor, L. T., (1999). Extraction of polyphenolic compounds from grape seeds with near critical carbon dioxide. *J. Chromatogr. A*, 849, 117-124.

Palma, M., Taylor, L. T., Varela, R. M., Cutler, S. J., Cutler, H. G., (1999). Fractional extraction of compounds from grape seeds by supercritical fluid extraction and analysis for antimicrobial and agrochemical activities. *J. Agric. Food Chem.*, 47, 5044-5048.

Pascual-Martı, M. C., Salvador, A., Chafer, A., Berna, A., (2001). Supercritical fluid extraction of resveratrol from grape skin of< i> Vitis vinifera</i> and determination by HPLC. *Talanta*, 54, 735-740.

Patras, A., Brunton, N. P., O'Donnell, C., Tiwari, B. K., (2010). Effect of thermal processing on anthocyanin stability in foods; mechanisms and kinetics of degradation. *Trends Food Sci. Techn.*, 21, 3-11.

Pekic, B., Kova, V., Alonso, E., Revilla, E., (1998). Study of the extraction of proanthocyanidins from grape seeds. *Food Chem.*, 61, 201-206.

Pinelo, M., Arnous, A., Meyer, A. S., (2006). Upgrading of grape skins: Significance of plant cell-wall structural components and extraction techniques for phenol release. *Trends Food Sci. Techn.*, 17, 579-590.

Pinelo, M., Fabbro, P. D., Manzocco, L., Nuñez, M. J., Nicoli, M. C., (2005). Optimization of continuous phenol extraction from< i> Vitis vinifera</i> byproducts. *Food Chem.*, 92, 109-117.

Rajha, H. N., Louka, N., El Darra, N., Hobaika, Z., Boussetta, N., Vorobiev, E., Maroun, R. G., (2014). Multiple Response Optimization of High Temperature, Low Time Aqueous Extraction Process of Phenolic Compounds from Grape Byproducts. *Food Nutr. Sci.*

Ramos, L., Kristenson, E. M., Brinkman, U. A., (2002). Current use of pressurised liquid extraction and subcritical water extraction in environmental analysis. *J. Chromatogr. A*, 975, 3-29.

Raventós, M., Duarte, S., Alarcón, R., (2002). Application and possibilities of supercritical CO2 extraction in food processing industry: an overview. *Food Sci. Technol. Int.*, 8, 269-284.

Revilla, E., Ryan, J. M., Martín-Ortega, G., (1998). Comparison of several procedures used for the extraction of anthocyanins from red grapes. *J. Agric. Food Chem.*, 46, 4592-4597.

Shrikhande, A. J., Race, E. J., Wightman, J. D., Sambueso, R. D., (2003).U.S. Patent No. 6,544,581. Washington, DC: U.S. Patent and Trademark Office.

Smith, R. M. (2002). Extractions with superheated water. *J. Chromatogr. A*, 975, 31-46.

Valle, J. D., Aguilera, J. M., (1999). ARTICLES-High pressure CO2 extraction. Fundamentals and applications in the food industry. *Food Sci. Technol. Int.*, 5, 1-24.

Valls, J., Millán, S., Martí, M. P., Borràs, E., Arola, L. (2009). Advanced separation methods of food anthocyanins, isoflavones and flavanols. *J. Chromatogr. A*, 1216, 7143-7172.

Vilkhu, K., Mawson, R., Simons, L., Bates, D., (2008). Applications and opportunities for ultrasound assisted extraction in the food industry—A review. *Innov. Food Sci. Emerg.*, 9, 161-169.

Wang, L., Weller, C. L., (2006). Recent advances in extraction of nutraceuticals from plants. *Trends Food Sci. Technol.*, 17, 300-312.

Wu, F., Wang, Z. B., Chen, W. Z., Zhu, H., Bai, J., Zou, J. Z., Su, H. B., (2004). Extracorporeal high intensity focused ultrasound ablation in the treatment of patients with large hepatocellular carcinoma. *Ann. Surg. Oncol.*, 11, 1061-1069.

Yang, Y., Hawthorne, S. B., Miller, D. J., (1997). Class-selective extraction of polar, moderately polar, and nonpolar organics from hydrocarbon wastes using subcritical water. *Environ. Sci. Technol.*, 31, 430-437.

Yilmaz, Y., Toledo, R. T., (2006). Oxygen radical absorbance capacities of grape/wine industry byproducts and effect of solvent type on extraction of grape seed polyphenols. *J. Food Comp. Anal.*, 19, 41-48.

INDEX

A

B

C

D

E

F

G

H

I

J

K

L

M

N

O

P

Q

R

S

V

W

X

Y

Z